ANTARCTIC GLACIAL HISTORY AND WORLD PALAEOENVIRONMENTS

The research ship 'R.S.A.' in a bukta near the SANAE base on Antarctica
Photo: Dept. of Transport, Pretoria

ANTARCTIC GLACIAL HISTORY AND WORLD PALAEOENVIRONMENTS

edited by
E.M. VAN ZINDEREN BAKKER
(Phil.Nat.D., D.Sc.h.c.)

INTERNATIONAL COUNCIL OF SCIENTIFIC UNIONS

SCIENTIFIC COMMITTEE ON ANTARCTIC RESEARCH

Proceedings of a symposium held on 17th August, 1977
during the Xth INQUA Congress at Birmingham, U.K.

A.A.BALKEMA / ROTTERDAM / 1978

The proceedings of the former two symposia organised by the 'SCAR Group of Specialists on Late Cainozoic Studies' have been published in:

PALAEOECOLOGY OF AFRICA, THE SURROUNDING ISLANDS AND ANTARCTICA

Volume V, 1969, 240 pp. (Cambridge Symposium, 24–27 July, 1968)
Volume VIII, 1973, 198 pp. (Canberra Symposium, 9–12 August, 1972)
A.A.Balkema, Cape Town

ISBN 90 6191 027 7
Published by A.A.Balkema, P.O. Box 1675, Rotterdam, Netherlands
Printed in the Netherlands

Foreword

The SCAR Group of Specialists on Late Cainozoic Studies, which was established under a different name at the IX SCAR meeting held at Santiago de Chile in 1966, organised its third symposium during the Xth INQUA Congress at Birmingham in August 1977. The two former symposia were held in Cambridge (1968) and Canberra (1972) and the proceedings of these meetings are available in published form. The last symposium was devoted to the influence the Antarctic glaciation had on world palaeoenvironments, a theme which has so far not been studied in a multidisciplinary context. The invited participants have succeeded in giving a valuable up to date assessment of our knowledge in this field.

SCAR is most grateful to Professor van Zinderen Bakker for all his efforts in organising this symposium and his work in editing the proceedings and arranging its publication.

T. Gjelsvik, president

Scientific Committee on Antarctic Research
Norsk Polarinstitutt, Oslo, Norway

Table of contents

Proceedings of a symposium held on 17th August, 1977 during the Xth INQUA Congress at Birmingham, U.K.

Introduction

As a consequence of our classical training and orientation it has become customary to view at our planet from the Northern Hemisphere and to study glaciations, their causes and far reaching effects mainly in the Quarternary. It is, however, very rewarding to go further back in time and to observe the Earth from a different angle as this will give more perspective to our ideas on the fascinating history of the Globe.

The concerted efforts of the scientists of the different SCAR countries have in recent years provided a wealth of new information which makes it possible to perceive correlations on a world wide scale. Much more light has been shed on the origin and evolution of the enormous ice masses of East and West Antarctica. Great advances have especially been made in the study of ocean sediments which have resulted in extremely important, continuous records of environmental changes. Cores which can be calibrated and compared using microfossils, isotope techniques, measurements of former magnetism and temperatures offer invaluable opportunities for correlating events in the different oceans of the World. All these studies have provided information for a better understanding of the oceanic and atmospheric circulation systems and are revealing the influence of astronomical variations on the energy budget of the Earth. This approach will eventually lead to more insight in the climatic relationship of the two hemispheres and on the possible evolution of climate in future.

The consequences of these processes have been discussed for the continents of the Southern Hemisphere and also for some of the islands in the Southern Ocean. The incomplete record of the terrestrial evidence and the complicated nature of the processes do not yet allow a detailed comparison of all the events on land and in the oceans but an integrated picture is slowly emerging. Within a few years a further evaluation should be made of the advances in these fields, so that we can assess how the contributions of the different disciplines fit in the fascinating jigsaw puzzle of our knowledge of the history of the Earth.

The time available for the symposium was not sufficient to cover the many subjects which should receive attention so that some contributions could not be presented personally. It is most unfortunate that the valuable information on the Pacific and Indian Oceans was not prepared in time to be included in this volume.

The Scientific Committee on Antarctic Research deserves our gratitude for its encouragement and support of the work of our Group of Specialists. We are deeply indebted to the Executive of INQUA which offered its hospitality by making the first day of its Tenth Congress at Birmingham available for our meeting. Our deepfelt thanks are due to the Transantarctic Association which provided valuable financial assistance for the publication of this volume. The co-operation of the Steering Committee and the authors made it a pleasant task to edit these proceedings.

E.M. van Zinderen Bakker Sr
Convenor, SCAR Group of Specialists on Late Cainozoic Studies.
Institute for Environmental Sciences, University of the Orange Free State,
Bloemfontein, South Africa

* 1 *

Comparison of Antarctic and Arctic climate and its relevance to climatic evolution

H. Flohn

Meteorological Institute, University of Bonn,
Auf dem Hügel 20, 53 Bonn 1, W. Germany

Manuscript received 15th October 1977

CONTENTS

ABSTRACT

Averaged over the whole year, the temperatures of the layer 300–900 mb (ca 3–9 km) in interior Antarctica are 11–12°C colder than those in the Arctic. As a consequence, the circumpolar vortex in the upper troposphere and lower stratosphere above the Antarctic is much deeper than above the Arctic, the intensity of the mid-latitude westerlies is higher and the tropical Hadley circulation of the southern hemisphere extends, during the greater part of the year, beyond the equator, thus suppressing the northern Hadley cell. The physical causes of this marked asymmetry of the global atmospheric circulation will be discussed.

Arctic and Antarctic climatic evolution could be essentially independent, as indicated by recent development. The coincidence of major Milankovich episodes can be interpreted as resulting from the seasonal parallels between the development of drifting ice of the Subantarctic and summer anomalies of the essentially continental climate of the Subarctic.

The role of possible Antarctic surges for the climatic evolution in the last 40×10^6 years will be discussed, as well as the role of the slow drift of Antarctica onto its isolated polar position.

1. TEMPERATURES AND RADIATION BUDGET

One of the basic features of global atmospheric circulation (and that of the wind-driven oceanic mixing layer) is its asymmetry with respect to the equator – a fact which is hardly covered satisfactorily in most textbooks. One of the fundamental parameters of the circulation is the thermal Rossby number Ro_T

$$Ro_T = U_T / r\Omega$$

where $U_T = \partial u / \partial z . \Delta z$ is the vertical shear of zonal wind u (or the 'thermal wind') of a layer Δz, r = earth's radius and Ω = angular speed of earth's rotation. This dimensionless number indicates the thermal zonal wind, depending on the temperature difference between Equator and Pole, in units of the rotation speed of the earth's equator (464 ms^{-1}).

Since the Antarctic ice-dome surface reaches an average altitude of about 3 000 m, the lower layers of the Arctic atmosphere have to be omitted for comparison. Thus we compare Antarctic and Arctic temperature soundings above 700 mbs (Amundsen-Scott station at 2 800 m and 90° S, with an average pressure of 681 mb). Results of the drifting T 3 station ('Fletcher's Island') have been taken as representative for the Arctic; differences to the drifting 'North Pole' stations of the USSR are negligibly below 300 mbs. In this case the Arctic low-level inversions are disregarded, while the (even stronger) Antarctic inversion is included: this error results in slightly lower average temperatures above the Antarctic. However the use of the thermal wind equation necessitates restriction to an isobaric layer. Taking the surface data from the South Pole as representing the 700 mb level, we obtain the following pressure-averaged temperatures (Fig. 1, Table 1) for the layer 300/700 mbs (about 3–9 kms). If we select the layer 300/680 mbs instead 300/700 mbs, the temperature difference North Pole–South

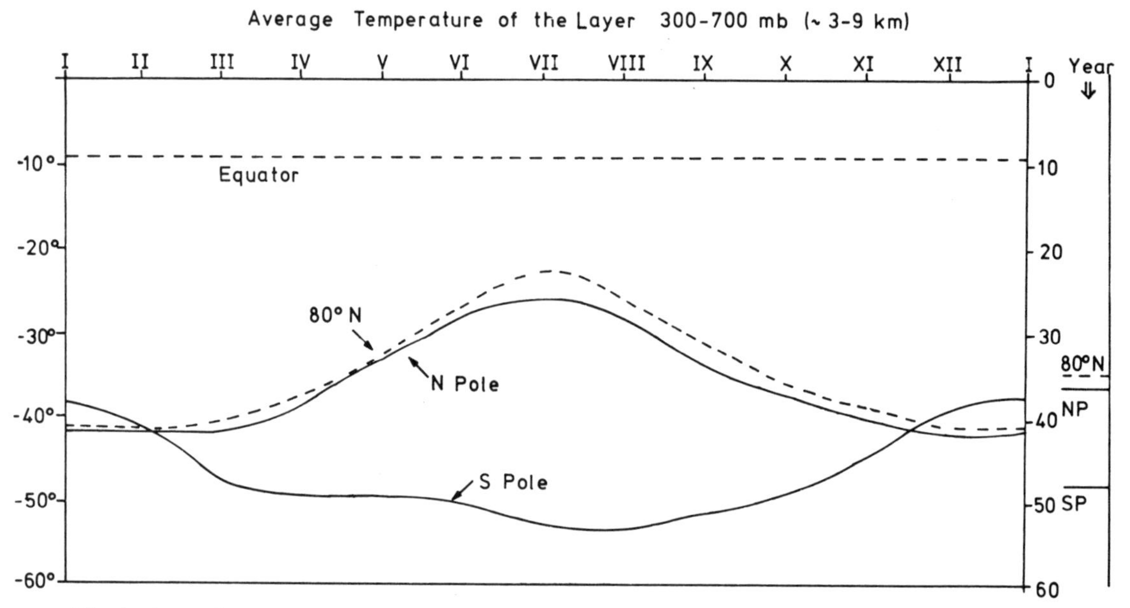

Fig. 1. Annual temperature trend (300/700 mb) for both Poles, 80 ° N and Equator (Flohn, 1967)

Pole decreases only by 0.7°C. For comparison the average of 7 stations near the equator (3° S–7° N) has been selected; this is practically identical with the Standard Tropical Atmosphere, without any appreciable seasonal trend. These averages comprehend 6–10 years and can be taken as representative (with errors of the order 0.5° C). At an annual average the upper and middle troposphere above the Arctic is about 11° C warmer than above the Antarctic. This difference is small during the northern winter, but most accentuated during northern summer, where it reaches nearly 27° C. Then the thermal Rossby number reaches, at the southern hemisphere, more than 250 percent of the value at the northern hemisphere; even in the annual average more than 140 percent. The kinetic energy of the Southern Westerlies is about 60 percent larger (Lamb, 1959) than that of the Northern Westerlies. This stronger circulation of the southern hemisphere pushes the meteorological equator towards the Northern Hemisphere, especially during northern summer, while during northern winter (January) the circulation intensity of both hemispheres is about equal (Fig. 2). Its zonally-averaged annual position can be estimated to about 6° N (Flohn, 1967); at the Atlantic (where maritime data along 25° W had been investigated with a meridional resolution of 1° Lat.) it varies between 1° N (February–March) and 11° N (August) (Fig. 2).

At surface stations, this difference may even be larger (Table 2). Comparable stations near 3000 m altitude in the central parts of the ice domes of Antarctica and Greenland – i.e. below a strong

Table 1
Average temperatures of Middle Troposphere (300/700 mb layer)

Region	Extreme months		Year
Equator E			—8.6
North Pole N	—41.5 (I)	—25.9 (VII)	—35.9
South Pole S	—52.7 (VII)	—38.3 (I)	—47.7
E–N	32.9 (I)	17.3 (VII)	27.3
E–S	29.7 (I)	44.1 (VII)	39.1

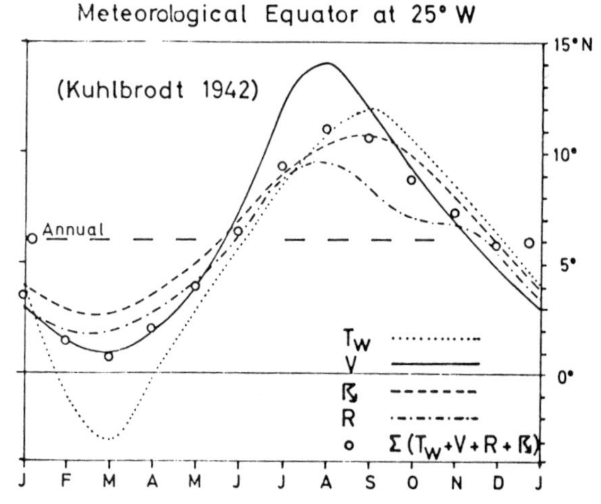

Fig. 2. Annual displacement of meteorological equator along 25 ° W (Atlantic) (Flohn, 1967, data after Kuhlbrodt, 1942)

Table 2
Climatic surface data, Antarctic–Arctic (Arctic 6 months shifted)

Location	Height (m)	Temperature (° C) Su (12–2)	Wi (6–8)	Year	Absolute Extremes		Cloudiness (%) Su	Wi	Period
South Pole	2800	—32.3	—58.2	—49.3	—15	—81	52	38	1957–66
Vostok (78° S)	3488	—36.8	—67.0	—55.6	—21	—88	38	33	1957–66
Eismitte (71° N)	3000	—13.9	—39.7	—28.8	—3	—65	65	59	1930/1, 49–51
McMurdo (78° S)	24	—5.2	—25.6	—17.4[1]	+6	—51	68	52	1957–66
Arctic Ice Drift	2	—1.0	—33.7	—19.2[2]	+6	—53	90	51	1957–61[3]
Orcadas (61° S)	4	0.0	—10.1	— 4.4	+12	—40	93	79	1903–68
Ivigtut (61° N)	30	+9.0	— 4.6	+ 1.8	+23	—23	65	63	1931–56

1. 5 years 1902–12 also — 17.4
2. Fram Drift 1893–96 also — 19.2
3. Several records, together 117 months

surface inversion – show differences up to 25–30°C, while the climate in the oceanic parts, mainly controlled by the drifting sea-ice with its polynyas, is nearly the same. The harsh climate in the oceanic latitude belt 50–65° S has often been compared to that of similar latitudes in the Northern Hemisphere: Bouvet Island (54° S), is nearly completely ice-covered, while Helgoland (54° N) is a summer bathing resort.

These differences are partly reflected in the average surface temperatures of the hemispheres and the earth (Table 3). However, the small area of the polar region suppresses the large differences, and during winter the large continents render the Northern Hemisphere average lower than those of the oceanic southern hemisphere.

The physical reason of this difference is to be found in the different radiation budget of an isolated ice-covered Antarctic continent and of nearly (85 percent) land-locked Arctic ocean (Table 4). A decisive factor is the high albedo of the Antarctic ice which reflects more than 80 percent of the high incoming radiation: see Vostok in comparison with the mostly snowfree station Oasis. The effective long-wave radiation E–C is a consequence of the low water vapour content above the Antarctic, to be compared with the relative moist Arctic atmosphere, where during the summer melting period a nearly permanent low stratus cover contributes substantially to the atmospheric counter-radiation C.

Due to the high latitude, the heat and radiation budget of both Arctic and Antarctic atmosphere are negative: a permanent (annual) loss to space must be balanced by an advective influx of sensible (and, to a lesser degree, latent) heat (Fig. 3). This can easily be formulated by a stationary heat balance equation of an atmospheric column, where the divergence of the vertical fluxes of radiation and heat is equal to the convergence of the horizontal heat transport. Preliminary values of the main terms of the interior Arctic and Antarctic are given from the data in Vol. 14 of *World survey of climatology*, putting the extra-terrestrial solar radiation = 100 (Fig. 3). Several important differences should be underlined:

a) As a consequence of its low cloudiness the reflected short-wave radiation (wavelength $< 3\ \mu m$) at the Antarctic surface (R_{sf}) is about twice as large than at the Arctic; in contrast to that the reflected radiation from clouds (R_{cl}) is only about one half.

b) Near the surface, the turbulent fluxes of sensible and latent heat are in both regions small and unimportant; in the Arctic these fluxes (mainly evaporation from the polynyas between the sea-ice floes) are directed upwards into the atmosphere, while in the Antarctic obvious downward fluxes predominate.

c) In both regions, the negative radiation budget at the top of the atmosphere is maintained by horizontal advection from temperate latitudes. This advective flow (mainly sensible heat) is larger in the

Table 3
Average surface temperatures of Hemispheres and Earth (van Loon, 1972)

Region	Extreme months		Year
N. Hemisphere	8.0 (I)	21.6 (VII)	15.0
S. Hemisphere	10.6 (VII)	16.5 (I)	13.4
Earth	12.3 (I)	16.1 (VII)	14.2

Table 4
Radiation Budget in both Polar Zones (data after Gavrilova, 1963; in Watt m^{-2})

Location	(S + H)	a	(S + H)(1 — a)	(E — C)	Q
Summer (2 months) (7–8):					
Vostok (78° S, 3 488 m)	432	81%	82	71	11
Oasis (66° S, 28 m)	288	18	236	66	170
Central Arctic (75° N)	274	72	74	16	58
Winter (4 months) (12–3):					
Vostok	0		0	14	— 14
Oasis	8.2	(41)	4.8	32.4	— 27.6
Central Arctic	0		0	26	— 26
Year:					
Vostok	143	83	26	34	— 8
Oasis	118	24	91	41	50
Central Arctic	90	76	22	21	— 1

S = direct solar radiation
H = diffuse sky radiation
a = surface albedo (%)
E = outgoing terrestrial radiation
C = atmospheric infrared radiation
Q = radiation balance: Q = (S + H)(1 — a) — (E — C)

Arctic due to more frequent meridional circulation patterns of the Northern Hemisphere. In comparison to that, oceanic advection (as operating across the thin floating sea-ice) into the Central Arctic is rather small.

d) Due to higher temperature and water vapour content, long wave infrared (> 3 μm) radiation at the top of the atmosphere (E_0) from the Arctic is about twice that from the Antarctic, equivalent to the effective heat loss to space.

2. CLIMATIC IMPLICATIONS

The global climatic consequences of this hemispheric asymmetry of the polar climates are far-reaching. Assuming that the baroclinic instability criterion as formulated by Smagorinsky (1963) – see Henning (1967) – can be adapted to describe the difference between the (dynamically stable) Hadley circulation of the Tropics and the (unstable) extra-tropical Ferrel vortex (Flohn, 1964; Korff & Flohn, 1969), the boundary φ_s between these two parts of the global atmospheric circulation – which coincides with the position of the subtropical jet as well as with that of the subtropical anticyclonic belt – can be derived from this criterion (Fig. 4). We obtain:

$$\operatorname{ctg} \varphi_s = -\frac{r}{h} \frac{\partial\Theta/\partial y}{\partial\Theta/\partial z}$$

with r = radius of the earth, h = scale of the atmosphere (~ thickness 250/750 mb), Θ = potential temperature, y(z) = meridional (vertical) component. Now the meridional temperature gradient $\partial\Theta/\partial y$ is proportional to the thermal wind $\partial u/\partial z$ and the vertical temperature gradient $\partial\Theta/\partial z$ is rather constant (~ 6.5°C km), except surface inversions, depending largely on the water vapour content of the atmosphere. This means that within present conditions $\operatorname{ctg}\varphi_s$ depends mainly on the thermal Rossby number.

This simple, but fundamental approach has been tested with the seasonal and hemispherical variation of actual data (Korff & Flohn, 1969). The correlation between the 24 monthly meridional temperature gradients and the simultaneous position of the subtropical anticyclonic belt on both hemispheres (Fig. 4) is 0.85. Taking into account a time-lag between large-scale temperature anomalies and the atmospheric response, the correlation between $\partial\Theta/\partial y$ and φ_s of the following month reaches 0.93. In the annual average the two subtropical anticyclonic belts are situated at latitudes 37° N and 31° S. The asymmetry of the position of equivalent climatic zones in both hemispheres can thus be interpreted as caused by the temperature difference between Antarctic and Arctic. In turn, this is a result of the different heat budget of a massive ice-dome and a thin floating sea-ice with many polynyas.

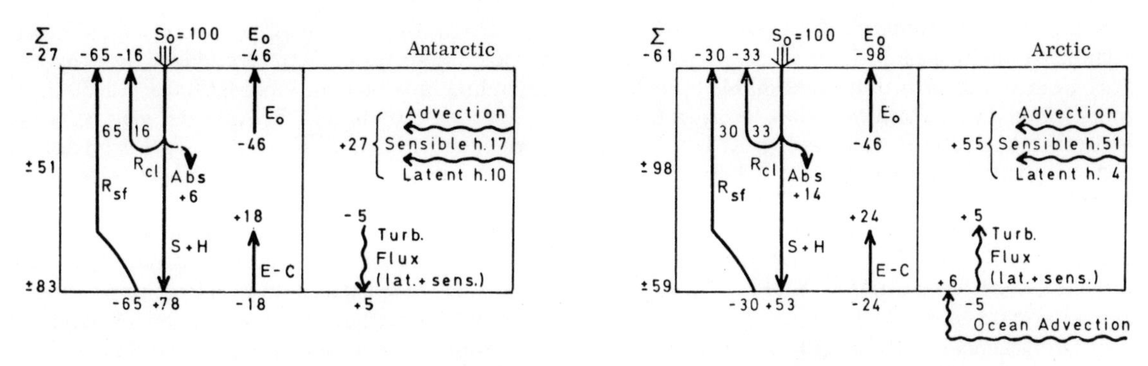

Fig. 3. Preliminary heat and radiation budget for Antarctic (Schwerdtfeger, 1970, partly revised) and Arctic (Vowinckel & Orvig, 1970)

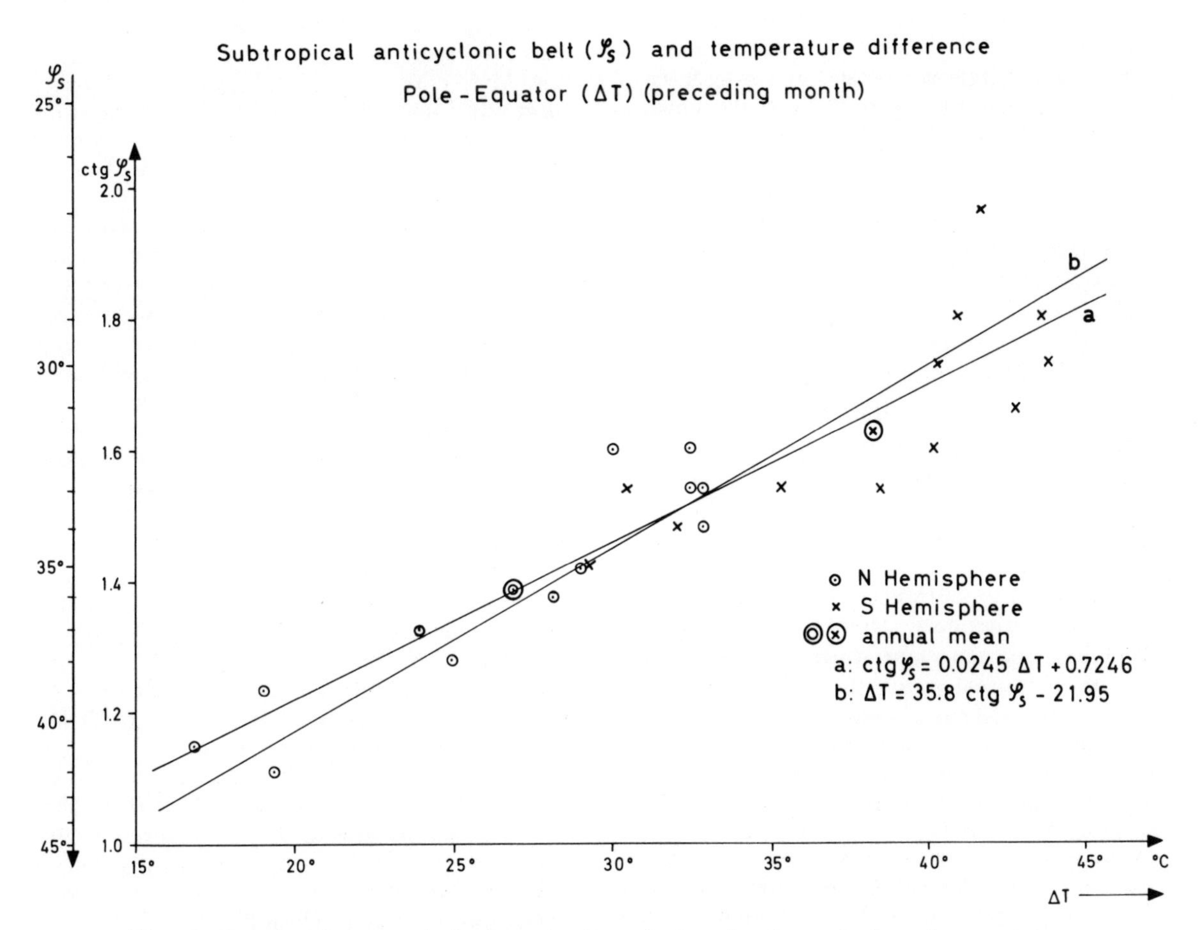

Fig. 4. Relation between latitudes of subtropical anticyclones (φ_S) and temperature difference Equator – Pole (ΔT 300/700 mb) of each month, Northern and Southern Hemisphere (Korff & Flohn, 1969)

Some further climatic consequences of the asymmetry shall be reminded:

a) Coexistence of two distinguishable baroclinic zones within the southern westerlies, i.e. a distinction between a subtropical jet-stream (near lat. 30° S) and a polar (circum-Antarctic) jet-stream (lat. 50–60° S) (see van Loon, 1972: Chapter 5 and 8);

b) Displacement of the equatorial oceanic rain-belt (identical with the main Intertropical Convergence Zone) to lat. 0–12° N (Fig. 2), with an annual average of 6° N;

c) Displacement of the zonally averaged equatorial pressure trough to 6° N, in connection with an encroachment of H of the southern trades across the equator. As a consequence, the seasonal temperature variations of many stations in Lat. 0–5° N are characterized by an advective minimum in northern summer/southern winter ('Sudanese' type). Furthermore, this encroachment causes the extension of oceanic upwelling – as produced by a divergence of the wind-driven Ekman-drift, from the equator southward (Flohn, 1972).

d) Occurrence of an Equatorial Counter Current in the upper mixing layer of Pacific and Atlantic in the belt 2–8° N, embedded between the two branches of Northern and Southern Equatorial Current driven by the atmospheric Trades.

3. PALAEOCLIMATIC ROLE OF THE HEMISPHERIC CLIMATIC ASYMMETRY

The climate of the early Tertiary (and probably also of the greater part of the Mesozoic era) was characterized by an ice-free climate over the whole globe. The above-mentioned relation between meridional temperature gradient and φ_s has been used (Flohn, 1964) to estimate φ_s from a fairly realistic estimate of the polar temperatures. From the available isotope temperatures at surface and bottom of tropical oceans the meridional surface temperature gradient was estimated to be 16°C. In this case it was assumed that the sea surface temperature in a polar ocean is about equal to the (observed) bottom temperature in low latitudes, i.e. that the abyssal circulation of the ocean was similar to the actual circulation. From these estimates, a reduction of the thermal Rossby number to about 50% of its present value (0.02–0.03) is obtained, i.e. a position of the subtropical anticyclones between lat. 50 and 60°; as a conservative estimate an annual average near lat. 50° should be preferable. Assuming solar radiation and earth's rotation to be constant within narrow margins, a reasonable climatic interpretation would be a (seasonally shifting) tropical Hadley circulation with convective summer rainfall controlling about 70–80% of the global surface. The remaining small polar caps should have been dominated by a Ferrel-type polar vortex much weaker than now, with travelling disturbances extending in winter into mid-latitudes (40–50°).

The long evolution of the Antarctic glaciation has been described by J.P. Kennett, A.T. Wilson & J.H. Mercer (this volume). From the view-point of climatology, the following stages are important:

1. During Eocene, local glaciation at Antarctica;
2. Near Eocene–Oligocene boundary (38 MY ago), wide-spread glaciation with freezing conditions at sea-level, development of sea-ice and formation of cold bottom water penetrating the world's oceans;
3. Late Oligocene: after separation of Tasmania and the South Tasman Rise from Antarctica and the opening of the Drake Passage south of South America, formation of a baroclinic circum-Antarctic ocean current, isolating the Antarctic continent from any significant meridional exchange with warm subtropic waters;
4. Middle Miocene (12 MY ago) rapid development of a complete Antarctic ice-cap (still near the melting point), probably correlated with a period of increased volcanic activity (Kennet & Thunell, 1975).
5. Late Miocene (5 MY ago, i.e. contemporaneous with the repeated desiccation of the Mediterranean, see Hsü et al., 1973; 1974; 1977) ice thickness increase to the present volume or even larger, most probably caused by the transition into a cold, slow-moving ice (Hoinkes, 1961). In this case a positive mass budget of 1 cm/a – at present 3–5 cm/a estimated – could build an ice-dome with an average thickness of 2 200 m in only (!) 2×10^5 years, in equilibrium with an eustatic sea-level drop of 78 m;
6. Late Pliocene (about 2.5 MY ago) development of Northern Hemisphere ice-sheets;
7. Following this stage (Y. Herman, 1977), after the first continental deglaciation, large volumes of fresh water from melting ice-bergs, ice-shelves and increased river runoff resulted in the formation of a shallow low-saline surface water layer (with low density) at the top of the Arctic ocean.

This sequence of events has, as a consequence, caused – for a period of more than 20, perhaps of 35 MY – a marked asymmetry of climate: glaciated Antarctic versus ice-free Arctic. A tentative climatic interpretation of these facts – derived from direct observations – may be given now:

Stage 1: wide-spread winter snow-cover.

Stage 2: at first during winter complete snow-cover and seasonal sea-ice, which may have persisted during summer only along some parts of the coast. The formation of cold bottom-water leads not only to a world-wide deep-sea fauna crisis, but also, *very* gradually, to a world-wide cooling of surface waters and air, as observed during the second half of the Tertiary.

Stage 3: climatic isolation of the Antarctic continent in a polar position, in strong contrast to the Arctic Ocean with its broad connections with the Tethys Ocean via Atlantic and Western Siberia. Formation of a deep cyclonic Antarctic vortex in the atmosphere, with baroclinic westerlies (similar to actual conditions), probably correlated with a weak surface anticyclone with easterly winds, along the coast with a katabatic component towards sea.

Stage 4: due to the high albedo of the Antarctic continent (slightly less than now only during the summer melting period), the intensity of the Southern Hemisphere circulation is about equal to actual conditions, while at the Northern Hemisphere conditions of an ice-free (acryogenic) circulation still prevail.

Stage 5: with the transition to a permanent cold ice, further increase of summer surface albedo and baroclinity, i.e. intensity of westerlies. Eustatic lowering of the ocean surface increases continentality of the earth's climate, Local glaciation at subarctic latitudes (e.g. Alaska) began between stages 4 and 5.

Stage 6: the late development of Northern Hemisphere glaciation must be attributed to other causes, such as a period of strongly increased explosive volcanism (as observed in the deep-sea drill-cores, see Kennett & Thunell, 1975, similar to stage 4) and the closure of the Central American isthmus.

Stage 7: Only in such a permanent low-saline top layer of the Arctic ocean the present permanent sea-ice cover could have developed: this is related to the evolution of the actual large-scale climatic boundary conditions (and those in the interglacial periods).

The first stage must have been the development of seasonal sea-ice with a thickness near 1 m (instead of 3 m of multi-annual ice); then the present permanent Arctic sea-ice (about 8.10^6 km^2) has developed. Its two decisive climatogenic effects are: increase of the surface albedo from about 0.10 to 0.60–0.70 (melting season) or 0.80 (snow-covered) and inhibition of most of the fluxes of sensible and latent heat from ocean to air, restricted now only to polynyas. Lower evaporation has caused the dry climate of all coasts of the interior Arctic Ocean, most notably in northeastern Greenland and at the Queen Elizabeth Islands with its partially unglaciated mountains.

During this very long (38–3 MY ago) time a hemispheric circulation asymmetry much stronger than the present one must have been developed: an isolated glaciated continent in the south, a warm ocean with broad meridional openings to the tropics in the north. The temperature contrast between Arctic and Antarctic may have reached, during this time, about 20° C – it would be quite interesting to look for evidence of strong climatic anomalies (especially aridity) in the equatorial region. Kozur has produced some evidence on a position of the meteorological equator even farther north (15° N) than now about 30 MY ago. His assumption of a nearly permanent position of all continents cannot be too far from reality during the second half of the Cainozoic. This marked asymmetry diminished about 9 MY ago when in Alaska mountain glaciers developed, indicating increased baroclinity also at the northern hemisphere.

Obviously, the Antarctic ice-dome and its much shallower predecessors belong to the most ancient boundary conditions of world's climate – being about in their present position now for more than about 25 MY. Since that time the air above the Antarctic must have been, as a consequence of the radiation and heat balance of extended snow and ice fields, much colder than above the Arctic. There are only short exceptions: the 15 (or more) Pleistocene ice-ages (each lasting, at its maximum, probably not much more than 10 000 years, as in the case of the Würm–Wisconsin glaciation) during which even larger ice-domes with an area of about 26.10^6 kms developed in subarctic (not arctic) latitudes and produced a greater baroclinity of the Northern Hemisphere. Available calculations with circulation models (Alyea, 1972; Williams et al., 1974; Saltzman, 1975; Gates, 1976) indicate that even then the high-tropospheric southern westerlies remained stronger than the (intensified) northern westerlies. Rognon and Williams (1977) have indicated, that during the Middle Würm (45 000–25 000 BP) the Arctic Convergence Zone may have been displaced about 5° Lat. towards N, as well as the intertropical rains in the Sahel belt. This distinct asymmetry was after 24 000 BP, replaced by a more symmetrical distribution of temperature and zonal circulation on both hemispheres, now with a narrowing of the ITCZ region and a reduction of the summer monsoon belt. This strengthening of the Hadley circulation coincides with a strong development and extension of equatorial upwelling at both Atlantic and Pacific (cf. Gates, 1976), i.e. with low evaporation. See also Appendix III.

Evidence of arid tropical continents, of a drastic reduction of the equatorial rain-forests of the Congo and Amazon basin and of the formation of fossil phosphate sediments (from Guano) at Pacific islands like Nauru and Malden is quite consistent with an intensified Hadley circulation, which was situated more symmetric to the equator than today.

It had been generally accepted in recent decades that the simultaneous occurrence of the last glaciation of northern, equatorial and southern latitudes (except Antarctica!) is inconsistent with a leading role of orbital elements (here precession). However this interpretation is based on an unjustified generalization of the radiation conditions of continental glaciations. In fact, the mass budget of northern *continental* glaciers depends largely on the ablation processes during summer (Hoinkes, 1961): cool, wet summers with frequent new snow and high albedo promote ice advances; warm, dry summers lead to ice retreat. However, in the *oceanic* climate of the southern hemisphere the situation is different. In contrast to the Arctic – where about two thirds of the drift-ice area is multiannual ice – the Antarctic drift-ice is largely a seasonal phenomenon, with only 15 percent of the ice being multiannual. While in the Arctic not more than about 4×10^6 km^2 are forming during winter and melting during summer, the same figure for the Subantarctic is 18×10^6 km^2. Here the formation of sea-ice is controlled by winter conditions: with decreasing winter temperatures at the Antarctic continent, from where strong katabatic winds blow on the ocean freezing its surface layer, the formation of seasonal sea-ice is increasing. When the perihelion occurs near the solstice of the southern hemisphere, the Antarctic receives, during its summer, about 7 percent more radiation than the Arctic. In other words, the *winter* insolation of the *southern* hemisphere reaches its minimum exactly simultaneous with the minimum *summer* insolation of the *northern* hemisphere. Thus the orbital term e sin π, at times of high eccentricity e, produces simultaneous periods of increasing snow and ice formation on both hemispheres, when the perihelion occurs in December/January.

From a climatological view-point, the observed dominant role of the 100 000 year cycle (Hays et al., 1976) cannot be correlated with a simultaneous variation of e. It is much more likely (Berger, in press; Wigley, 1976) that it is a consequence of the near coincidence of two observed main periods of precession (π_1 with ~ 19 000 years, π_2 with ~ 23 000 years), where the 'beat frequency' is near 100 000 years and produces a strong amplitude modulation of the precession signal. See also Appendix II.

The exceptional situation of some mountain glaciation in the exterior tropical belt (Heine, 1974) around Lat. 20°, with their peak glaciation not before 14 000–12 000 BP, i.e. during one of the stadials before the Alleröd, when the precipitation increased sharply, is not well understood until now. Also in other areas recent radiocarbon data signalize more interesting deviations from a hitherto over-simplified picture. Much evidence for a simultaneous occurrence of volcanic explosions and the beginning of glacier advances has been collected by Bray (1974, 1976, 1976, 1977). It should be pointed out, that the coincidence of many severe earthquakes at the subduction zones and of volcanic eruptions in the world's rift system is a consequence of plate tectonics: obviously these events occur discontinuously in time and are therefore clustering, as in recent years.

Results of model simulations of ice-age circulation – as impressive as they are – should be interpreted with caution: none of the available models is able to simulate the present-day climate with sufficient regional detail, because of their deficiencies in dealing simultaneously with so many, mostly non-linear interactions between soil, vegetation, sea, ice and atmosphere. However, such mocels are indispensable for a rational judgement of the physical climatogenic processes.

4. RECENT EVOLUTION AND OUTLOOK INTO THE FUTURE

During the last 20–30 years, stations along the Antarctic coast and in high southern latitudes have experienced slight warming (Limbert, 1974; Salinger & Gunn, 1975; Damon & Kunen, 1976; Coughlan, 1975) in contrast to the more marked cooling in the arctic and subarctic latitudes (Borzenkova et al., 1976) (Fig. 5). Above the Antarctic itself, summer temperatures have been warming, while winters are getting cooler (von Loon & Williams, 1977). This different evolution of recent climate in both hemispheres can be interpreted as a short-lived regulatory phenomenon within the climatic system with its subsystems operating at very different time-scales. Damon has proposed instead, that the Southern Hemisphere warming is due to the CO_2 effect, which is, at the Northern Hemisphere, overwhelmed by the cooling of man-made particulate matter. However, most authors now agree that the role of low-level man-made aerosol leads mainly to warming instead of cooling, due to high absorption coefficients in the visible and near-infrared portion of the solar spectrum. On the other side, it can be argued that this cooling at the Northern Hemisphere is partly due to increased volcanic activity (Flohn, 1974), at least since 1963

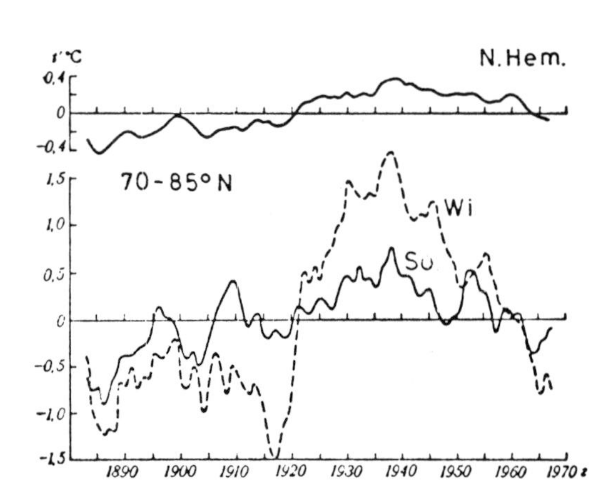

Fig. 5. Temperature trend in the Arctic (75°–87° N, Borzenkova, 1976)

(Agung eruption), and would have been more intense without the role of a 13% increase of CO_2. Indeed, recent discussion of the possibility of large-scale man-made climatic effects has led to a remarkable consensus among most specialists, that CO_2 and other infrared-absorbing man-made gases may inevitably lead to a substantial warming at a time-scale, which is extremely short from a geological point of view.

Since our knowledge of the physical causes of the recent climatic fluctuations is limited, such questions must remain open. Nevertheless it must be clearly stated, that – in the time-scale between 10 and 10^4 years – the climatic trends of Arctic and Antarctic can be of opposite sign. Since the temperature fluctuations in high northern latitudes have been – at least during the instrumental period (Borzenkova et al., 1976), but most probably during the last millenium – more accentuated (by a factor 3–5) than those in low latitudes, the role of both polar regions in controlling the earth's climatic history can hardly be overestimated. This is mainly a consequence of the positive feedback mechanism between albedo, snow cover and temperature (Kellog, 1973).

Looking at the future of the next 10^3–10^4 years, the Antarctic ice-dome will certainly remain the most constant climatic factor, even if the Arctic sea-ice should completely disappear, as foreshadowed by some simplified models already for the next century. In the geologic past, all latitudes between about 65° S and the North Pole have been subject to parallel and apparently simultaneous climatic fluctuations at a time-scale of 10^4 years and above. It has been frequently suggested – with admittedly scanty evidence – that the Antarctic ice-dome is out of phase with these (orbitally controlled) fluctuations. A large-scale warming of the Southern Hemisphere with a higher water-vapour content and a reduced area of seasonal subantarctic drift-ice will almost certainly increase the snow accumulation along the margins and even in the interior of Antarctica. This may be true also during a drastic man-made 21st century warming. But even in the (not expected) case of melting, the annual net loss of the Antarctic ice volume cannot be more than, at a maximum, 5 cm/a, because of the high albedo and the intense (negative) heat content of this vast ice body. Thus the time necessary for melting of the East Antarctic ice-dome must be at least (!) some 40 000 years, equivalent to an eustatic sea level rise of less than 2 mm/a (at present 1.2 mm/a).

In this context the possibility of large-scale Antarctic surges has to be considered seriously (see Appendix I and IV). In several papers, Hughes (1973; 1975; 1977) has discussed evidence and mechanism of West-Antarctic ice surges; if man-made warming reaches the expected level, it may drastically increase to an ice-volume not observed since the late Tertiary. Such macro-surges at a scale of 10^6 km^3 may occur at time distances of the order 10^5 years, i.e. they are probably rare events contributing to, but not solely responsible for the climatic history of the Pleistocene. There is evidence that an equilibrium stage of the ice-dome as prerequisite of a macro-surge has still not been reached. Under such circumstances it may be safely stated that the East Antarctic ice-dome will remain constant, not only during the next century but during the next millenium and probably much longer. A macro-surge from West-Antarctica, however, cannot be excluded, especially not after an increased snow-load caused by man-made warming. As a consequence the climatic asymmetry as described above will further control the global circulation mainly because the East Antarctic ice-dome can be considered as stable.

APPENDIX

I. For obvious reasons it is quite difficult to obtain convincing evidence for the occurrence of large-scale surges in the sense of Wilson, with a magnitude of 1–2×10^6 km^3 (Hollin, 1976). Such macro-surges are rare events and are difficult to verify because of their relatively short duration. The above-mentioned estimate of their magnitude would be equivalent to a eustatic sea-level rise of 2.8–5.6 m. The comparatively high speed of formation and decay of Northern Hemisphere glaciations in a time-span of 5–8 000 years – Barbados III, building of Würm III after 25 000 BP, decay of Würm III between 14 500 BP and 6 500 BP – indicates a short duration of such a surge-initiated flood. This time-scale suggests an annual accumulation of ~ 30 cm water equivalent (which is not unreasonable, with summer temperatures around 0° C) over an area of 12.10^6 km^2 (40 percent of the glaciated area of the northern continents); then in one year $3.6 = 10^{12}$ m^3 could be stored in form of ice, or 10^6 km^3 in about 280 years. Thus the duration of a flood of the above-mentioned scale would be in the order of 300–500 years only.

Such estimates are of some importance when discussing the possibility of a West-Antarctic macro-surge in a foreseeable future, perhaps after a drastic CO_2-produced warming of the whole atmosphere causing a drastic reduction also of the Antarctic winter sea-ice, which could certainly lead to greatly increased snowfall at the margins of the Antarctic ice-cap (National Science Foundation).

II. Hays et al. (this volume) have indicated that within the general synchroneity of climatic change in the last 0.7 MY from Arctic to the Subantarctic there exists a delay of $\sim 3\,000$ years in the response of the global ice volume compared with the Subantarctic sea surface temperature variations. This delay can be interpreted as caused by the necessary duration of the building (and decay) of Northern continental ice-domes (see Appendix I) in contrast to the immediate action at the drifting Subantarctic sea-ice. In turn, this effect may be also responsible for the earlier evolution of the Southern Hemisphere climatic optimum (starting about 11 000 BP, maximum 9400 ± 600 BP, see Mercer, this volume, simultaneous with the beginning of the Younger Dryas Stadial and with the cool Preboreal).

III. The occurrence of asymmetric climatic features near the equator – especially the great role of equatorial upwelling in the Atlantic and Pacific, together with the increasing coastal upwelling along the West coast of South America (Humboldt Current) – at 18 000 BP indicates the probability that the Southern Hemisphere has remained still cooler than the heavily glaciated Northern Hemisphere. If during winter the Subantarctic drift-ice has reached Lat. 45–50° S (J.D. Hays in this Volume), then the global asymmetry during northern summer must have been still serious. This latitude must be compared with an extension of polar water with wintertime sea-ice in the northern Atlantic up to about Lat. 43° N, similarly the Laurentide ice-lobe, but with a distinctly more northerly position of permanent snow in Siberia and of polar water and sea-ice in the Pacific, i.e. over about 170° longitude. At both hemispheres, the regional climatic difference between East and West coasts must have been increased. This may explain the marked difference between glaciated Marion Island (45–47° S) and unglaciated Falkland Island (52–53° S) during the last glaciation, in spite of the strong temperature gradient of the Atlantic surface waters. Apparently the position of the Antarctic Convergence has remained orographically fixed in the vicinity of Drake passage, but swings freely in other longitudes.

IV. Medium-scale West-Antarctic surges (or better: intensified calving) with an ice-volume of $1\text{–}2 \times 10^5$ km^3, as found by A.T. Wilson (this Volume), may be much more frequent, at a time scale of $10^2\text{–}10^4$ years. Lamb (1967) and the author (1969) have discussed some evidence in the 19th century, notably around 1840 and 1895, with periods of marked climatic anomalies reflected in a cool Arctic and in a drier climate in some tropical continents. It can by no means be excluded that such events contribute much to the variability of the climate at a time-scale $10^2\text{–}10^4$ years, which is attributed, by many climatologists (Bray 1974; 1976; 1976; 1977) to discontinuous phases of intense eruptive volcanism. Their eustatic effect will be comparatively small (10^5 km^3 equivalent to 28 cm) but not negligible – it may remain most probably undetected before the time of instrumental observations. Their climatic effect (Flohn, 1974) – especially via large-scale decrease of sea-surface temperatures – could be rather worldwide, at a time-scale of one or several decades.

DISCUSSION

C. Lorius: Could you say a few words about the role of possible Antarctic surges on the climatic evolution.

H. Flohn: In several papers T. Hughes (1973, 1975, 1977) has investigated the mechanism of large-scale surges in the Ross Ice Shelf area. They seem to be rare events, with a time distance of several 10^5 years. Unfortunately no investigation has been made in the Weddell Sea area, which is apparently more productive. Hollin (1977) has presented new local evidence for an abrupt eustatic sea-level rise at the end of the last glaciation.

Apparently large-scale surges (ice volume $1\text{–}2.10^6$ km^3) are *possible*, with a sea-level rise of 3–6 m; more detailed studies of their time-scale and their time-distance are needed. Evidence is now available of at least 6–8 abrupt coolings of hemispherical scale since about 230 000 years (Holstein interglacial). Their interpretation from the view-point of physical climatology is ambiguous: either by large-scale Antarctic surges, and or (preferably) by clustering volcanic events (Flohn, in print, 1977), both at irregular time distances.

REFERENCES

Alyeae, F.N. 1972. *Numerical simulation of the Ice-Age paleoclimate.* Colorado State University, Atmos. Sci. Paper 193.

Berger, A. Support for the astronomical theory of paleoclimatic change (submitted for publication in *Nature*).

Borzenkova, I.I. et al. 1976, Izmeneniye tempertur yi; Wozducha Severnogo Poluschariya za period 1882–1975. *Meteor. i. Gidrol.* (7): 27–35.

Bray, J.R. 1977. Pleistocene volcanism and glacial initiation. *Science* 197 (1977): 251–254; see also *Nature* 254 (1974): 679–680, 260 (1976), 414 and l.c. 262 (1976): 300.

Coughlan, M.J. 1975. Paper read at the conference on climate and climatic change, Monash Univ. Dec.

Damon, P.E. & Kunen, St. M. 1976. Global cooling? *Science* 193: 447–453.

Flohn, H. 1964. Grundfragen der Paläoklimatologie im Lichte einer theoretischen Klimatologie. *Geol. Rundschau* 54: 504–515.

Flohn, H. 1967. Bemerkungen zur Asymmetrie der atmosphärischen Zirkulation. *Ann. Meteor.* N.F. 3: 76–80.

Flohn, H. 1969. Ein geophysikalisches Eiszeit-Modell. *Eiszeitalter und Gegenwart* 20: 204–231.

Flohn, H. 1972. Investigations of equatorial upwelling and its climatic role. In: *Studies in physical oceanography – A tribute to George Wüst on his 80th birthday*, Gordon & Breach, New York & London, Vol. 1: 93–102.

Flohn, H. 1974. Instabilität und anthropogene Modifikation des Klimas. *Ann. Meteor.* N.F. 9: 25–31.

Flohn, H. 1974. Background of a geophysical model on the initiation of the next glaciation. *Quat. Res.* 4: 385–404.

Gates, W.L. 1976. The numerical simulation of Ice-Age climate with a global general circulation model. *J. Atmos. Sci.* 33: 1844–1873.

Hays, J.D., Imbrie, J. & Shackleton, H.J. 1976. Variations in the Earth's orbit: Pacemaker of the Ice-Ages. *Science* 1121–1131.

Heine, K. 1974. Bemerkungen zu neueren chronostratographischen Daten zum Verhältnis glazialer und pluvialer Klimabedingungen. *Erdkunde* 28: 303–312.

Heirtzler, J.R. The evolution of the southern oceans. In: L.Q. Quam (ed.), *Research in the Antarctic*. Amer. Assoc. Adv. Sci., Washington, DC, Publication 93: 667–684.

Henning, D. 1967. Zur Interpretation eines Zirkulationskriteriums. *Arch. Meteor. Geophs. Biokl.* A 16: 126–136.

Herman, Y. 1977. *X. INQUA congress, Birmingham, Abstracts* p. 203.

Hoinkes, H. 1961. Die Antarktis und die geophysikalische Erforschung der Erde. *Naturwiss.* 48: 354–374.

Hollin, J.T. 1977. Thames interglacial sites, Ipswichian sea levels and Antarctic ice surges. *Boreas* 6: 33–52.

Hsü, K.J. et al. Late Miocene desiccation of the Mediterranean. *Nature* 242 (1973): 240–244; cf. also *Naturwiss.* 64 (1974): 137–142; *Nature* 267 (1977): 399–403.

Hughes, T. 1973. Is the West Antarctic ice sheet disintegrating? *J. Geophys. Res.* 78: 7884–7910.

Hughes, T. 1977. West Antarctic ice streams. *Rev. Geophys. Space Phys.* 15 (1977): 1–46; also l.c. 13 (1975): 502–526.

Kellogg, W.W. 1973. Climatic feedback mechanism involving the polar regions. In: *Climate of the Arctic*, 24th Alaska Science Conference, Fairbanks: 111–116.

Kennett, J.P. & Thunell, R.C. 1975. Global increase in Quaternary explosive volcanism. *Science* 187: 497–503.

Korff, H.Cl. & Flohn, H. 1969. Zusammenhang zwischen dem Temperaturgefälle Äquator-Pol und den planetarischen Luftdruckgürteln. *Ann. Meteor.* N.F. 4: 163–164.

Kozur, H. 1976. Paläontologische, paläogeographische und paläoklimatologische Kriterien der Globaltektonik. Franz. Kossmat Symposium, *Nova Acta Leopoldina* N.F. 224: 413–472.

Lamb, H.H. 1959. The southern westerlies: a preliminary survey, main characteristics and apparent associations. *Quart. J. Roy. Meteor. Soc.* 85: 1–23.

Lamb, H.H. 1967. On climatic variations affecting the far South. *WMO Techn. Note* 87: 428–453.

Limbert, D.W.S. 1974. Variations in the mean annual temperature for the Antarctic Peninsula, 1904–1972. *Polar Record* 17: 303–306.

National Academy of Science. 1975. *Understanding climatic change.* A program for action, Washington, D.C. 239 p.

Rognon, P. & Williams, M.A.J. 1977. Late Quaternary climatic changes in Australia and North Africa: A preliminary interpretation. *Palaeogeography, Palaeoceanography, Palaeoecology* 21: 285–327.

Salinger, M.J. & Gunn, J.M. 1975. Recent climate warming around New Zealand. *Nature* 256: 396–398.

Saltzman, B & Vernekar, A.D. 1975. A solution for the Northern Hemisphere zonation during a glacial maximum. *Quat. Res.* 5: 307–320.

Schwerdtfeger, W. 1970. The climate of the Antarctic. *World survey of climatology*, Vol. 14: 253–355.

Smagorinsky, J. 1963. General circulation experiment with the primitive equations (Appendix B). *Monthly Weather Review* 91: 159–162.

Van Loon, H. et al. 1972. Meteorology of the Southern Hemisphere. *Meteor. Monogr. Am. Meteor. Soc.* 35.

Van Loon, H. & Williams, J. 1977. The connection between trends of mean temperature and circulation at the surface; Part IV. Comparison of surface changes in the Northern Hemisphere with the upper air and with the Antarctic in winter. *Monthly Weather Review* 105: 636–647.

Vowinckel, E. & Orvig, S. 1970. The climate of the North Polar Basin. *World survey of climatology*, Vol. 14: 129–252.

Wigley, T.M.L. 1976. Spectral analysis and the astronomical theory of climate change. *Nature* 264: 629–631.

Williams, J., Barry, R.G. & Washington, W.N. 1974. Simulation of the atmospheric circulation using the NCAR global circulation model with Ice-Age boundary conditions. *J. Appl. Meteor.* 13: 305–317.

* 2 *

Glacial history of New Zealand and the Ross Dependency, Antarctica

A. T. Wilson

School of Science, University of Waikato, Hamilton, New Zealand

Manuscript received 18th November 1977

CONTENTS

ABSTRACT

The glacial history of the New Zealand sector of the Southern Hemisphere is reviewed. The history of the Last Glacial in New Zealand seems to be broadly similar to that of Europe and North America, except that New Zealand appears to have emerged somewhat earlier than Europe and considerably earlier than North America. Possible reasons for this are discussed. In the New Zealand sector it appears that the latitudinal temperature gradient increased at 38° S latitude with the glacial cooling in northern North Island being only 2° Celsius, which was considerably less than the 5 tot 6°C cooling in southern New Zealand.

The glacial history of the New Zealand sector of Antarctica is reviewed and it is shown that glaciations in Antarctica are more complicated than in the temperate regions of the world and not simply related to climate cooling. It is concluded that there are four separate superimposed sets of glacial events ultimately related to one another and to events in the rest of the world. Some of the glacial advances are clearly out of phase with the major glacials of late Cainozoic time, and the fascinating question is whether the Antarctic glacial system is being driven by the climate of the rest of the world or vice versa.

THE GLACIAL HISTORY OF NEW ZEALAND

Introduction

The glacial periods of the Pleistocene have left spectacular geomorphological features on the land forms of New Zealand in the form of glacially cut valleys and moraines. The southwest corner of the South Island is called Fiordland and consists of fiords as magnificent as those of Norway. The Canterbury Plains, with an area of 8 000 km^2 have been built from glacial debris during the cold periods of the Middle and Late Pleistocene. These gravels are many hundreds of metres thick. The loess deposited during glacial periods provide much of the parent material for the agricultural soils of central and southern New Zealand.

Early glacial events

Ross Glaciation: This glacial event was first recognised as glacial beds in a conglomerate sequence more than 30 m thick near the town of Ross on the

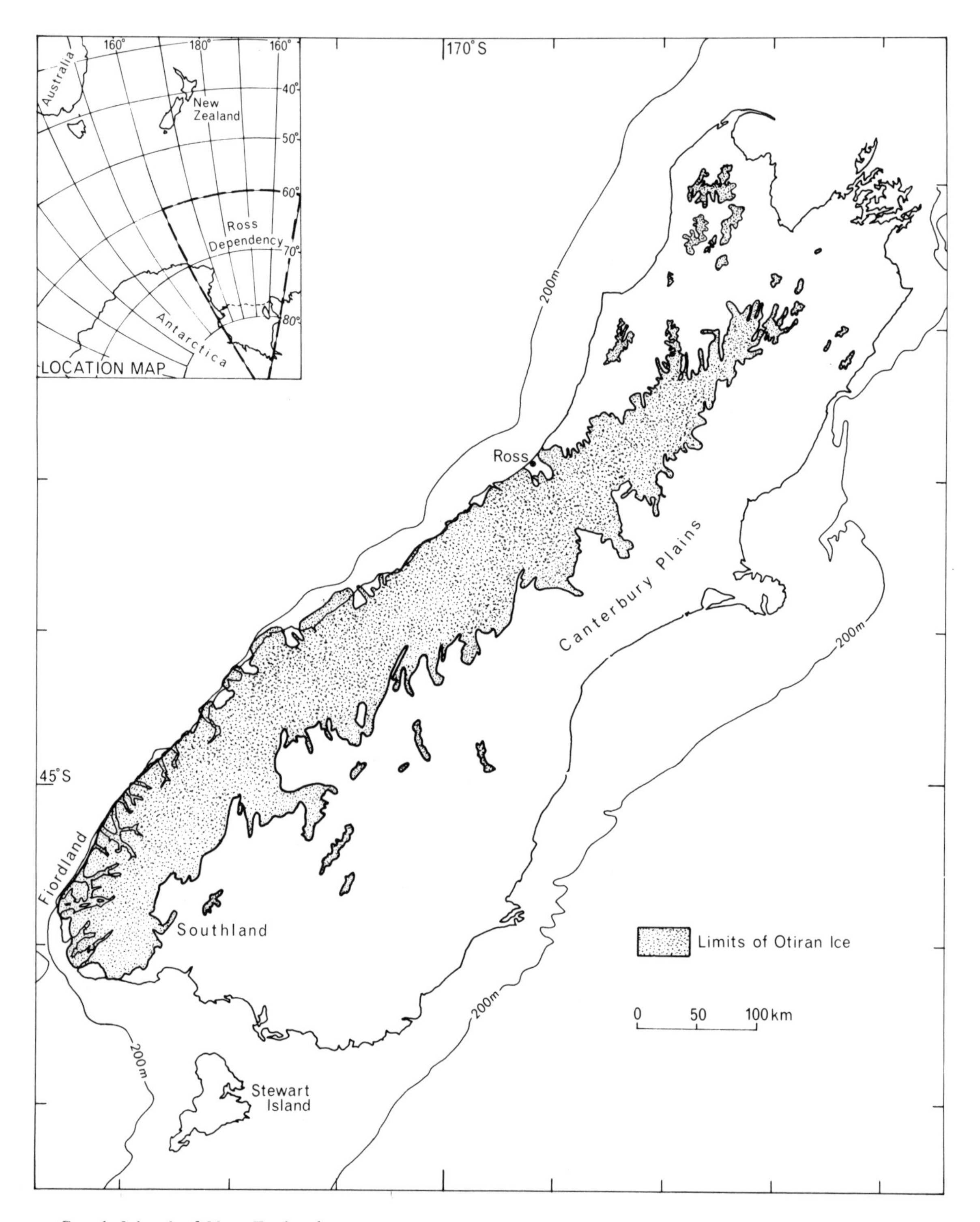

a. South Island of New Zealand

Fig. 1. Map of New Zealand showing distribution of ice and forest during the height of the last major glaciation (Otiran)
(Distribution of ice based on N.Z. Geological Survey, Miscellaneous Series Map 6: Quaternary Geology, South Island. D.S.I.R., Wellington, New Zealand, 1973).

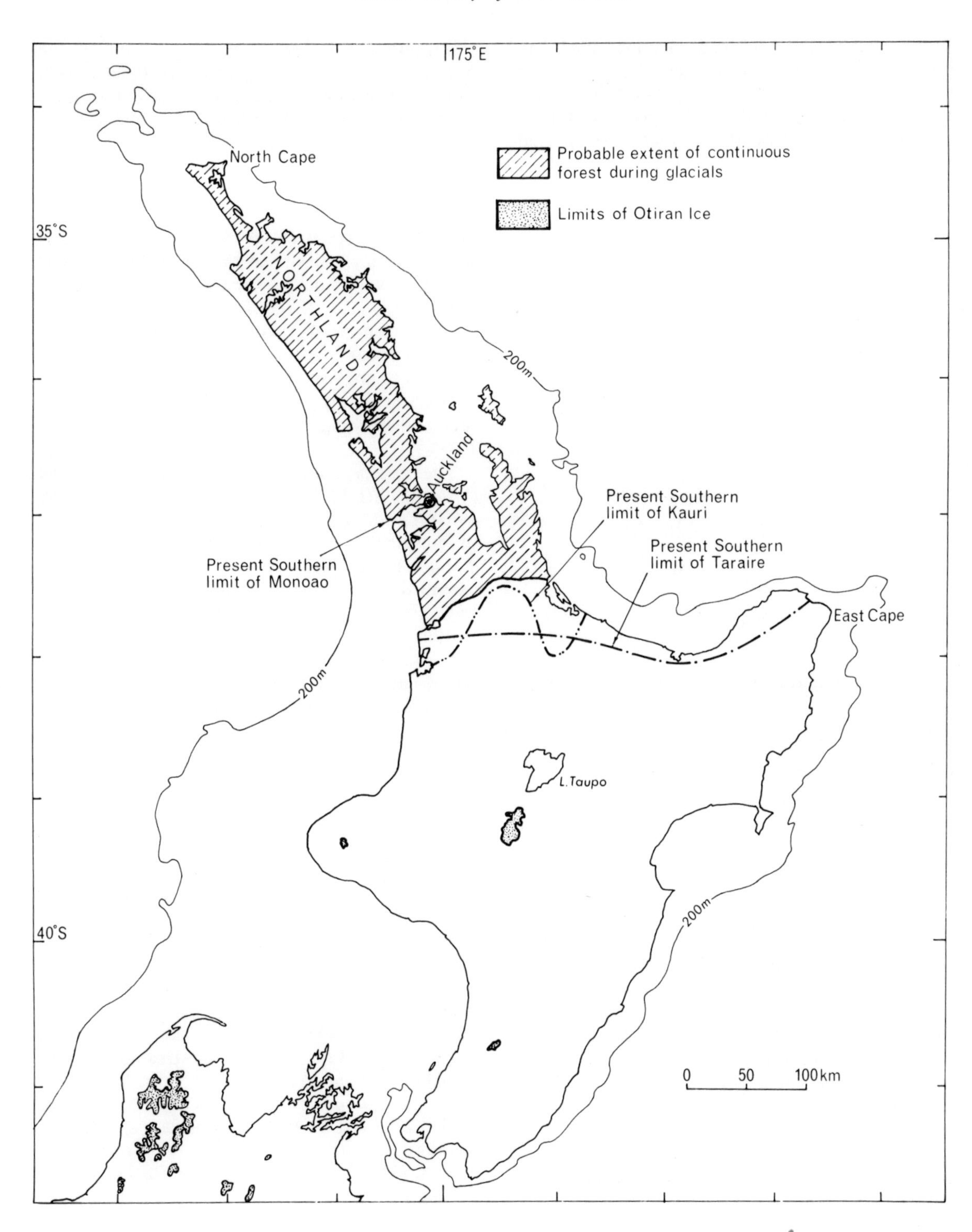

b. North Island of New Zealand

West Coast of the South Island (Fleming, 1956). The beds have been described by Gage (1961) and more recently by Bowen (1967). The considerable degree of weathering suggests that the Ross Glaciation is considerably older than the four later glacials and is generally believed to be early Pleistocene or Late Pliocene in age.

Late Pleistocene glacial history

The abundant evidence in the forms of moraines has enabled the mapping of the extent of the ice during the last few glaciations.

Five glaciations are recognised in the South Island of New Zealand. They are (after Suggate, 1965):

Glaciation	*Interglacial*
Otira	
	Oturi
Waimea	
	Terangi
Waimaunga	
	Waiwhero
Porika	
Very long interval	
Ross	

The four more recent glacial events are generally believed to be the equivalents of the classical four major late Pleistocene glaciations recognised in Europe and North America. The evidence is not as clear as it is in the Northern Hemisphere and, except for the latter stages of the Otiran, no dating control is available. Even the existence of the Waimea is a matter of controversy (Soons, 1966).

Material for radiocarbon dating is much less abundant than in Europe and North America. However the last two advances of the Last Glacial have been dated by radiocarbon and it is evident that their ages correspond to advances in the Northern Hemisphere, supporting the rather obvious meteorological conclusion that the glacial periods are broadly synchronous in both the Northern and Southern Hemispheres. Other dating techniques have been inapplicable, and the next advance in this field will probably come from ocean cores off the coast of New Zealand.

For details of the sequences of deposits from successive advances and retreats of glaciers from the various areas, see:

Waimakariri Basin	Gage (1958)
Hurunui Valley	Powers (1962)
Waiau Valley	Clayton (1968)
Rakaia Valley	Soons (1963)
Lake Pukaki area (Mackenzie Basin)	Speight (1963)
Upper Clutha Valley	McKellar (1960)

Suggate (1965) has given an account of the Nelson-Marlborough area, North Westland, as well as a general review of the northern half of the South Island. Other small areas are described by Fitzharris (1967) and Rains (1967).

Climate of New Zealand during the glacial periods

Willett (1950) reviewed the evidence of Pleistocene snow lines and concluded that New Zealand was some 6° C cooler during the glacial periods than at present.

The pollen record is complicated by the need to interpret interaction between *Nothofagus* and podocarp-mixed angiosperm forest. So far it has not been possible to distinguish clearly climate change other than the warming at the end of the Otiran (Moar, 1971).

The vegetational changes have been studied (Wardle, 1962; Molloy, 1969; Moar, 1971; McGlone, 1973). During glacial times there was a general lack of forest in the South Island except for refuges in the extreme north and south and probably also on the extended coastal plains exposed by lowered sea level. The northern part of the North Island above latitude 38° S was probably forested throughout the last glaciation, but the southern part was not. Afforestation of the southern area began about 14000 years BP (McGlone, 1973).

Temperature gradients during interglacial times – The glacial climate of Northland

The view currently held by most workers in the late Pleistocene is (following Willett, 1950) that the whole of New Zealand was some 5 to 6° C cooler during the glacial periods. The range of mean temperature from North Cape to Southland is also 5 to 6° C. The botanical implication is that all plants presently growing in Northland should be capable of growing and reproducing in the southern part of New Zealand. In fact, there are many plants found only in New Zealand that are only found in the northern part of the North Island. Examples would include (see Allan, 1961):

Kauri *(Agathis australis)* – found from North Cape (34° 22′ S lat.) to Maketu/Kawhia (38° S lat.)

Taraire *(Beilschmiedia tarairi)* – found from North Cape (34° 22′ S lat.) to East Cape/Raglan (37° 50′ S lat.).

Litsea calicaris – found from North Cape (34° 22′ S lat.) to 38° S lat.

Tawari *(Ixerba brexioides)* – Hokianga (35° 30′ S lat.) to a little south of 38° S lat.

Monoao *(Dacrydium kirkii)* – found from Hokianga (35° 30′ S lat.) to Auckland city (lat. 37° S).

Toatoa *(Phyllocladus glaucus)* – found from lat. 35° S to lat. 38° S.

Monoao *(Dacrydium kirkii)* is perhaps a good case to illustrate the problem. Under present conditions Monoao does not grow well in Northland although at times (presumably warmer) in the Holocene, extensive forests have existed as far south as lat. 37° 50′ S (A.T. Wilson, unpublished work). It can be found growing at present in favoured sites as far south as Auckland city (lat. 37° S). The question of how such a species could survive the glacial periods presents a problem. If under present conditions Monoao has difficulty surviving as far south as Auckland city it seems unreasonable to invoke survival based on refugia and micro climate. A very simple-minded interpretation of the botanical data would suggest that the temperature of the most northern part of Northland at the time of low glacial sea level (see Fig. 1b) was never lower than the present temperature of Auckland.

The implication is that whereas the temperatures in the southern part of the North Island and the South Island dropped by some 6°C during glacial times the temperature of northern Northland dropped by no more than 2°C. This would also imply a steepening of the latitudinal temperature gradient at latitude 38° S with a consequent increased atmospheric circulation.

The hypothesis presented above is rather radical but is the only evidence we have at present for the glacial climate in Northland during glacial times. Interestingly enough, recent work on ocean cores, particularly in the Tasman Sea to the west of northern New Zealand, give strong evidence that the above hypothesis is in fact correct (pers. com. T.C. Moore, 1977 (CLIMAP Project); see also CLIMAP, 1976). There is clearly a need for further work on this problem.

Isotopic evidence

The lack of datable material and the difficulty of determining past climate in non-glaciated areas led New Zealand workers to seek alternative techniques. Hendy and Wilson (1968) developed a technique for obtaining past climate information from cave formations (speleothems). Fig. 2 gives a climate curve for the last 35 000 years of the Waitomo area. The dating at present is limited to that attainable by C-14 dating techniques. Unfortunately New Zealand speleothems are unusually low in Uranium which makes Uranium/Thorium dating difficult. Strictly the isotopic data from speleothems gives temperature, only if the isotopic composition of the precipitation is known or can be inferred. This depends on latitudinal temperature gradients which, as suggested above, may change dramatically during glacial periods.

Neoglacial events in New Zealand

There is little doubt that the glaciers in New Zealand have advanced from time to time during the Holocene – see for exampe Burrows (1977). Wilson et al. (1973) have established, by isotope measurements on speleothems, that the temperature curve for New Zealand for the last 1 000 years is broadly similar to that of England, and that New Zealand was subjected to the Little Ice Age and the Medieval Warm Period. Work is progressing in southern New Zealand by continuing studies of glacial moraines. In northern New Zealand studies of wood in the extensive peat swamps (Wilson, unpublished work) is enabling the southern limit through time of climatically sensitive plant species, such as Monoao, to be determined.

Relationship to other parts of the world

It is clear from evidence from the last two advances of the Otiran Glaciation, and from general meteorological considerations, that the glacial events in New Zealand are broadly in phase with those in the Northern Hemisphere.

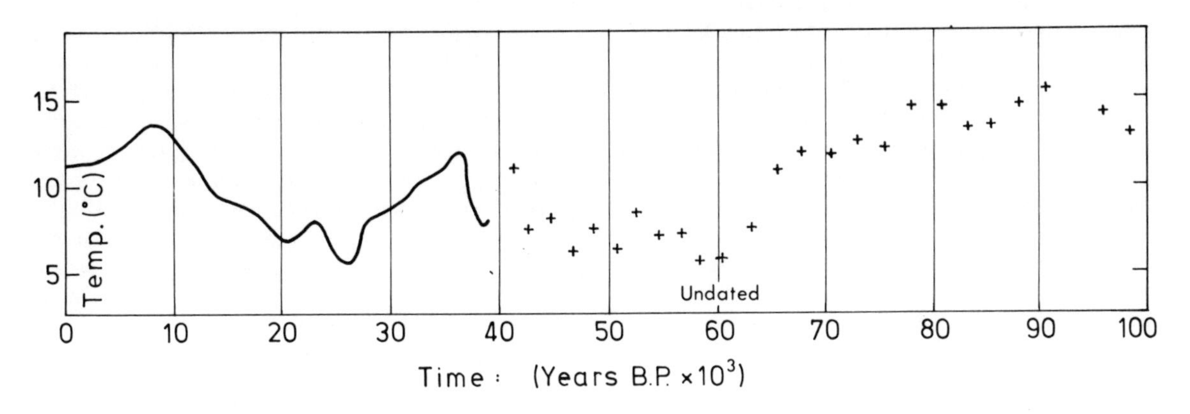

Fig. 2. Climate curve for New Zealand as derived from isotope measurements on speleothems (after Hendy & Wilson, 1968)

Isotopic and palaeobotanical evidence suggests that New Zealand emerged from the Last Glacial period rather earlier than Europe or North America. This is probably because the climate of New Zealand was dominated by events at high southern latitudes. For example, during the glacial periods the sea ice in the Southern Ocean must have come very close to New Zealand. This in itself would have caused considerable cooling due to the regional albedo change. The tree line in southern South Island is currently at 1 000 metres yet only 650 km to the south of Campbell Island the shrub line *(Dracophyllum)* is at only 150 m.

The data from New Zealand suggests that the last glaciation may have ended first at high latitudes in the Southern Hemisphere. Work in Australia (J. Bowler pers. com., 1975) suggests that the climate break there was also very early and certainly well before that in Europe and North America.

When it is considered that from a palaeobotanical point of view the break in climate came in New Zealand 14 000 years BP, in Europe 10 000 years BP, and in southwestern North America 8 000 years BP (Van Devender, 1976), it suggests that too much emphasis has been placed on dates and events from atypical areas whose climate is dominated by large ice sheets rather than the more typical areas of the earth. It suggests also that the driving force for the glacial periods is probably at high southern latitudes.

GLACIAL HISTORY OF THE ROSS DEPENDENCY, ANTARCTICA

Causes of glaciations in polar deserts

In the non-polar areas of the earth, glaciations are usually caused by the lowering of the summer snow line, which leads to the accumulation of ice to form glaciers. This is usually induced by a climatic cooling. In Antarctica, where summer temperatures never rise above 0°C for more than a few days each summer, glaciation is controlled by ice budgets. Ice can be removed by flowing away from the area of interest, or by evaporation. To illustrate the first, consider the lowering of sea level which could ground a floating ice shelf and prevent wasting of an ice sheet by iceberg formation. Since ice can no longer be lost as rapidly as before, the lowering of sea level will lead to the extension of the ice sheet (Hollin, 1962). For example, in the Ross Dependency at times of low sea level, we have the grounding of the Ross Ice Shelf with the consequent extension of the West Antarctic ice sheet.

An example of the second phenomenon is the local alpine glaciers. These depend for their nourishment on the local precipitation on to their névés. If precipitation is reduced – caused for example by an increase of sea ice, or the filling of McMurdo Sound with an ice sheet as described above – these will retreat. If the reverse happens, they will advance.

The Ross Dependency

In Ross Dependency we have a coastal mountain range (the Transantarctic Mountains) damming the side of the East Antarctic ice sheet (Fig. 3). At points along the mountain chain, ice from the East Antarctic ice sheet breaks through and flows down valleys to join the floating Ross Ice Shelf. The Ross Ice Shelf is also fed from the West Antarctic ice sheet. Precipitation on the surface of the Ross Ice Shelf is a very important source of nourishment: so much so, that, at the carving line, only ice formed from precipitation on the ice shelf itself is found in the icebergs that carve from the front: all the glacier ice is lost by melting from the bottom of the ice shelf.

Evidence of very early glacial history in the Ross Dependency

The Ross Dependency abounds in fossil cirques. For example, the large cirques cut into the ridges above the Wright Valley in the McMurdo Oasis region. Present theories on the origin of cirques imply that these cirques were cut by temperate ice. Simple plate tectonic considerations suggest that it has been millions of years since temperatures in this region could have been warm enough for temperate alpine glaciation. These considerations, together with palaeomagnetic data, suggest that the Victoria Land region of Antarctica was subjected to a temperate glaciation in the Eocene and Lower Oligocene (Wilson, 1973). It was probably at this time that the major valley systems were cut.

Investigations of the Late Pleistocene glacial history of the Ross Dependency

Studies of the glacial history of the Ross Dependency were initially hampered by a failure to realise that glaciations in the Antarctic were controlled by different mechanisms to those in more temperate areas. Initially it was attempted to force the more obvious glacial deposits into the classical glacial periods of North America and Europe. Wilson (1962; 1967) pointed out that on geochemical evidence a through glacier could not have passed through the Wright Valley during Wisconsin time. Hollin's (1962) analysis of the effect of sea levels on an ice sheet provided the mechanism for glacial advances from McMurdo Sound. Wilson (1960), and Denton et al. (1970; 1971) proposed that the glacial history of the Ross Dependency should be divided into three separate glacial sequences: local

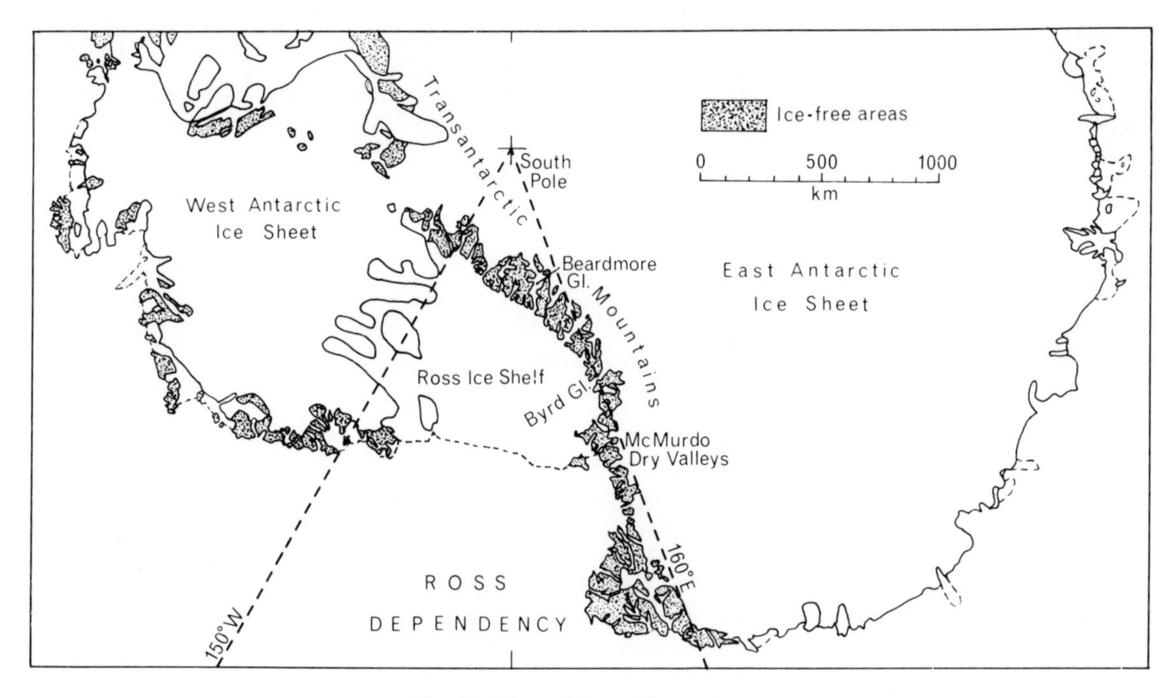

Fig. 3. Map of Ross Dependency

alpine glaciations, Polar Plateau fed glaciations and glaciers from an ice sheet in McMurdo Sound.

There have been two principle lines of attack for studying past glaciations in the Antarctic.

1. The classical approach of mapping moraines.

2. Studying lakes and lake deposits in order to gain information on past climate. This is complementary to the first approach since, as stated above, glacial events in Antarctica are usually caused by climatic events other than a simple climatic cooling.

The first approach is bedevilled by the fact that most Antarctic glaciers are of the polar type whose bases are frozen to the underlying material – all movement taking place within the glacier itself. Such glaciers do no cutting and the moraine that they carry is largely rock debris that has fallen on to their surfaces from the surrounding mountains. Thus, past glacial advances sometimes leave only a thin veneer of material occupying a small percentage of the surface and are recognisable only if exotic rock types are present. For example, the Kenyte in the lower Taylor Valley enables one to map glacial advances from McMurdo Sound. It is recognisable only because Kenyte is a very unusual rock not found in the Taylor Valley. In some cases, for example, the Darwin Mountains and the Erehwon Dry area, a past glacial advance is marked by a wall of rock at some distance from the glacier front. One is often quite unable to distinguish between the surfaces on either side of these moraines. If the rock wall were not present the glacial advance would have left no recognisable record. In some cases the rock wall moraine has been laid across an old stream bed, and the stream bed inside the moraine wall has survived intact the advance and retreat of the glacier. A particularly good example of this phenomenon is the stream bed from Nussbaum Riegel that is intercepted by rock wall moraine in front of the Canada Glacier in the Taylor Valley, McMurdo Oasis.

Despite the remarkable resistance of land surfaces to erosion by polar glaciers in the Antarctic there are large deposits of glacial till which can be shown by U/Th dating to have been formed during the last few hundred thousand years (Hendy et al., 1977). How can this paradox be resolved? It is the author's opinion that in the Antarctic these large till deposits are always associated with past bodies of liquid water, and that all glacial deposits of significant volume in the Antarctic are the result, not only of glacial advance, but of a glacial advance into or through a lake (or the ocean). In Antarctica, the lakes are usually perennially ice covered. The ice is very transparent to light because it consists of large vertically orientated crystals formed as a consequence of continuous growth due to ablation from the surface during summer and crystal growth during winter. Sufficient light energy penetrates the ice to warm the underlying water. If the lakes are filled with fresh water the bottom waters can be warmed as high as 4°C. This in time melts the underlying frozen ground. Many present Antarctic lakes contain saline water which can be solar heated and temperatures can rise as high as 26°C (Wilson &

Wellman, 1962). Any water entering the lake at temperatures above 0°C will finger to their appropriate density level. The physics of Antarctic lakes has been reviewed by Wilson (1967). Clearly if a glacier advances through such a lake the unfrozen material can be readily moved by an advancing glacier.

Aragonite and gypsum are deposited in the lakes and these can be dated using U/Th dating techniques (Hendy et al., 1976). It is this latter development that has provided the dating control so vital to the building of any glacial chronology.

Pleistocene glacial history of the Ross Dependency

The glacial history of this portion of Antarctica is becoming clear, at least in broad outline. There appear to be four quite separate but ultimately related superimposed sets of glacial events.

1. Glaciations from the East Antarctic ice sheet. It appears that at intervals of 100 000 years (probably towards the end of each interglacial) the East Antarctic ice sheet thickens (Hendy et al., 1977) and the flow of ice in the outlet glaciers feeding into the Ross Shelf increases. Ice flows over the rock thresholds and into the McMurdo Dry Valleys.

There are currently two theories on the behaviour of the East Antarctic ice sheet during Cainozoic times. Scott (1905) suggested that the ice sheets thicken due to increased precipitation during warmer (interglacial) periods. On this model the ice sheet would slowly respond to glacial vs interglacial climate, thickening during interglacials and thinning during glacials. Wilson (1964) presented the hypothesis that the East Antarctic ice sheet was unstable against surging,and built up in interglacial times eventually to collapse, causing the glacial period.

In both views the East Antarctic ice sheet is seen to rise during interglacial times, and this is supported by the fact that the present ice budget of the West Antarctic ice sheet is strongly in credit (Giovinetto et al., 1966).

These glaciations appear to have happened at least in each of the past few interglacials and probably all through the Quaternary and late Tertiary times. During this time the land has been slowly rising with respect to sea level and each successive glacial advance has moved a shorter distance down the valleys. Fiords become valleys with through glaciers, and valleys with through glaciers become dry valleys. Till bodies are laid down as a result of the glacier advancing through the unfrozen material under a lake or the sea. Type examples of the sequence are as follows:

- Ferrar Valley - Presently a fiord occupied by the Ferrar Glacier which is floating in its lower reaches.
- Taylor Valley - Upper Taylor Valley occupied by the Taylor Glacier which only penetrates part way through the Valley.
- Wright Valley - a very small amount of Polar Plateau ice penetrates the valley via the Airdeveronsix Ice Falls, but except for local alpine type glaciers the valley is ice free.

The East Antarctic ice sheet glaciations leave evidence of two types. If they enter a dry valley which does not contain a lake of liquid water then we have a cold based glaciation with very little moraine recording the event. Examples are the upper Wright and Victoria Valleys. If, on the other hand, they enter a dry valley which contains a lake of liquid water, they are able to erode the underlying unfrozen material and can lay down till bodies which record their advance. Such a case is the Taylor Valley where advances at 100 000, 200 000 and 300 000 years have been recorded (Hendy et al., 1977).

2. Glaciations from the West Antarctic ice sheet.

a) At each lowering of world sea level, the West Antarctic ice sheet expanded on to the continental shelf. This 'backed up' the outlet glaciers which have been flowing through the Transantarctic mountains into the Ross Ice Shelf. In some cases the expanded West Antarctic ice sheet flowed into the dry areas. In such cases they can, being wet based, erode and also push quantities of unfrozen marine sediments into the valleys. Examples of this can be found in the lower ends of the Waikato Valley in the Browns Hills Dry Area and the Taylor Valley in the McMurdo Oasis.

b) It is suggested (Wilson in this volume) that each rise in sea level renders the West Antarctic ice sheet unstable and that periodic surging results. These surges have produced moraines by pushing unfrozen material from the bottom of the sea into the Brown Hills Dry area.

3. Alpine glaciers. The alpine glaciers advance and retreat according to the precipitation on their névés. This is, in turn, controlled by the distance between their névés and the open sea. They have advanced and retreated according to the position of the Ross Ice Shelf and size of the West Antarctic ice sheet.

It can be seen that the glacial history of the Ross Dependency is complicated on the one hand by having ice advances controlled by different mechanisms (the thickness of East Antarctic ice sheet and world sea levels). On the other hand, because of tectonic uplift, a large number of glacial events are recorded, some of great antiquity. The interrelation of the glacial events in Antarctica with events in the rest of the world is indeed a fascinating study. The glacial history of the Antarctic is of considerable importance in that it is probable that at least some global climatic events have causes related to processes taking place in Antarctica.

REFERENCES

Alan, H.H. 1961. *Flora of New Zealand*, Vol. 1, Govt. Printer, Wellington, New Zealand, 1085 pp.

Bowen, F.E. 1967. Early Pleistocene glacial and associated deposits of the West Coast of the South Island, New Zealand. *N.Z. J. Geol. and Geophys.* 10: 164–181.

Burrows, C.J. 1977. Late Pleistocene and Holocene glacial episodes in the South Island, New Zealand and some climatic implications. *N.Z. Geographer* 33: 34–39.

CLIMAP. 1976. The surface of the Ice-Age Earth. *Science* 191: 1131–1137.

Clayton, L. 1968. Late Pleistocene glaciations of the Waiau Valleys, North Canterbury. *N.Z. J. Geol. and Geophys.* 11(3): 753–767.

Denton, G.H., Armstron, R.L. & Stuiver, M. 1970. Late Cenozoic glaciation in Antarctica: The record in the McMurdo South Region. *Antarctic J. U.S.* 1: 15–21.

Denton, G.H., Armstrong, R.L. & Stuiver, M. 1971. The Late Cenozoic glacial history of Antarctica. In: Turekian, K.K. (ed.), *The Late Cenozoic Glacial Ages*, Yale Univ. Press, New Haven. 267–306.

Fitzharris, B.B. 1967. Some aspects of the Quaternary geomorphology of the mid-Waiau and Monowai Valleys, South Island, New Zealand. *Proc. 5th N.Z. Geography Conf.*: 181–189.

Fleming, C.A. 1956. Quaternary geochronology in New Zealand. *Congress Internat. Quat., Roma-Pisa, 1953*, Actes IV: 925–930.

Gage, M. 1961. On the definition, date, and character of the Ross glaciation, Early Pleistocene, New Zealand. *Trans. Roy. Soc. N.Z.* 88: 631–638.

Giovinetto, M.B., Robinson, E.S. & Swithinbank, C.W.M. 1966. The regime of the western part of the Ross Ice Shelf drainage system. *J. Glaciology* 6 (43): 55–68.

Hendy, C.H. & Wilson, A.T. 1968. Palaeoclimatic data from speleothems. *Nature* 219: 48–51.

Hendy, C.H., Healy, T.R., Rayner, E.R. & Wilson, A.T. 1967. A chronology for Late Pleistocene events in the Taylor Valley, Victoria Land, Antarctica. *Dry Valley Drilling Project Bull.* 6: 9 p.

Hendy, C.H., Healy, T.R., Rayner, E.M., Shaw, J. & Wilson, A.T. 1977. Late Pleistocene glacial chronology of the Taylor Valley, Antarctica, and the global climate. *Quat. Res.* (in press).

Hollin, J.T. 1962. On the glacial history of Antarctica. *J. Glaciology* 4: 173–196.

McGlone, M.G. 1973. Vegetation changes in the Otiran and Aranuian. In: Mansergh, G.D. (ed.), *New Zealand Quaternary: an introduction*, IX INQUA Congress, Christchurch: 13–19.

Moar, N.T. 1971. Contributions to the Quaternary history of the New Zealand flora 6. Aranuian pollen diagrams from Canterbury, Nelson and North Westland, South Island. *N.Z. J. Botany* 9: 80–145.

Molloy, B.P.J. 1969. Evidence for Post-Glacial climate changes in New Zealand. *J. Hydrology (N.Z.)* 8: 56–67.

Powers, W.E. 1962. Terraces of the Hurunui River. *N.Z. J. Geol. and Geophys.* 5: 114–129.

Rains, R.B. 1967. The Late Pleistocene glacial sequence of the High Peak Valley, Canterbury. *N.Z. J. Geol. and Geophys.* 10(4): 1145–1158.

Scott, R.F. 1905. Results of the national Antarctic expedition, 1, Geographical. *Geographical J.* 25: 353–373.

Soons, J.M. 1963. The glacial sequence in part of the Rakaia Valley, Canterbury, New Zealand. *N.Z. J. Geol. and Geophys.* 6: 735–736.

Soons, J.M. 1966. A review of Suggate, 1965. *N.Z. Geographer*, 22(1): 100–102.

Speight, J.G. 1963. Late Pleistocene historical geomorphology of the Lake Pukaki area, New Zealand. *N.Z. J. Geol. and Geophys.* 6(2): 160–188.

Suggate, R.P. 1965. Late Pleistocene geology of the northern part of the South Island, New Zealand. *N.Z. Geological Survey Bull.* n.s. 77 p.

Van Devender, T.R. 1976. The biota of the hot deserts of North America during the last glaciation: the Packrat Midden record. *Abstr. Am. Quat. Assoc. Meeting, Oct. 1976.*

Wardle, P. 1962. Evolution and distribution of the New Zealand flora, as affected by Quaternary climates. *N.Z. J. Botany* 1: 3–17.

Willett, R.W. 1950. The New Zealand Pleistocene snow line, climate conditions, and suggested biological effects. *N.Z. J. Science and Technology,* 32B: 18–48.

Wilson, A.T. & Wellman, H.W. 1962. Lake Vanda, an Antarctic lake. *Nature* 196: 1171–1173.

Wilson, A.T. 1964. Origin of ice ages: an ice shelf theory for Pleistocene glaciation. *Nature* 201: 147–149.

Wilson, A.T. 1967. The lakes of the McMurdo Dry Valleys. *Tuatara* 15: 152–164.

Wilson, A.T., Hendy, C.H. & Rennolds, C.P. 1973. New Zealand temperatures during the last millenium. *Abstr. IX INQUA congress, Christchurch*: 406–407.

Wilson, A.T. 1973. The great antiquity of some Antarctic landforms – Evidence for an Eocene temperate glaciation in the McMurdo region. In: E.M. van Zinderen Bakker Sr (ed.), *Palaeoecology of Africa* 8, Balkema, Cape Town.

* 3 *

Aspects of the early evolution of West Antarctic ice

David J. Drewry
Scott Polar Research Institute,
Cambridge CB2 1ER, U.K.

Manuscript received 10th November 1977

CONTENTS

ABSTRACT

The inference of Palaeogene ice masses of considerable extent in West Antarctica conflicts with interpretations of palaeoenvironmental conditions and a consideration of glacio-marine processes responsible for any ice-rafting into the southeast Pacific Ocean.

The earliest ice-rafted sediments recovered by the Deep Sea Drilling Project off Antarctica are from the Ross Sea and include pebbles mainly of metasedimentary lithology. Barrett suggested a provenance in sub-ice Marie Byrd Land, and from this Hayes and Frakes argued for the formation of an extensive ice sheet in West Antarctica in the Oligocene. Radio echo sounding along the front of the southern Transantarctic Mountains, oxygen-isotope and sedimentological studies suggest that DSDP results are better explained by development of vigorously calving tidewater valley and outlet glaciers, possibly located in the Whitmore-Thiel-Transantarctic Mountains area during the Oligocene rather than establishment of a large ice sheet – ice shelf environment in Marie Byrd Land.

Conditions suitable for the growth of a marine ice sheet in West Antarctica did not obtain until Late Miocene times (4–5 MY BP) when a further severe cooling of Antarctic waters occurred. The first unequivocal evidence for the presence of grounded West Antarctic ice is provided by a marked unconformity in Ross Sea stratigraphy at this time. That an ice mass discharging from Marie Byrd Land extended to the continental break-of-slope is demonstrated by the regional extent of the unconformity, terrestrial geologic evidence of expanded conditions and $\delta^{18}O$ determinations which impute ice volumes greater than today.

INTRODUCTION

The Antarctic ice sheet may be divided into two principal but unequal portions. In East Antarctica the larger 10.2 M km^2 ice mass is grounded above sea level over much of its area and is constrained by the surrounding narrow, though deep continental shelf. Occupying Ellsworth and Marie Byrd Lands (1.6 M km^2) the West Antarctic ice sheet is grounded mainly below sea level and possesses floating ice shelf extensions into both the Ross and Weddell Seas.

The early history of these ice masses is still poorly known. The relative chronology between growth of ice in East and West Antarctica is also problematical and details of the evolution of the marine ice cover in Marie Byrd Land remain unclear. In the following discussion several types of data are examined (results of the Deep Sea Drilling Project in Antarctic and Subantarctic waters, recent radio echo sounding and a consideration of glaciomarine sedimentation) in an attempt to resolve some of these current problems of early ice sheet history principally in West Antarctica.

EARLY CAINOZOIC

Mercer (1973; 1976), Drewry (1975; 1976) and Kennett (1977) have reviewed the growing body of palaeoclimatic information accruing from palaeobiological and palaeo-oceanographic studies in and around Antarctica. The data indicate a distinctive cooling trend discernable in the Southern Ocean in Late Cretaceous and Palaeogene times with warm to cool temperate environments, supporting widespread vegetation cover around Antarctica (Thomson & Burn, 1977).

The results of DSDP Leg 35 in the Bellingshausen Sea provide little support for intense cold until Late Oligocene–Late Miocene, although no sequences of Eocene to Mid-Oligocene strata were recovered (Craddock & Hollister, 1976). Oxygen-isotope data from South Tasman rise and Campbell Plateau yield no evidence consistent with substantial ice accumulation until the Mid-Late Miocene (Shackleton & Kennett, 1975). Cainozoic hyaloclastite sequences from large strato-volcanoes in northern Marie Byrd Land have been investigated by LeMasurier (1972) and provide additional evidence of glacial activity although for the Palaeogene their palaeo-glacial significance is minimal. LeMasurier considers sub-glacial eruption as the primary mechanism in hyaloclastite formation: lack of interbedded sediments would apear to favour sub-glacial rather than submarine eruption (LeMasurier, 1976). Since an ice overburden in excess of several hundred metres is required for development of hyaloclastite textures LeMasurier argues that the northern Marie Byrd Land sequences must have been erupted beneath an 'ice sheet'. This is a simplification as ice bodies of required thickness and size can be provided without invoking extensive ice sheets. In the Canadian Arctic and in Iceland, Scandinavia and Patagonia ice caps several thousand km^2 in extent attain depths in excess of 1 000 m. Even valley glaciers can achieve thicknesses of this order (e.g. Aletsch, Kaskawulsch, Fedchenko Glaciers). Interpretation of an ice sheet environment rather than local ice fields cannot be clearly demonstrated and must therefore remain uncertain. The oldest volcanic sequences reported by LeMasurier were originally ascribed to the Eocene and Oligocene. Re-analysis has shown the Turtle Peak and Bowyer Butte material to be of Miocene age (10–20 MY BP – LeMasurier & Rex, in press). The oldest hyaloclastites are now those from the USAS Escarpment and are dated at 26–28 MY BP.

'Eltanin' piston cores from the southeast Pacific Ocean have been examined by Geitzenauer and others (1968) and Margolis & Kennett (1971) (see inset Fig. 2). Their studies led them to believe that during Early–Mid Eocene times West Antarctica was glaciated to a degree sufficient to allow calving of debris-rich icebergs into a proto-Southern Ocean. Such views, clearly divergent with other palaeoenvironmental indicators discussed above, are based upon the 'glacial' interpretation of quartz grain surface micro-textures, percentage coarse sand and low species diversity amongst planktonic formanifera within the cores. Several appraisals, critical of these results (Mercer, 1973; 1976; this volume; Drewry, 1975) have demonstrated that it is difficult to provide unambiguous evidence for Early Cainozoic conditions as cold or colder than today, as necessary for extensive ice sheet development in West Antarctica.

A brief examination of the processes of ice rafting, extending some of Mercer's (1973) intimations, further demonstrates a serious objection to the model proposed by Margolis and Kennett which requires debris to be transported to and beyond the present iceberg limit (1 700–3 000 km from the coast – see Fig 2 inset). The discussion below is taken from a more detailed analysis (in preparation) of glacio-marine sedimentation.

Glacio-marine sedimentation

Debris-charged icebergs calve principally from outlet glaciers (those constrained by mountains and ice streams) rather than ice shelves where melting may clean the ice of sediment prior to calving. Such 'dirty' icebergs are usually small in size (i.e. <150 m thickness, $\leqslant 1$ km length). Debris is concentrated in the lowermost layers of the ice mass (usually $\leqslant 1\%$ ice thickness) with an upward decreasing concentration of solids. The release of sediment is controlled by heat fluxes at the iceberg/water interface. Heat flux from the sea is governed by 1) water temperature, and 2) water flow characteristics principally velocity of the berg relative to the water body. Weeks and Campbell (1973) provide the following expression for iceberg melt:

$$M_b = \left[\frac{LV^{0.8}\Delta T}{X^{0.2}}\right] \quad (1)$$

where

L = latent heat of fusion of ice
V = relative free stream flow velocity
ΔT = temperature difference between berg and water
X = length of iceberg.

Suggested values for ΔT and V are 5°C and 0.1 m s^{-1} respectively. These are based upon a lower limit for high latitude surface water temperatures during the Palaeogene as indicated by $\delta^{18}O$ results (Shackleton & Kennett, 1975) and an 'effective' value of V being the product of significant current gradients over the immersed depth of the berg and a velocity less than half the mean rate of West Wind Drift (Tchernia, 1974). We observe (Fig. 1 and Table 1) that smallish icebergs (i.e. those typical of outlet glaciers) would be rapidly 'cleaned' of any basal debris. These results are supported by the findings of Hult & Ostrander (1973) who note that the average life of icebergs around Antarctica today is only a few

Table 1

Time (a) and distance (km) to melt out all basal debris in iceberg, initial length 2 km, drift velocity 5 km day^{-1}

Initial debris layer thickness (m)	Melt rate (m a^{-1}) 5 a	5 km	10 a	10 km	20 a	20 km	50 a	50 km
1	0.2	370	0.1	185	0.05	92	0.02	37
5	1.0	1 850	0.5	925	0.25	462	0.1	185
10	2.0	3 700	1.0	1 850	0.5	925	0.2	370
20	4.0	7 400	2.0	3 700	1.0	1 850	0.4	740
50	10.0	18 500	5.0	9 250	2.5	4 625	1.0	1 825

years and that there is a significant diminution in their size on approaching the edge of the sea ice zone. Complications to the melting and debris-release history may be introduced by berg framentation and overturning.

Radial, straight-line drift assumed above is unreasonable, berg trajectories being influenced by oceanic currents. Present-day tracks show a strong east-west movement close to the coast in a 100–150 km zone dominated by the East Wind Drift. Movement is irregular with twisting motions due to the constraining sea ice (Tchernia, 1974). Bergs which after a year or two are eventually pushed further north towards the edge of the pack (400–900 km from the coast) come under the influence of the dominant Southern Ocean current (West Wind Drift). Thus we might anticipate considerably longer periods in which bergs are held close to shore when the bulk of any included basal debris would be melted out. Icebergs finally swept east and north into warmer waters would experience enhanced melt rates. Longevity in this case would simply be a function of berg size at breakout. Since debris is primariy restricted to the small bergs their ability to retain debris beyond a few hundred kilometres from the Antarctic coast may be seriously questioned.

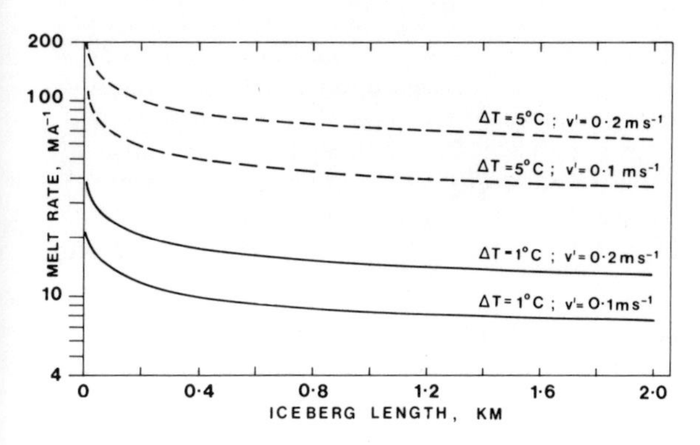

Fig. 1. Melt rates for small icebergs (⩽2 km length) for various values of ΔT and V (see Eq. 1)

The application of such present day glacio-oceanographic patterns to the Early Cainozoic is not without difficulty. While it is certain that ongoing crustal separation (New Zealand and Campbell Plateau and Australia from Antarctica and opening of the Drake Passage) would have seriously influenced oceanic flow adjacent to Antarctica, nevertheless Frakes & Kemp (1973) postulate a predominant west–east component to a proto-Southern Ocean circulation at this time. DSDP abyssal plain sites in the Bellingshausen Sea (Sites 322 and 323) confirm inception of a protocircum-Antarctic current in the Early Cainozoic (Craddock & Hollister, 1976; Drewry, 1976 (Fig. 6)) with final impediments to flow through the Drake Passage being removed by the Early Miocene (Barker et al., 1977).

Such considerations appear to seriously challenge the hypothesis of ice rafting of debris into the southeast Pacific Ocean during Early Tertiary times. A unique 'glacial' origin for 'Eltanin' core materials of Palaeogene age is, therefore, thought highly questionable (see also Mercer, this volume). In view of negligable supporting evidence and much that apparently refutes it the contention of Early Cainozoic continental glaciation of Antarctica must be considered decidedly unconvincing.

MID–LATE CAINOZOIC

Much of the discussion relating to early origins of extensive ice cover in West Antarctica turns on the relative timing of large scale ice sheet events between East and West Antarctica. The Ross Sea embayment provides a crucial area for evaluating the relative chronology of these ice masses.

The oldest sediment containing clear evidence of ice on Antarctica comes from the Ross Sea region where the earliest ice rafted deposits recovered from JOIDES sites 270–273 possess Late Oligocene–Early Miocene ages. No marine investigations have however been conducted off the Queen Maud Land coast, and the glacial history of this region is little known. The dating of the basal part of site 270 has recently been questioned (Weaver & Anderson, in press). While it is clear that caution should be exercised under conditions of poor fossil control, the combination of several techniques, yielding consistent results, nevertheless provides a degree of confidence in the reported ages (i.e. palaeomagnetic stratigraphy to 233 m depth (20.7 MY BP) with extrapolation to base on constant sedimentation rates (25 MY BP); a K-Ar age on basal greensand glauconite (26 MY BP) and sparse bethonic forams (Oligocene or younger)).

Glacial origin of the sediments on the Ross Sea continental shelf is recognised from the very poor sorting and scattered exotic clasts (many of them striated), including a dropstone. Deep water sediment (Site 274) shows a significant coarse sand fraction in otherwise very fine grained materials (Piper & Brisco, 1975; Hayes & Frakes, 1975). Pebbles extracted from the Ross Sea cores are mainly of metasedimentary lithologies (e.g. greywacke, argillite, phyllite, schist, gneiss, hornfels and granofels), though a few are granitic (Barrett, 1975). Barrett assumed a southern source and suggested sub-ice Marie Byrd Land since the exposed Transantarctic Mountains, bordering the Ross Ice Shelf, comprise a basement of predominantly granitic rocks overlain by Beacon strata with extensive dolerite sills.

Recent radio echo sounding results (Figure 2) allow some refinement of possible source areas for Barrett's pebbles. Isostatically adjusted bedrock in Marie Byrd Land suggests a preglacial surface mostly below sea level (accepting the likelihood of some 100–200 m erosion during the last few million years and possible tectonic effects). Areas of extensive exposed terrain would have flanked the head of the Ross Sea embayment, mainly in northern Marie Byrd Land along the Transantarctic Mountains, an extended Ellsworth Mountain block and a zone between the Whitmore, Thiel and Horlick Mountains. This latter area may also have extended eastwards towards the scattered nunataks of the Ellsworth–Thiel Mountains ridge. Terrain, where determined, constitutes very rough mountains cut by steep-sided troughs up to 2 km deep and 20 km wide (Rose, in press). The Whitmore Mountains support a poorly known sequence of folded Palaeozoic strata exhibiting low grade metamorphism and intruded by granites of Mesozoic age (Bushnell & Craddock, 1970; Webers, in press). A seismic refraction profile at 83.38° S 124.08° W indicates an upper 1.1. km thick layer possessing a P-wave velocity of 5.3 km s^{-1} overlying basement (Robinson, 1964; Bentley & Clough, 1972). Velocities of this magnitude correspond to Palaeozoic sediments, low rank metasediments or granite. Thus the Whitmore–

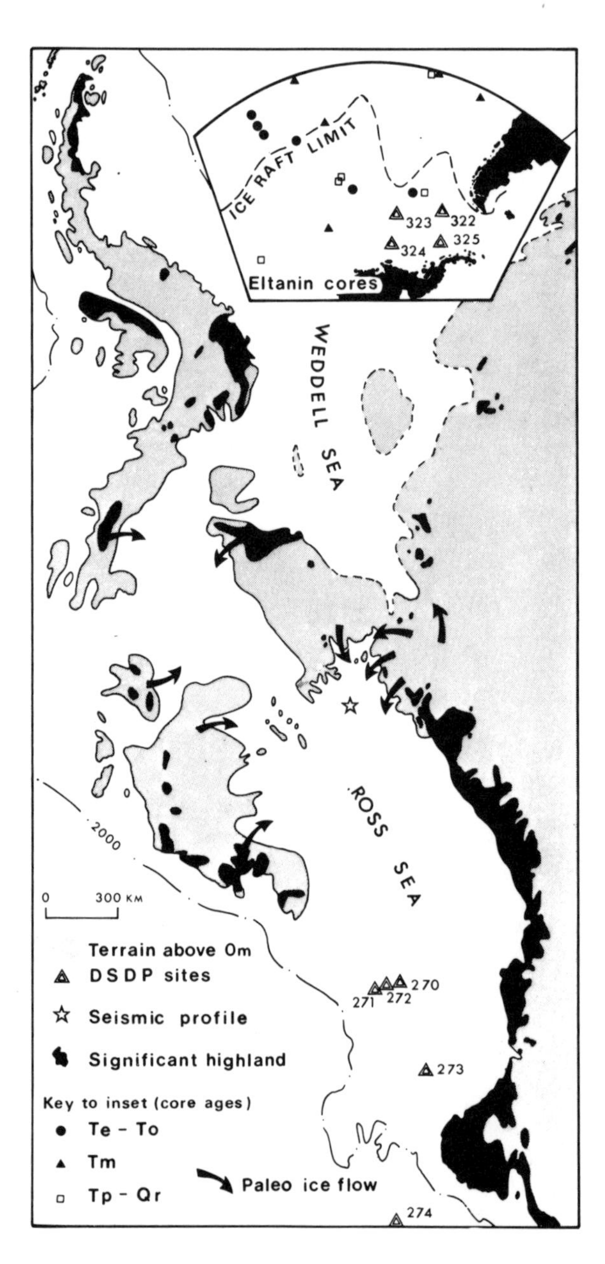

Fig. 2. 'Pre-glacial' terrain above sea level in the Ross and Weddell Sea embayments for approximately 10 MY BP. Based upon isostatically adjusted radio echo sounding data (Rose, in press), 'preglacial shoreline' from Bentley (1972) and areas of volcanic outcrop (Le Masurier & Rex, in press). Selected arrows suggest palaeo-ice discharge into an open Ross Sea from southern Transantarctic Mountains.

Inset. Location of 'Eltanin' piston cores with age of earliest 'ice-rafted' debris as reported by Margolis & Kennett (1971). Present limit of ice-rafting as shown.

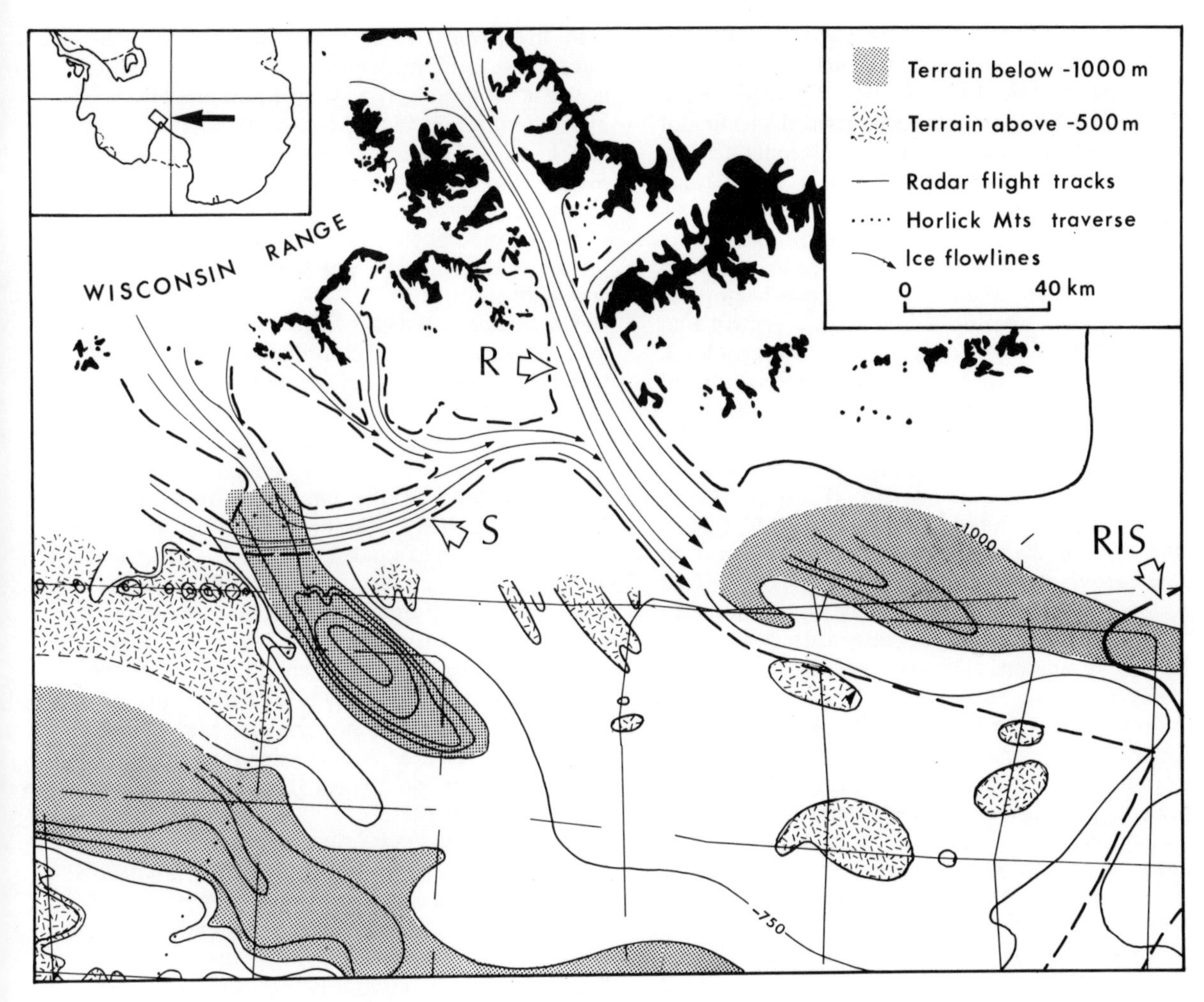

Fig. 3. Bedrock topography, principally from radio echo sounding, in the vicinity of the Horlick Mountains. southern Transantarctic Mountains. The Ross Ice Shelf (RIS), Reedy Glacier (R) and Shimizu Ice Stream (S) are marked. The edge of Marie Byrd Land ice streams (Rose, in press) are shown by a pecked line. Note strong discordance between current ice flow along Shimizu Ice Stream and the associated deep bedrock channel.

Thiel–Horlick Mountains zone could well have formed a major source of lithologies detected at Sites 270–272. The absence of diabase pebbles from the Late Oligocene–Mid Miocene core further strengthens the conjecture that the southern corner of the Ross Sea embayment, lacking extensive outcrops of such rocks, was a primary source of ice-rafted debris.

Little evidence is available to support the notion of Frakes (1975) and Hayes & Frakes (1975) that an extensive West Antarctic ice sheet–ice shelf system was developed in Late Oligocene times, and responsible for the glacio-marine sedimentation at Site 270. While oxygen-isotope studies from the Campbell Plateau and South Tasman Rise imply a rapid and substantial cooling in the Early Oligocene (Kennet and Shackleton, 1976), they provide no support until at least the Mid–Late Miocene for depletion of oceanic ^{16}O as required by a large fresh-water ice mass in Antarctica. In addition DSDP 270 has yielded palynomorphs which suggest the presence of cool-temperate vegetation, dominated by *Nothofagus*, in the central Ross Sea until the Late Oligocene–Early Miocene, contemporaneous with the earliest ice accumulations (Kemp & Barrett, 1975). Disappearance of this vegetation probably occurred during the Miocene as climate became more severe. Relatively high rates of sedimentation at Site 270 (Allis et al., 1975) and scattered forams and molluscs in deposits also argue for an open Ross Sea (Dell & Fleming, 1975) consistent with abundant ice rafting from melting bergs. It would seem clear that the ice masses involved in rafting debris into the central Ross Sea were of small dimensions, probably local highland ice fields nourishing vigorously calving tidewater glaciers and small ice shelves.

During the Mid–Late Miocene highland fringes of the Ross Sea must have witnessed thickening of the sea ice fields and extension of tidewater glaciers and ice shelves, principally along the western Transantarctic Mountains coast. Major expansion of ice into the Ross and Weddell Seas can probably only have been brought about in response to two main factors. 1) A substantial lowering in ocean temperatures close to freezing would enable growth and maintenance of large floating ice shelves at sea level. $\delta^{18}O$ data indicate that such conditions were fulfilled by Late Miocene times (Shackleton & Kennett, 1975). DSDP Leg 35 results confirm this pattern by showing that siliceous microfossils become dominant from the Mid-Miocene onwards. The Polar Front reached Site 322 in the Late Miocene. 2) Considerable influx of ice must have been contributed from an already enlarging continental ice sheet in East Antarctica, mostly through gaps in the southern Transantarctic Mountains, eastwards beyond the Horlick Mountains (Drewry, 1972; Rose, in press).

Recent radio echo sounding along the front of the mountains between 115° and 150° W shows several sub-ice bedrock channels issuing from the southern Transantarctic Mountains and clearly predating the present glacial regime. Two small troughs are present near the Ohio Range while two major channels occur at Reedy Glacier and the Shimizu ice stream. The latter feature is of principal interest and a map covering the area is depicted in Fig. 3. The bedrock trough displays a northwest trend, extends for at least 100 km and possesses an overdeepened character (reaching 2 060 m below sea level) with reversed gradient and seaward threshold. With isostatic adjustment the feature would constitute a well developed fjord issuing from East Antarctica through the Wisconsin Range. Fig. 3 also shows present ice flowlines. It can be seen that Shimizu ice far from following its bedrock channel flows westwards under the influence of ice discharging out of southern Marie Byrd Land. This striking discordance may be explained if the Shimizu trough owes its origin to glaciation prior to the evolution of a full-bodied West Antarctic ice sheet. East Antarctic ice of considerable proportions must have breached the Horlick Mountains in several places releasing ice via outlet glaciers and ice streams. The observed fjord-like overdeepening can best be explained by a decoupling ice tongue discharging into an open Ross Sea in a manner similar to that proposed for the formation of Skelton Inlet (Crary, 1966).

With continued climatic deterioration, West Antarctic ice would have spread out, partly grounded, partly as floating ice tongues and small ice shelves, first from the Ellsworth–Whitmore–Transantarctic Mountains axis finally reaching and combining with that from the highlands of northern Marie Byrd Land. Late Miocene ice is clearly demonstrated in the latter region by a striated pavement overlain by a tilloid in the Jones Mountains (Rutford et al., 1972), erosion of an andesite surface on Mount Sidley (Doumani, 1964) and 'sub-glacial' hyaloclastites (LeMasurier & Rex, in press). Enlargement of West Antarctic ice could have allowed grounding of an ice sheet below sea level in the Ross Sea embayment.

The first unequivocal evidence for the presence of West Antarctic ice of considerable extent is provided by a marked stratigraphic break in Late Miocene–Early Pliocene times in the Ross Sea sedimentary sequence. That ice extended to the continental break-of-slope is demonstrated by the regional extend of the unconformity (Houtz & Davey, 1975; Hayes & Frakes, 1975). It seems certain that the expanded ice sheet regime documented for the Late Miocene–Early Pliocene in the Ross Sea affected much of the Antarctic continent. In southern Victoria Land glacial geologic evidence relates to enlarged ice sheet conditions earlier than 4.2 MY BP (Denton et al., 1971; Fleck et al., 1972; Mayewski, 1975) and possible pre-Mid Miocene (Brady, in press). $\delta^{18}O$ determinations impute ice volumes greater than today (Shackleton & Kennett, 1975). JOIDES Site 274, at the continental margin, documents a massive influx of sediments at this time, while sedimentation decreased dramatically on the Ross Sea Shelf. In addition cores from the Southern Ocean reveal a substantial and contemporaneous northwards shift in the calcareous/siliceous boundary (Kemp et al., 1975). Numerous other widespread marine investigations suggest cold conditions peaking between 3.5 and 5 MY BP (several are quoted by Mercer (1976) and Kennett (1977), but in addition we may note Kennett & Brunner (1973); Thayer (1971) and Anderson (1972)).

The nature of events in parts of East Antarctica distant from the Ross Sea area remain uncertain. The presence at high elevation in the Prince Charles Mountains of lithified tills which Bardin (1974) correlated with the Sirius Formation (Mercer 1972) possibly suggests contemporaneous expansion of the ice sheet there.

CONCLUSIONS

Little or no support is provided for extensive Early Cainozoic glaciation in West Antarctica. Conditions necessary for glacio-marine sedimentation in the southeast Pacific Ocean as required by Margolis and Kennett (1971) are incompatible with glacio-oceanographic and palaeoenvironmental interpretations.

Climatic deterioration and Southern Ocean cooling progressed through Mid-Cainozoic times, possibly in a series of step-like declines. Such refrigeration allowed upland areas fringing the Ross Sea embayment to support small ice fields and glaciers which reached to sea level during the Late Oligocene and Early Miocene. Icebergs vigorously calving from these floating ice tongues, especially in the southeast corner of the embayment, enabled ice rafting of debris into the central Ross Sea region.

By Mid–Late Miocene times substantial ice sheets were developing in East Antarctica, discharging ice through gaps in the southern Transantarctic Mountains into an open Ross Sea. Other coastal upland areas of Marie Byrd Land supported growing ice fields (e.g. Jones Mountains) but a full-bodied West Antarctic ice sheet was not yet present.

Oceanographic conditions were favourable by the Early Pliocene for the coalescence of ice shelves and their subsequent grounding over the Ross Sea. Between 5 and 3.5 MY BP a West Antarctic ice sheet occupied the whole of the Ross Sea embayment.

ADDED NOTE

During recent field investigations in northern Marie Byrd Land LeMasurier (personal communication, December 1977) reports the discovery of tillite interbedded with hyaloclastites of Miocene age. This corroborates his interpretation of subglacial origin of these sequences. The inference of widespread, continental glaciation from any hyaloclastite, however, remains equivocal.

ACKNOWLEDGEMENTS

Radio echo sounding results reported here were gathered during joint US National Science Foundation–Scott Polar Research Institute–Technical University of Denmark operations in Antarctica during 1974–75. I would like to thank P.J.Barrett for stimulating discussions on Antarctic glacial history and for commenting on a draft of this paper.

REFERENCES

Allis, R.G., Barrett, P.J. & Christoffel, D.A. 1975. A palaeomagnetic stratigraphy for Oligocene and Early Miocene glacial sediments at Site 270, Ross Sea, Antarctica. *Initial reports of the DSDP* 28, U.S. Gov. Printing Off., Washington, D.C.: 879–884.

Anderson, J.B. 1972. *The marine geology of the Weddell Sea*, The Sedimentological Research Laboratory, Dept. Geol., Florida State Univ. Contribution 35. 222 pp.

Bardin, V.I. 1974. Geographic investigations on MacRobertson Land. *Soviet Ant. Exped. Info. Bull.* 8 (10): 540–543.

Barker, P.F. et al. 1977. Evolution of the southwestern Atlantic Ocean Basin: results of Leg 36, Deep Sea Drilling Project, *Initial reports of the DSDP* 36: 993–1014.

Barrett, P.J. 1975. Characteristics of pebbles from Cenozoic marine glacial sediments in the Ross Sea (DSDP sites 270–274) and the Southern Indian Ocean (Site 268). *Initial reports of the DSDP* 28: 769–784.

Bentley, C.R. 1972. Subglacial topography Plate 7. In: B.C. Heezen, M. Tharp & C.R. Bentley, (eds.), *Morphology of the Earth in the Antarctic and sub-Antarctic,* Antarctic Map Folio Ser. No 16.

Bentley, C.R. & Clough, J.W. 1972. Antarctic subglacial structure from seismic refraction measurements. In: R.J. Adie (ed.), *Antarctic geology and geophysics* Universitetsforlaget, Oslo: 683–691.

Brady, H.T. (in press). Late Neogene history of Taylor and Wright Valleys and McMurdo Sound, derived from diatom biostratigraphy and palaeoecology of DSDP cores. *Symposium on Antarctic geology and geophysics, Madison, August 1977.*

Bushnell, V.C. & Craddock, C. (eds.). 1969–70. *Geologic maps of Antarctica*, Antarct. Map Folio Ser., Folio 12.

Craddock, C. & Hollister, C.D. 1976. Geologic evolution of the southeast Pacific Basin. *Initial reports of the DSDP* 35: 723–744.

Crary, A.P. 1966. Mechanism of fiord formation indicated by studies of an ice-covered inlet. *Geol. Soc. Am. Bull.* 77: 911–930.

Dell, R.K. & Fleming, C.A. 1975. Oligocene–Miocene Bivalve mollusca and other macrofossils from sites 270 and 272 (Ross Sea), DSDP, Leg 28. *Initial reports of the DSDP* 28: 693–704.

Denton, G.H., Armstrong, R.L. & Stuiver, M. 1971. The late Cenozoic glacial history of Antarctica. In: K.K. Turekian (ed.), *Late Cenozoic glacial ages*, Yale University Press: 267–306.

Doumani, G.A. 1964. Volcanoes of the Executive Committee Range, Byrd Land. In: R.J. Adie, (ed.), *Antarctic geology*, North Holland: 666–675.

Drewry, D.J. 1972. Sub-glacial morphology between the Transantarctic Mountains and the South Pole. In: R.J. Adie (ed.), *Antarctic geology & geophysics*, Universitetsforlaget, Oslo: 693–703.

Drewry, D.J. 175. Initiation and growth of the East Antarctic ice sheet. *J. Geol. Soc. London* 131: 255–273.

Drewry, D.J. 1976. Deep sea drilling from Glomar Challenger in the Southern Ocean. *Polar Record* 18 (112): 47–71.

Fleck, R.J., Jones, L.M. & Behling, R.E. 1972. K-Ar dates of the McMurdo volcanics and their relation to the glacial history of Wright Valley. *Antarctic J. US* 7 (6): 245–246.

Frakes, L.A. 1975. Paleoclimatic significance of some sedimentary components at Site 274. *Initial reports of the DSDP* 28: 785–788.

Frakes, L.A. & Kemp, E.M. 1973. Palaeogene continental positions and evolution of climate. In: D.H. Tarling & S.K. Runcorn (eds.), *Implications of continental drift to the earth sciences*, Vol. 1, Academic Press, London: 539–558.

Geitzenauer, K.R., Margolis, S.V. & Edwards, D.S. 1968. Evidence consistent with Eocene glaciation in a south Pacific deepsea sedimentary core. *Earth and Planetary Sci. Letters* 4: 173–177.

Hayes, D.E. & Frakes, L.A. 1975. General synthesis, Deep Sea Drilling Project Leg 28. *Initial reports of the DSDP* 28: 919–942.

Houtz, R.E. & Davey, F.J. 1973. Seismic profiles and sonobuoy measurements in the Ross Sea, Antarctica. *J. Geophys. Res.* 78: 3448–3468.

Hult, J.L. & Ostrander, N.C. 1973. Antarctic icebergs as a global fresh water resource. *Rand. Corp. Report* R–1255–NSF. 84 pp.

Kemp, E.M. & Barrett, P.J. 1975. Antarctic glaciation and early Tertiary vegetation. *Nature* 258 (5535): 507–508.

Kemp, E.M., Frakes, L.A.W. & Hayes, D.E. 1975. Palaeoclimatic significance of diachronous biogenic facies, Leg 28, Deep Sea Drilling Project. *Initial reports of the DSDP* 29: 909–917.

Kennett, J.P. & Brunner, C.A. 1973. Antarctic late Cenozoic glaciation: evidence for initiation of ice rafting and inferred increased bottom water activity. *Geol. Soc. Am. Bull.* 84: 2043–2052.

Kennett, J.P. & Shackleton, N.J. 1976. Oxygen isotope evidence for the development of the psychrosphere 38 Myr ago. *Nature* 260 (5551): 513–515.

Kennett, J.P. 1977. Cenozoic evolution of Antarctic glaciation, the circum-Antarctic Ocean, and their impact on global paleoceanography. *J. Geophys. Res.* 28 (27): 3843–3861.

LeMasurier, W.E. 1972. Volcanic record of Cenozoic glacial history of Marie Byrd Land. In: R.J. Adie (ed.), *Antarctic geology and geophysics*, Universitetsforlaget, Oslo: 251–260.

LeMasurier, W.E. 1976. Interglacial volcanoes in Marie Byrd Land. *Antarctic J. U.S.* 11(4): 269–270.

LeMasurier, W.E & Rex, D.C. (in press). Volcanic record of glacial history in Marie Byrd Land and western Ellsworth Land: revised chronology and evaluation of tectonic interrelationships. *Symposium on Antarctic geology and geophysics, Madison, August 1977.*

Margolis, S.V. & Kennett, J.P. 1971. Cenozoic paleoglacial history of Antarctica recorded in deepsea cores. *Am. J. Sci.* 271 (1): 1–36.

Mayewski, P.A. 1975. *Glacial geology and Late Cenozoic history of the Transantarctic Mountains, Antarctica.* Report 56, Inst. Polar Studies, Ohio State Univ. 168 pp.

Mercer, J.H. 1968(b). Glacial geology of the Reedy Glacier area, Antarctica. *Geol. Soc. Am. Bull.* 79: 471–486.

Mercer, J.H. 1972. Some observations on the glacial geology of the Beardmore Glacier area. In: R.J. Adie (ed.), *Antarctic geology and geophysics*, Universitetsforlaget, Oslo: 427–433.

Mercer, J.H. 1973. Cainozoic temperature trends in the southern hemisphere: Antarctic and Andean glacial evidence. In: Van Zinderen Bakker (ed.), *Palaeoecology of Africa* 8. Balkema, Cape Town: 85–114.

Mercer, J.H. 1975. Southern Patagonia: glacial events between 4 m.y. and 1 m.y. ago. In: R.P. Suggate and M.M. Cresswell (eds.), *Quaternary Studies*, Roy. Soc. N.Z., Wellington: 223–230.

Mercer, J.H. 1976. Glacial history of southernmost South America. *Quat. Res.* 6: 125–166.

Piper, D.J.W. & Brisco, C.D. 1975. Deepwater continental margin sedimentation DSDP Leg 28. *Initial reports of the DSDP* 29: 727–756.

Robinson, E.S. 1964. *Geological structure of the Transantarctic Mountains and adjacent ice covered areas,* PhD dissertation, Univ. Wisconsin.

Rose, K.E. (in press). Radio echo studies of bedrock in Marie Byrd Land, Antarctica. *Symposium on Antarctic Geology and Geophysics, Madison, Wisconsin, August 1977.*

Rutford, R.H., Craddock, C. & White, C.M. 1972. Tertiary Glaciation in the Jones Mountains. In: R.J. Adie (ed.) *Antarctic Geology and Geophysics*, Universitetsforlaget, Oslo: 239–243.

Shackleton, N.J. & Kennett, J.P. 1975. Paleotemperature history of the Cenozoic and the initiation of Antarctic glaciation: oxygen and carbon isotope analyses in DSDP Sites 277, 279, 281. *Initial reports of the DSDP* 29: 743–755.

Tchernia, P. 1974. Etude de la dérive antarctique Est-Ouest au moyen d'icebergs suivis par le satellite "Eole". *C.R. Acad. Sc. Paris*, 278, Ser. B: 667–670.

Thayer, F. 1971. Benthic foraminiferal trends, Pacific-Antarctic Basic. *Deep-Sea Res.* 18 (7): 723–738.

Thomson, M.R.A. & Burn, R.W. 1977. Angiosperm fossils from latitude 70° S′. *Nature* 269 (5624): 139–141.

Webers, G.F. (in press). Geology of the Whitmore Mountains. *Symposium on Antarctic Geology and Geophysics, Madison, Wisconsin, August 1977.*

Weeks, W.F. & Campbell, W.J. 1973. Icebergs as a freshwater source: an appraisal. *J. Glaciology* 12 (65): 207–234.

* 4 *

Past surges in the West Antarctic ice sheet and their climatological significance

A.T. Wilson
School of Science, University of Waikato, Hamilton, New Zealand

Manuscript received 2nd November 1977

CONTENTS

ABSTRACT

Evidence is presented from both Antarctica and Australasia to show that the West Antarctic ice sheet has surged several times in the last 6 000 years. During these surges perhaps up to 3×10^5 km^3 of ice in the form of icebergs are delivered to the Southern Ocean in a period of a few hundred years. The dates of these surges correspond to the post-glacial cold periods and it is suggested that this may be the mechanism for at least some of the post-glacial climate fluctuations.

INTRODUCTION

The Antarctic ice sheet consists essentially of two parts - the larger East Antarctic ice sheet, which occupies the Antarctic Continent itself, and the smaller West Antarctic ice sheet, which occupies an archipelago of offshore islands. The bedrock under a large fraction of the West Antarctic ice sheet is below sea level now and would still be below sea level if the ice sheet was removed and complete isostatic rebound occurred. It is generally believed that at times of low glacial sea levels the West Antarctic ice sheet extended further out on to the continental shelf (Hollin, 1962; Wilson, 1970; Denton et al, 1971). At the end of the Wisconsin the rising sea level would have changed the situation of the West Antarctic ice sheet and it would have collapsed back to the interglacial situation. There is evidence that this did in fact occur (Denton et al., 1971). Hughes (1973) has suggested that the West Antarctic ice sheet is disintegrating at present. This paper presents evidence to suggest that the West Antarctic ice sheet partly collapsed as a result of the post-glacial rise in sea level, and is now in the situation of periodic surging followed by recovery and that it is presently in the recovery phase. It is suggested further that each collapse has surged vast numbers of icebergs into the Southern Ocean which changed the albedo of the earth to produce the post-glacial cold periods.

EVIDENCE FROM THE PAST CLIMATE OF THE MCMURDO DRY VALLEYS

For a number of years the author has been involved in studying the lakes in Victoria Land, Antarctica (Wilson, 1967). Part of the program has been to extract palaeoclimate information from lake levels, salt gradients and sediment compositions. This work has shown that the following major climatic events occurred in the McMurdo Dry Valley system:

For some period before 3 000 years PB all lakes were at high level.
3 000 years BP - lakes suddenly dropped to very small volume.
1 200 years PB - Lake Vanda rapidly filled.
800 years BP - Lake Bonney dropped to a level below present.
200 years BP–present. Lake Bonney's level has risen.

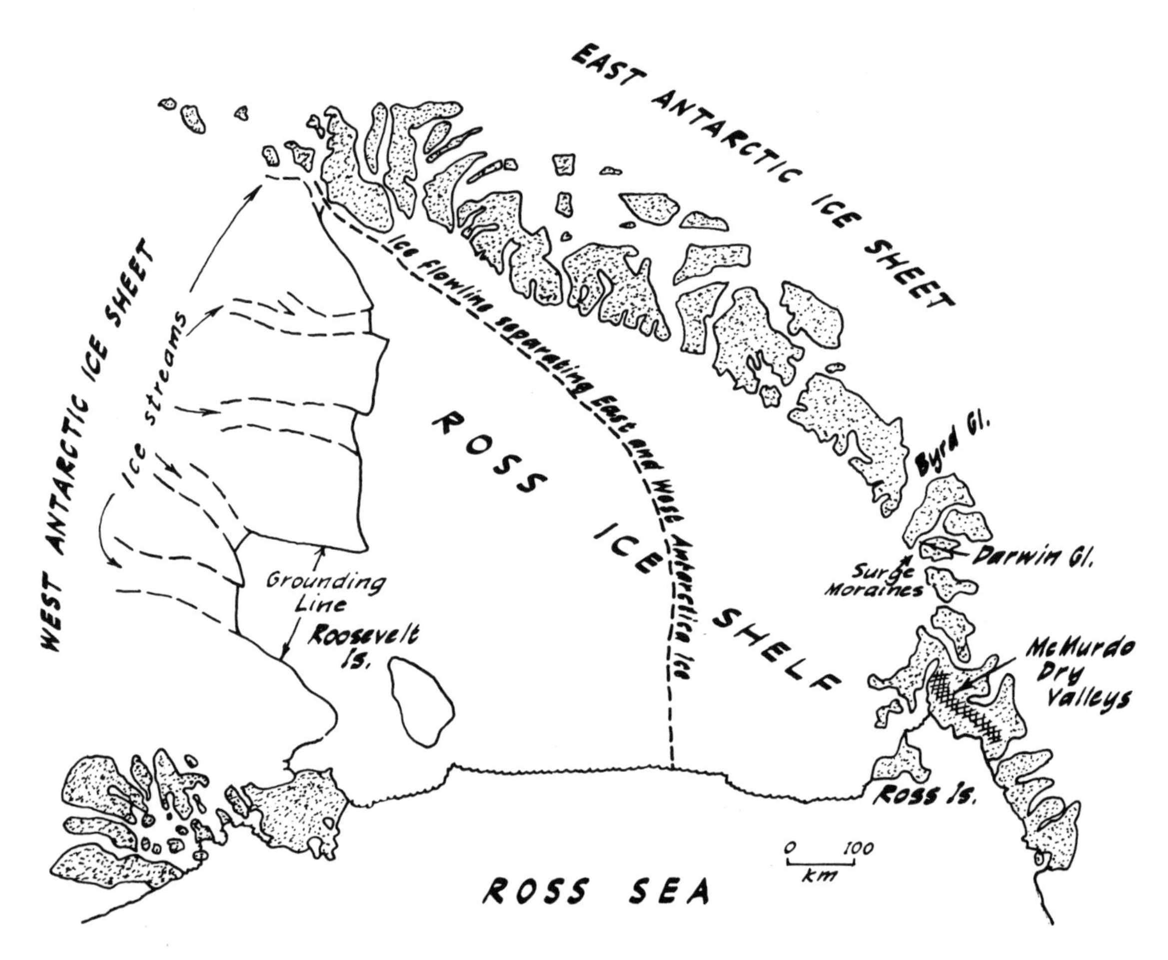

Fig. 1. Map of Ross Ice Shelf showing its relationship to the East and West Antarctic ice sheets

The McMurdo Oasis Dry Valley Area consists of a number of enclosed drainage basins. The lowest part of each is occupied by a saline lake. If we consider the evaporation/precipitation balance of the entire drainage system it can be seen that the net excess precipitation of a snow field will flow below the snow line as a glacier. If the surface area of the glacier is insufficient to balance the sublimation/precipitation budget, the glacier will advance further and further below the snow line. It will proceed toward a situation where the total positive net precipitation above the snow line is balanced by the total negative net precipitation below the snow line. Usually the glacier has pushed sufficiently far below the snow line so that some summer melting takes place. For a few days during the hottest part of the summer, in such cases, a stream flows away from the glacier snout and feeds a lake which occupies the lowest point of that particular enclosed drainage basin. The size of the lake is determined by the area needed to balance the evaporation/precipitation equation for that particular drainage area. If there is a net precipitation increase to the area the lake levels will rise and if there is a decrease in precipitation the lake levels will fall. The problem of immediate interest is what caused these lake levels to change so suddenly and so dramatically.

In most areas of the world precipitation delivered to a given area is controlled by its position relative to the open sea. A very attractive hypothesis is that the Ross Ice Shelf was much further to the south than its present position prior to 3 000 years ago. This would mean that there would have been more open sea closer to the snow fields supplying Lake Vanda. The hypothesis that the local alpine glaciers, as distinct from those fed from the polar plateau, are controlled by mean distance to the sea (i.e. the position of the Ross Ice Shelf) is further supported by evidence that the coastal regions to the south of the McMurdo Oasis area appear to have been more heavily glaciated until a few thousand years ago.

Thus we are forced to the conclusion that the precipitation history of the McMurdo Dry Valleys has changed suddenly at times in the past, presumably due to changing position of the front of the Ross Ice Shelf. In turn it is proposed that this has been pushed forward from time to time by a surging East Antarctic ice sheet. In particular there was a dramatic surge immediately prior to 3 000 C-14 years BP.

The only reasonable explanation seems to be that for some time prior to 3 000 years BP the front of Ross Ice Shelf was much further to the south, due perhaps to the collapse of the West Antarctic ice sheet as a result of the post-glacial rise in sea level. Some time immediately before 3 000 years BP the West Antarctic ice sheet surged and pushed the front of the Ross Ice Shelf rapidly forward to some point considerably to the north of its present position.

There was a retreat of the Ross Ice Shelf prior to 1 200 years BP which enabled more precipitation to enter the Wright Valley catchment. These changes are illustrated in Figure 2.

The rapid and dramatic fluctuations of the lake levels in the McMurdo Dry Valleys argue strongly in favour of a periodic surging of the West Antarctic ice sheet followed alternatively with a build-up of ice in the West Antarctic ice sheet rather than a continuous collapse as proposed by Hughes (1973).

GLACIAL THEORY OF ICE SHEETS GROUNDED BELOW SEA LEVEL

Weertman (1974) has made an analysis of the steady-state size of a two-dimensional ice sheet whose base is below sea level and which terminates in floating ice shelves. Under the assumption of perfect plasticity, it is found that an ice sheet placed on a bed whose surface was initially flat cannot exist if the depth of the bed below sea level exceeds a critical depth. If this depth is less than the critical level, the ice sheet extends out to the edge of the continental shelf.

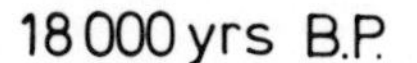

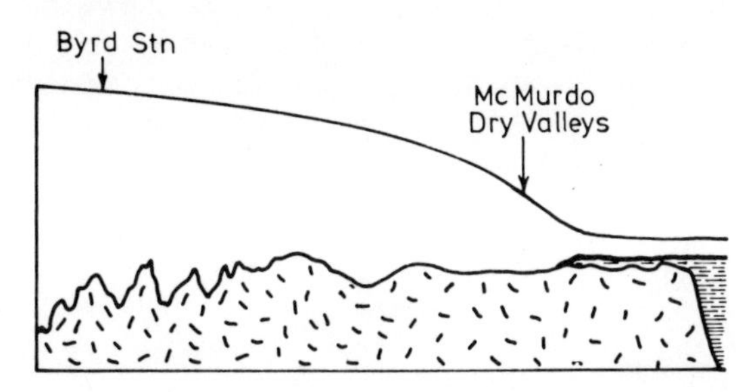

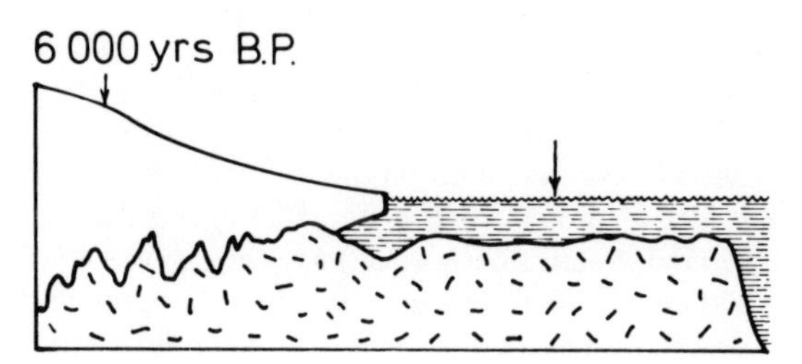

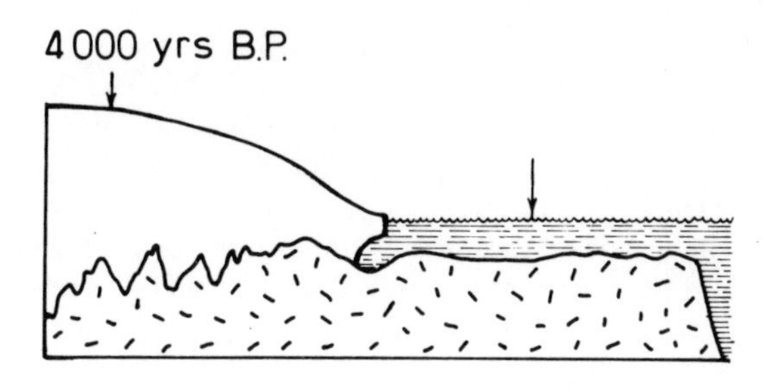

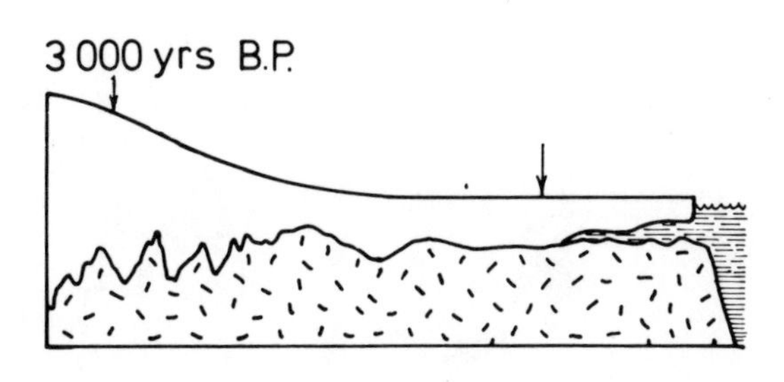

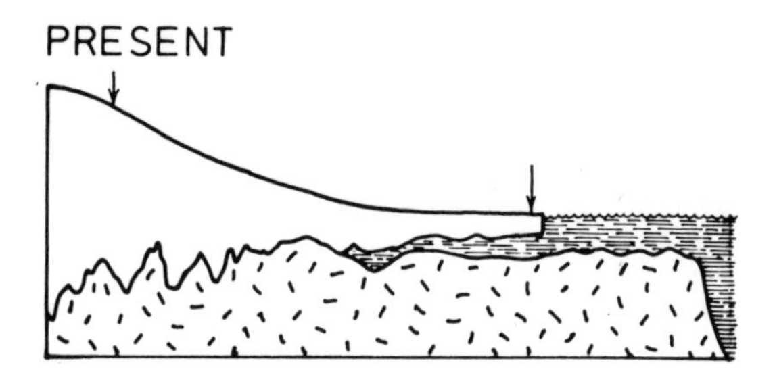

Fig. 2. Schematic representation of changes in West Antarctic ice sheet

18 000 years BP. End of last glacial period. Sea level low. West Antarctic ice sheet thicker and expanded on to continental shelf.

6 000 years BP. Sea level rises at end of glacial period and the West Antarctic Ice Sheet collapses back. Period of advance for local alpine glaciers. Lake Vanda at upper level.

4 000 years and 3 000 years BP. Example of a surge – West Antarctic Ice Sheet thickens due to increased precipitation and surges. Reduced precipitation to neves of local alpine glaciers, which retreat.

The hypothesis presented in this paper suggests that in interglacial times the West Antarctic ice sheet cannot extend to the edge of the shelf because the intermediate situation was unstable against periodic surging. This led to the rapid wasting of the ice sheet to icebergs. Evidence is presented in this paper to suggest that the range over which the surges take place and the period of surging is controlled by sea level.

EVIDENCE FROM THE WEST ANTARCTIC ICE SHEET ITSELF

This evidence comes from two main sources: The shape of the West Antarctic ice sheet itself and isotope profiles through the ice sheet.

The ice surface of the West Antarctic ice sheet is generally concave. Glacial theory (Nye, 1957) would predict a convex surface. Most ice sheets including the East Antarctic ice sheet have the convex profile predicted by Nye. The unusual concave shape of the West Antarctic ice sheet suggests that it is collapsing or has surged recently.

$\delta^{18}O$ and δD measurements of the ice core from Byrd Station yield information on the temperature at which the precipitation formed (Dansgaard, 1964). Johnsen et al. (1972) believe that the West Antarctic ice sheet has undergone considerable fluctuations during the last 100 000 years. By comparing the isotopic data of the ice core from Byrd Station with the Greenland ice core from Camp Century, Johnsen et al. (1972) concluded that during the period 4 000 years BP to 3 000 years BP the ice surface of the West Antarctic ice sheet dropped by 300–500 metres at the site of Byrd Station.

Turning to the details of these data as shown in Fugure 3 we can see evidence for several lowerings of the ice level. These occurred, interestingly enough, just at the onset of the post-glacial cold periods (marked with arrows) and it is an attractive hypothesis to suggest that the surging of 400 metres of ice from the West Antarctic ice sheet in the form of icebergs into the Southern Ocean would change the albedo of the Earth sufficiently to cause the neoglacial periods. Looking at Fugure 4 evidence can be seen for similar surges in the West Antarctic ice sheet during glacial times. These appear to occur at less frequent intervals due perhaps to the increased size of the ice sheet – which in turn is the result of the lower sea level during glacial times.

DIRECT FIELD EVIDENCE OF SURGES

The author has made a search for direct geomorphological evidence for the surges in the West Antarctic ice sheet suggested in this paper. It was apparent that the surging ice sheet had not reached the McMurdo Sound region.

A search was made to the south of Ross Island for a 'dry area' on the coast where moraines might be preserved. These moraines would be expected to be very young (<6 000 years BP). The searched-for moraines were found in the summer of 1975/76 in a dry area in the Brown Hills immediately north of where the Darwin Glacier joins the Ross Ice Shelf (see Fig. 1).

There were four ice-cored moraines of successively older ages, all very recent and at least in the author's experience quite unique in Victoria Land. These

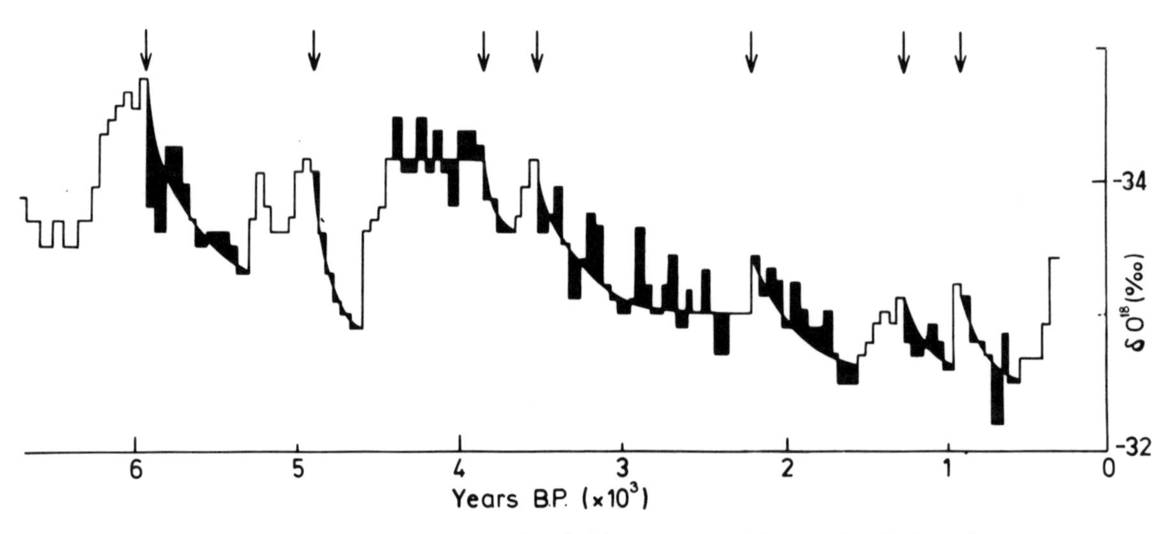

Fig. 3. Isotopic data from Byrd Core (after Johnsen et al., 1972). Shading is that of present author to suggest periods of possible surging.

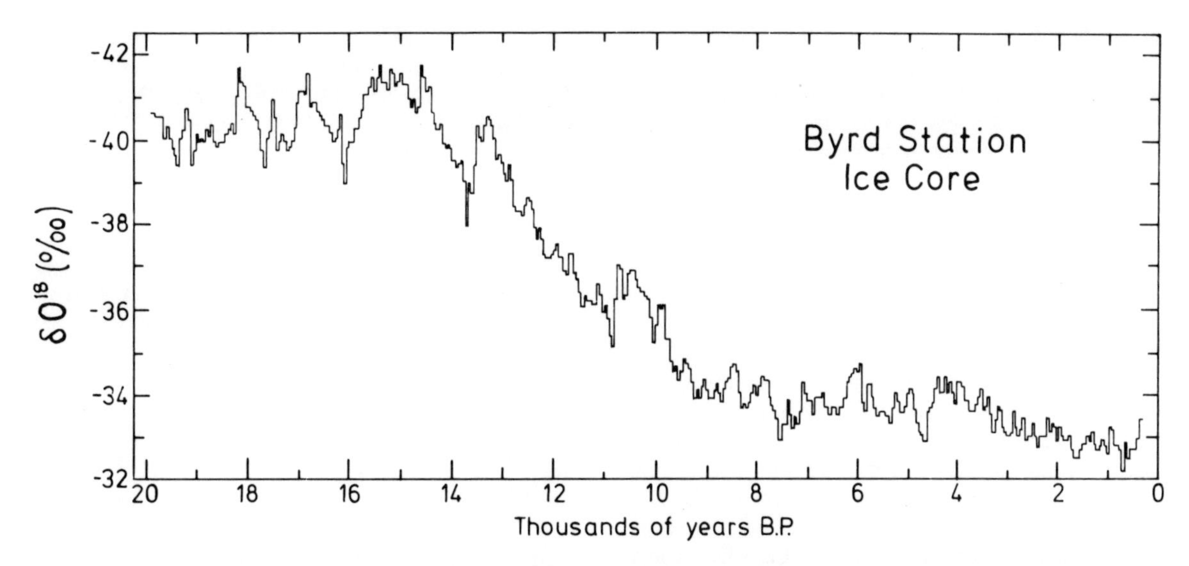

Fig. 4. Isotope data from Byrd Core (after Johnsen et al., 1972). Shows periods in glacial times when surging may have taken place.

moraines consisted of glacially scratched boulders (not common in the moraines of the polar glaciers of Antarctica) ice-cored and ablating rapidly and hence must be young.

The ice was sampled for isotopic analysis (Table 1) which showed it to be much too depleted in deuterium to have come from the adjacent Darwin Glacier. The Darwin Glacier rather fortunately is only partly supplied by the Polar Plateau, the rest being supplied by alpine glaciers formed from local precipitation, and hence is much less depleted in deuterium than the polar plateau.

Table 1
Isotopic composition of ice cored moraines from Brown Hills

	δD (SMOW)
Ice-cored moraines (shore of Brown Hills)	
a) South of Lake Wilson	— 332‰
b) North of Lake Wilson	— 321‰
Darwin Glacier Ice (adjacent to Brown Hills Dry Area)	— 284‰
Ice in West Antarctic ice sheet	
Byrd core – glacial	— 350‰
– interglacial	— 250‰
East Antarctic ice sheet (Surface at South Pole)	— 400‰

EVIDENCE FROM HIGH SEA LEVELS

The hypothesis presented in this paper suggests that at each surge of the West Antarctic ice sheet large quantities of fresh water would be supplied rapidly to the Southern Ocean in the form of melting icebergs. The area of the West Antarctic ice sheet is about 1.2% of the area of the earth's ocean below 30° S latitude. If the interpretation of the Byrd Station ice core by Johnsen et al. is taken, then the surface of the West Antarctic ice sheet dropped 300–500 metres in a period of a few hundred years between 3 000 and 4 000 years BP. An instantaneous transfer of 200 metres of ice into the ocean below 30° S for example, would raise the level by 2.5 metres. In actual practice the 3×10^5 km^3 of water would be added as icebergs perhaps 300 metres thick during a period of a few hundred years. As the icebergs melted they would have reduced the salinity of the upper 300 metres of the Southern Ocean and raised its surface by perhaps 2 metres. These levels would be maintained until mixing with the other oceans of the world would have reduced the salinity differences and the level would have dropped over a period of a hundred years or so to a level only a few centimetres above pre-surge times. Surges of the West Antarctic ice sheet could explain some of the controversy on post-glacial sea levels.

Most New Zealand and many Australian geologists believe that there was a high post-glacial sea level of 2 metres (Fleming, 1973). But this view is not popular in more northern latitudes.

The position of the sea level during the late Quaternary is a subject of much controversy. Of the period since the late Wisconsin glacial maximum, the last 6000 years are especially controversial, despite the fact that it is for this short period that we have the most data. It is suggested the reason for this controversy might be that there were periods when there were sea level rises in one part of the world's oceans but not in other, and that this was caused by a surge from the West Antarctic ice sheet adding over a short period of time large quantities of fresh water and reducing for a time the salinity of the Southern Ocean.

The literature on past sea level studies is vast and conflicting. If we restrict ourselves to the Pacific sector there is extremely good evidence that the tropical Pacific sea levels were at least near their present level for the last 4000 to 5000 years (see for example Curray et al., 1970). To the south in New Zealand and Australia there is extremely good and well-dated evidence that there were high sea levels at 3500 years BP and 2400 years BP.

Perhaps the best evidence comes from Schofield's work (Schofield, 1960; 1973). The best-dated of these are at Miranda (Gulf of Thames, New Zealand), where a very pronounced 2 metre high sea level is dated at 3900 ± 90 BP. This date is on shell, which if corrected would yield a date of 3500 years BP. This high sea level is found at many places in New Zealand and Australia and has been reported from South America.

Sea level studies are plagued with distinguishing between eustatic sea level changes and tectonic changes, but two facts emerge. The West Antarctic surge hypothesis would help resolve some of the apparent controversy in sea level studies. The post-glacial sea level rises in New Zealand appear to occur at the same time as the isotope and palaeoclimatic data from Antarctica suggest the West Antarctic ice sheet surges took place.

THE PRESENT SITUATION – EVIDENCE FROM THE RISP PROGRAM

Data gathered by the Ross Ice Shelf Project (RISP) from the southeast quadrant of the Ross Ice Shelf, Thomas (1976) indicate that, near the grounding line, the ice shelf is growing thicker by almost 1 metre year^{-1}. This thickening rate implies an advance of 1 kilometre year^{-1} of the grounding line and suggests that the West Antarctic ice sheet is presently thickening after a surge. This evidence favours the surging model for the West Antarctic ice sheet, rather than a continuous collapse model as proposed by Hughes (1972).

CONCLUSION

Field evidence for the Ross Dependency shows conclusively that the West Antarctic ice sheet extended further onto the Continental Shelf in glacial times and has recently collapsed back, probably as a result of the post-glacial rise in sea level. Hughes (1973) has proposed that the West Antarctic ice sheet is continuing to collapse. In this paper evidence of many kinds is presented to support the hypothesis that the West Antarctic ice sheet is presently in a quasi-equilibrium situation with periodic surges followed by recovery. Evidence is presented to suggest that these surges change the level of the Southern Ocean by up to 2 metres and may have caused at least some of the neoglacial periods.

DISCUSSION

H. Flohn: What amount of ice should be injected into the world's ocean in each of your repeated surges (7 in 6000 years)? How long could the duration of each 'surge' last?

A.T. Wilson: Something of the order of 1 to 3×10^5 km^3 over a period of a hundred years or so.

H. Flohn: Starting from the evolution of a mass budget, I could agree with 'surges' of the order 1–2.10^5 km^3, equivalent to a sea level rise of 30–60 cm, spread over periods of the order of several decades or centuries. To avoid misunderstanding, one should describe such prolonged phenomena better as periods of intense calving or similar. They have been described by H. Lamb for ~1840 and ~1895; they certainly may contribute to global climatic fluctuations in a time-scale of 10^2–10^4 years.

J.M. Bowler: Is it not perhaps risky appealing to post-glacial sea level changes in southern Australia to support the thesis of Antarctic surging. The question of Holocene high sea levels in Australia remains a vexed one with current opinion favouring shelf deformation by water loading. Moreover, post-glacial high levels also exist around the south coast of Ireland for which an explanation by Antarctic surging and sea water dilution cannot be invoked. Why appeal to one explanation for Australian coasts and another for Ireland?

A.T. Wilson: Because there is probably more than one phenomena involved. I think we may be talking about different types of sea level change. I am talking about rises and falls in sea level during time intervals of centuries e.g. the Gulf of Thames, New Zealand. You are talking of changes of sea level relative to the land which took place over thousands of years.

C. Lorius: I am not convinced by your interpretation of the $\delta^{18}0$ Byrd ice core as indicating

Fig. 5. Ice-cored moraines on the coast of Brown Hills dry area

Fig. 6. Close-up of ice-cored moraines on coast of Brown Hills showing rapid ablation

periodic surges in the West Antarctic ice sheet. Our measurements of the total gas content (which indicates past elevations) show a rather regular decrease of ice thickness of the order of a few hundred metres for the last 10 000 years in the Byrd station area.

A.T. Wilson: Your evidence clearly favours the Collapse Hypothesis of Terry Hughes.

J. Mercer: You used the presence of striated clasts in a moraine above Darwin Glacier as evidence for a surge, because of rarity of striations in Antarctica. But striated clasts are, in fact, very common locally: for example, on Buckley Island, upper Beardmore Glacier.

A.T. Wilson: Striated clasts are only formed when rock moves against rock. This in my experience is not common in the cold polar glaciers of Antarctica where most of the movement takes place in the ice itself. I believe it will only happen in Antarctica when unfrozen material is being moved or when the energy dissipated by the glacier is sufficient to raise the ice to the pressure melting point.

J.T. Andrews: The Little Ice Age is dated about 1 600 and 1 700 A.D. On your diagram your last surge occurred about 1 000 B.P. Are these data compatible with your surge neoglaciation association. I would assume a surge would lead to a major cooling very shortly after the surge, i.e. the Little Ice Age should be dated ~900 BP.

A.T. Wilson: It depends on what you call the Little Ice Age. There was a marked cooling of the earth that began about 650 years ago in Europe and a little earlier in Iceland. The dates you give are for the coldest part of this cold period. We have a date from Lake Bonney of 700 years B.P., and Gill has a high sea level in Australia at 800 years B.P.

D.J. Drewry: Could you not explain the change in the precipitation regime of lakes in the McMurdo Sound area by changes in sea ice cover in the Ross Sea? Fast ice could be locked to the Victoria Land coast to a distance of several hundreds of kilometres, thus providing the required changes in distance from open water.

A.T. Wilson: Yes you could and it would be nice if this were the case as then our lakes would give us a record of past sea ice cover. However the change was so rapid and complete at 3 000 years B.P. that I feel it was something more catastrophic.

R.H. Thomas: It is not clear why the West Antarctic ice sheet, after collapsing, should establish a temporary equilibrium and then surge again. The stability of a marine ice sheet is determined by sea depth at the grounding line: increase in sea depth causes retreat; decrease causes advance. Presumably isostatic rebound following initial collapse would render the ice sheet more, rather than less, stable.

A.T. Wilson: It is not suggested that it forms a temporary equilibrium but that in advancing towards the edge of the continental shelf it becomes unstable against a surge.

REFERENCES

Curray, J.R., Shepard, F.P. & Veeh, H.H. 1970. Late Quaternary sea level studies in Micronesia: Carmarsel expedition. *Geol. Soc. Am. Bull.* 81: 1865–1880.

Dansgaard – 1964 – Stable isotopes in precipitation. *Tellus* 16: 436–468.

Denton, G.H., Armstrong, R.L. & Stuiver, M. 1971. The Late Cenozoic glacial history of Antarctica. In: K.K. Turekian (ed.), *The Late Cenozoic glacial ages*, Yale University Press, New London Coun.: 267–306.

Fleming, C.A. 1975. The Quaternary record of New Zealand and Australia. In: R.P. Suggate & M.M. Cresswell (eds.), *Quaternary Studies – Selected Papers from IX INQUA Congress, Christchurch, New Zealand*, De. 1973, Roy. Soc. N.Z. Bull. 13, Wellington, New Zealand.

Hollin, J.T. 1962. On the glacial history of the Antarctic. *J. Glaciology* 4: 173–95.

Hughes, T. 1972. Is the West Antarctic ice sheet disintegrating? *J. Geophys. Res.* 78: 7884–7910.

Johnsen, S.J., Dansgaard, H.B., Clausen, H.B. & Langway Jr, C.C. 1972. Oxygen isotope profiles through the Antarctic and Greenland ice sheets. *Nature* 235: 429–432.

Nye, J.F. 1957. The distribution of stress and velocity in glaciers and ice sheets. *Proc. Roy. Soc. Ser. A.* 239: 113–133.

Schofield, J.C. 1960. Sea level fluctuations during the last 4,000 years as recorded by a chenier plain, Firth of Thames, New Zealand. *N.Z. J. Geol. and Geophys.* 3: 467–485

Schofield, J.C. 1973. Post Glacial sea levels of Northland and Auckland. *N.Z. J. Geol. and Geophys.* 16: 359–366.

Thomas, R.H. 1976. Thickening of the Ross Ice Shelf and equilibrium state of the West Antarctic ice sheet. *Nature* 259: 180–183.

Weertman, J. 1974. Stability of the junction of an ice sheet and and ice shelf. *J. Glaciology* 13: 3–11.

Wilson, A.T. 1967. The lakes of the McMurdo Dry Valleys. *Tuatara* 15: 152–164.

Wilson, A.T. 1970. The McMurdo Dry Valleys. In: M.W. Holdgate (ed.), *Antarctic Ecology*, Academic Press, London, Vol. 1: 21–30.

* 5 *

Cainozoic evolution of circumantarctic palaeoceanography

James P. Kennett
Graduate School of Oceanography, University of Rhode Island, Kingston, R.I. 02881, U.S.A.

Manuscript received 12th October 1977

CONTENTS

ABSTRACT

Deep-sea drilling in the Antarctic region (DSDP Legs 28, 29, 35 and 36) has provided much new data about the development of circum-Antarctic circulation through the Cainozoic. The development of this circulation has had profound effects on the total global oceanic circulation and climatic change.

During the Palaeocene (t = ~65 to 55 MY ago), Australia and Antarctica were joined. In the Early Eocene (t = ~55 MY ago), Australia began to drift northwards from Antarctica forming an ocean, although circum-Antarctic flow was blocked by the continental South Tasman Rise and Tasmania. During the Eocene (t = 55 to 38 MY ago) the Southern Ocean was relatively warm and the continent largely non-glaciated. Cool temperate vegetation existed in some regions. By the Late Eocene (t = ~39 MY ago) a shallow water connection had developed between the southern Indian and Pacific Oceans over the South Tasman Rise.

The first major climatic threshold was crossed 38 MY ago near the Eocene–Oligocene boundary, when substantial Antarctic sea-ice began to form. This resulted in a rapid temperature drop in bottom-waters of about 5° C and a major crisis in deep-sea faunas. Thermohaline oceanic circulation was initiated at this time much like that of the present day. The resulting change in climatic regime increased bottom-water activity over wide areas of the deep ocean basins creating much sediment erosion especially in western parts of oceans. A major (~2000 m) and apparently rapid deepening also occurred in the calcium carbonate compensation depth (CCD). This climatic threshold was crossed as a result of the gradual isolation of Antarctica from Australia and perhaps the opening of the Drake Passage.

By the Middle to Late Oligocene (t = ~30 to 25 MY ago), deep-seated circum-Antarctic flow had developed south of the South Tasman Rise as this separated sufficiently from Victoria Land, Antarctica. Major reorganization resulted in Southern Hemisphere deep-sea sediment distribution patterns. During the Early Miocene calcareous biogenic sediments began to be displaced northwards by siliceous biogenic sediments with higher rates of sedimentation reflecting the beginning of circulation related to the development of the Antarctic Convergence. In the Middle Miocene (t = 14 to 11 MY ago) at about the time of closure of the Australian–Indonesian deep-sea passage, the Antarctic ice-cap formed and has remained a semi-permanent feature exhibiting some changes in volume. The most important of these occurred during the latest Miocene (t = 5 MY ago) when ice volumes increased beyond those of the present day. This event was related to global climatic cooling; a rapid northward movement of about 300 km of the Antarctic Convergence and a eustatic sea-level drop that may have been partly responsible for the isolation of the Mediterranean basin.

The Quaternary in the Southern Ocean marks a

peak in activity of oceanic circulation as reflected by widespread deep-sea erosion, very high biogenic productivity at the Antarctic Convergence and resulting high rates of biogenic sedimentation, and maximum northward distribution of ice-rafted debris.

INTRODUCTION

Until rather recently, little was known about the long-term palaeoceanographic and climatic record of the Southern Ocean and Antarctica. Knowledge was mostly restricted to the latest Cainozoic based on studies of piston-core sequences and from accessible geomorphological evidence in areas near the periphery of the Antarctic continent which are not covered by ice. Since December 1972, this however dramatically changed when the first of four deep sea drilling legs of the Deep Sea Drilling Project (DSDP) commenced activities in present-day Antarctic and Subantarctic regions (Table 1). During these expeditions 26 sites were drilled, these being equally distributed throughout the present-day Antarctic and sub-Antarctic water-masses (Figure 1). The initial reports for each of these legs (Leg 28, Hayes, Frakes et al., 1975; Leg 29, Kennett, Houtz et al., 1975; Leg 35, Craddock, Hollister et al, 1976; Leg 36, Barker, Dalziel et al., 1977) form the basis for our recent advances in understanding of palaeoclimatic, palaeoglacial and palaeoceanographic evolution of the region. Major Southern Ocean palaeoceanographic trends as developed in these studies are summarized here.

The Southern Ocean region is important for several reasons: it is the seat of formation of much of the world's abyssal and deep waters resulting from Antarctic sea-ice formation; it forms a critical component of the thermohaline circulation of the world ocean through the production of bottom waters and through upwelling of intermediate waters near the Antarctic Convergence; it is a region of formation of deep ocean currents that may cause widespread deep-sea erosion; it represents the primary mixing arena between the present-day oceans affecting in turn chemical budgets of each ocean and biogeographic patterns, it is the seat of production of much of the present-day biogenic silica due to upwelling of nutrient-rich waters near the Antarctic Convergence; it represents an important site of ice-rafted terrigenous sediments from Antarctica; and the Southern Ocean is intimately linked with the climatic regime of the continent. Each of the environmental characteristics of the present-day Antarctic Ocean has not been permanent throughout the Cainozoic but is the result of a continuing dynamic evolution of the high latitude environmental regime. The deep-sea sedimentary sequences so far obtained have provided information that has either enabled the timing of the development of these features or placed certain constraints on our knowledge of this timing. Useful summaries on various aspects of Antarctic environmental development include: Hayes & Frakes (1975); Kemp et al. (1975); Kennett et al. (1974; 1975); Craddock & Hollister (1976); Barker et al. (1977) and Drewry (1975; 1976). The chronology applied to the geological epochs in this discussion are after the scheme established by Berggren (1972a).

Table 1

Data on the four DSDP expeditions to the Southern Ocean

DSDP Legs		
28.	Dec., 1972–Feb., 1973:	10 Antarctic Sites, Australasian Sector.
29.	March–April, 1973:	8 sub-Antarctic sites, Australasian Sector.
35.	Feb.–March, 1974:	4 Antarctic Sites, S.E. Pacific Sector.
36.	April–May, 1974:	4 sub-Antarctic Sites, Falkland Plateau Area.

EVIDENCE FOR PALAEOCEANOGRAPHIC HISTORY

Geophysical evidence

The primary control over palaeoceanographic change is continental and plate tectonic motions which are determined principally from geophysical evidence. The geophysical data places certain constraints on the interpretation of the timing of particular palaeocirculation events such as the development of passageways as continents move apart. The circum-Antarctic current developed during the Cainozoic as a result of the isolation of Antarctica, by the development of the Drake Passage between South America and the Antarctic Peninsula and the formation of a continuously deep seaway between Australia–Tasmania and East Antarctica.

Geophysical reconstructions of the southwestern Pacific and the Australian–Antarctic regions have been based on magnetic anomaly patterns. The oldest identifiable magnetic anomalies in the southwest Pacific region are approximately 80 MY old and occur along the southwest margin of the Campbell Plateau (Christoffel & Falconer, 1972).

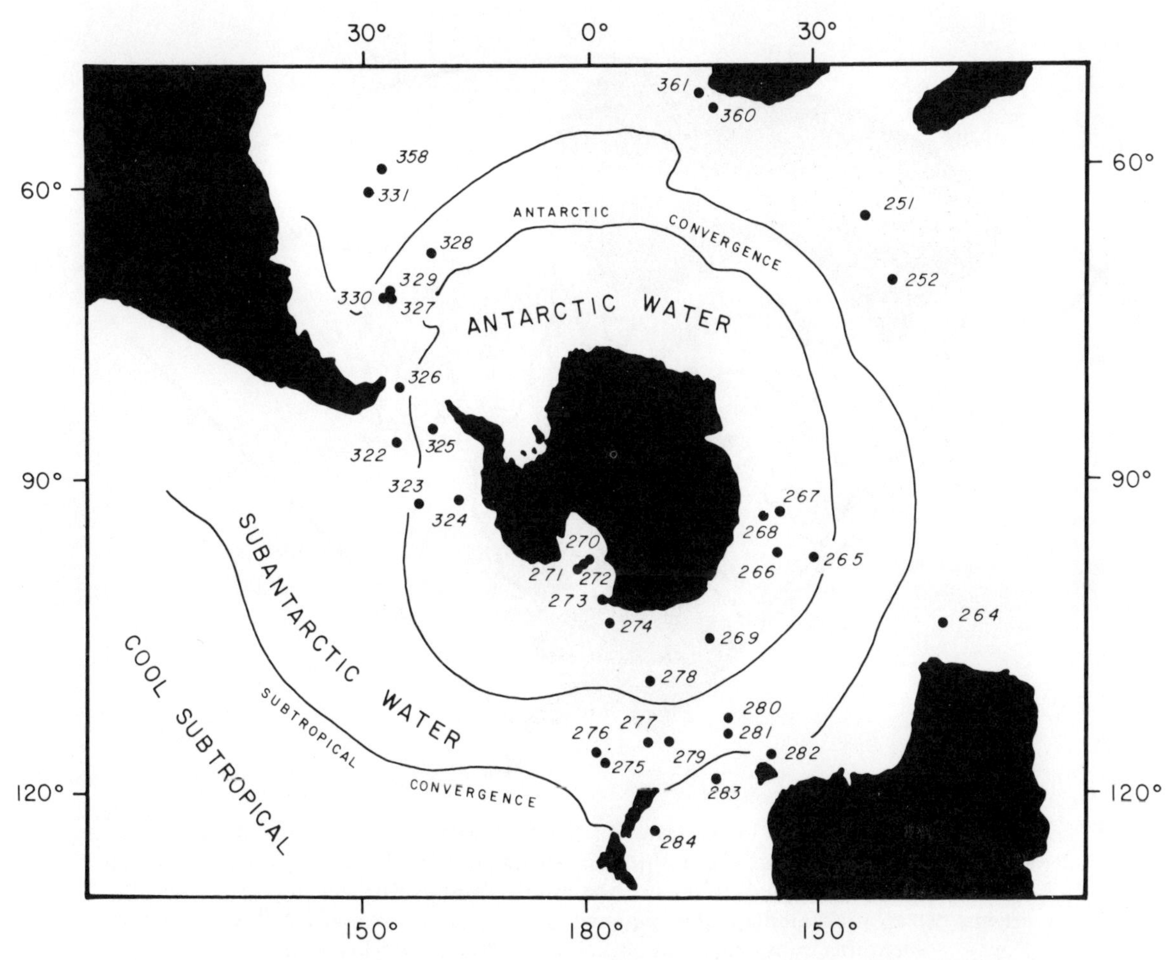

Fig. 1. Location of D.S.D.P. cores in Southern Ocean region and present-day distributions of the Antarctic, sub-Antarctic and cool subtropical water masses each respectively separated by Antarctic and Subtropical Convergences

Sea floor of approximately the same age is thought to occur on both margins of the Tasman Sea beyond Anomaly 32 (Hayes & Ringis, 1973). Anomalies adjacent to New Zealand record the separation of New Zealand from Australia and Antarctica 60–80 MY ago, forming the Tasman Sea between Australia and New Zealand, while Australia remained firmly connected with Antarctica. Approximately 55 MY ago, spreading ceased in the Tasman Sea region and Australia detached from Antarctica and commenced drifting northwards towards its present position (Weissel & Hayes, 1972; Deighton et al, 1976), forming a deep ocean seaway between the continents (Figure 2). Although this spreading commenced during the Early Eocene, a complete deep ocean passageway between the two continents did not develop until well after initiation of spreading because the continental South Tasman Rise remained in close connection with Victoria Land, Antarctica forming a shallow-water barrier to circum-Antarctic flow. The reconstructions (Figure 2) show that the South Tasman Rise would not have cleared Victoria Land sufficiently enough for deep oceanic circulation to develop at least until the time of the Eocene-Oligocene boundary (t = 38 MY ago) and can be expected to have occurred at some time during the Oligocene (t = 38 to 22 MY BP; Kennett et al., 1975; Deighton et al., 1976). The region south of Tasmania is structurally complicated by numerous subparallel fracture zones (Hayes & Conolly, 1972) and thus the existing geophysical evidence is by itself not of sufficient resolution to provide the timing of the opening of the deep seaway south of the South Tasman Rise. Furthermore, the bathymetric history of such developing oceanic seaways is also critical in controlling palaeoceanographic development because A.L. Gordon (in: Hayes & Frakes, 1975) has considered that any topographic high with an average depth of about 1 500 m or less would serve as an effective barrier against significant circum-polar

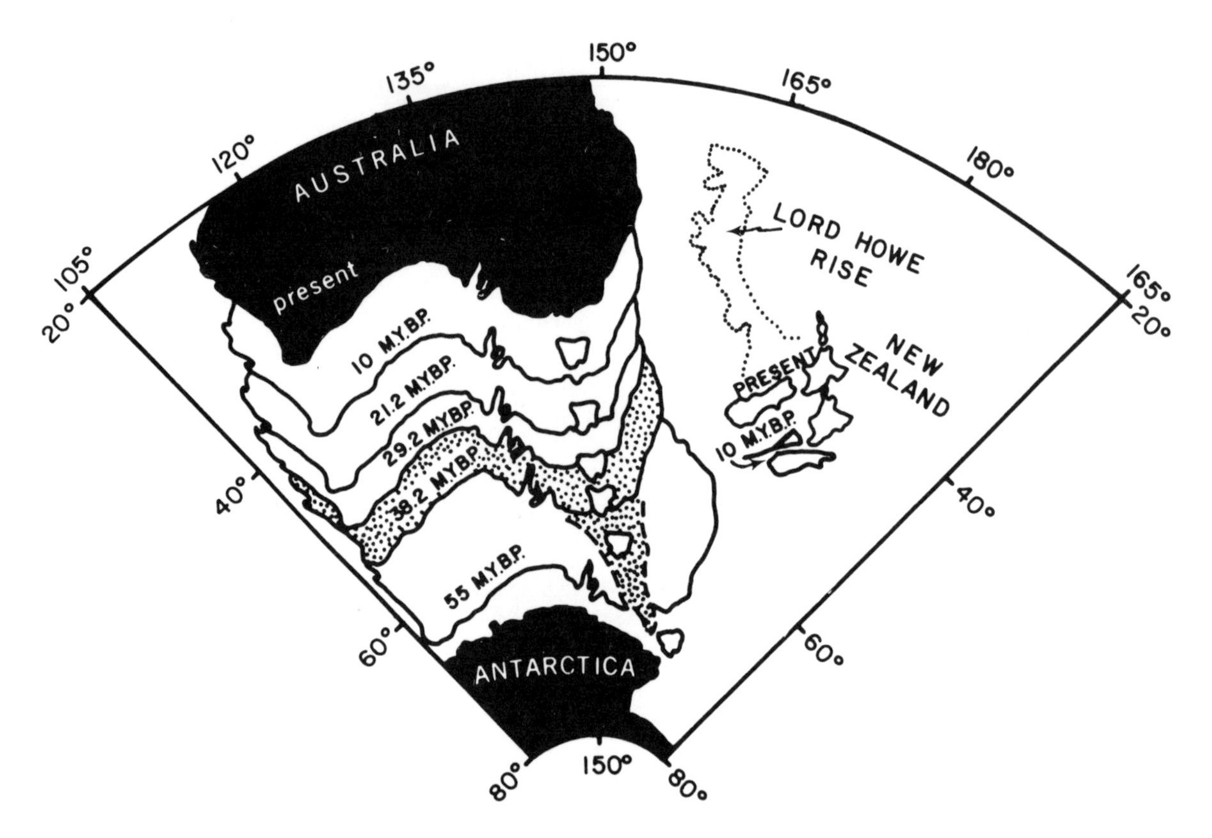

Fig. 2. Successive positions of Australia relative to Antarctica as Australia moved northwards during the Cainozoic. The position of Australia at 38 MY ago (Eocene–Oligocene boundary) is shown (stippled) to include the South Tasman Rise which is of continental crust and which prevented the development of deep circum-Antarctic flow well after spreading commenced (modified after Weissel & Hayes, 1972)

current flow. This information is lacking in the South Tasmanian region. As a result of these constraints imposed by the geophysical and structural data, the sedimentological evidence in the region becomes important in assisting to resolve the palaeoceanographic history.

The opening of the Drake Passage represents the other primary structural event involved in the development of Southern Ocean Circulation. The existing sea-floor magnetic anomaly data in this region does not appear to be of sufficient resolution to precisely define the time of the opening of the Drake Passage to circum-polar circulation, but the possible identification of magnetic anomaly 6 age ocean crust in the Drake Passage (Herron & Tucholke, 1976; Figure 3) suggests that the passage was formed no later than about 21 MY ago (earliest Miocene). Barker & Burrell (1976) consider that the major phase of opening commenced about 30 MY ago (Middle Oligocene) although a shallow gap may have existed earlier. Thus the geophysical data seems to indicate opening of the Drake Passage region between 30 and 22 MY ago (Middle to Late Oligocene) although it is not clear if shallow barriers prevented deep-water circulation (Tucholke & Houtz, 1976).

Changes in sediment facies

In addition to the geophysical evidence which forms the basis for tectonic reconstructions, primary data for palaeoceanographic interpretations obtained is from the sedimentary record. The evolution of sedimentary facies forms a basis for palaeoenvironmental analysis of the oceanic realm, providing information such as the availability and sources of terrigenous sediments, the relative importance of biogenic sedimentary processes and the nature of the deep-sea environment where erosion has not removed the record. One of the clearest and best documented examples of the study of changing sedimentary facies in the Southern Ocean region in response to palaeoceanographic development is for the SW Pacific Basin, the Tasman Basin and the southeast Indian Ocean (Andrews et al., 1975; Hampton, 1975; Kennett et al., 1975). In this region most of the

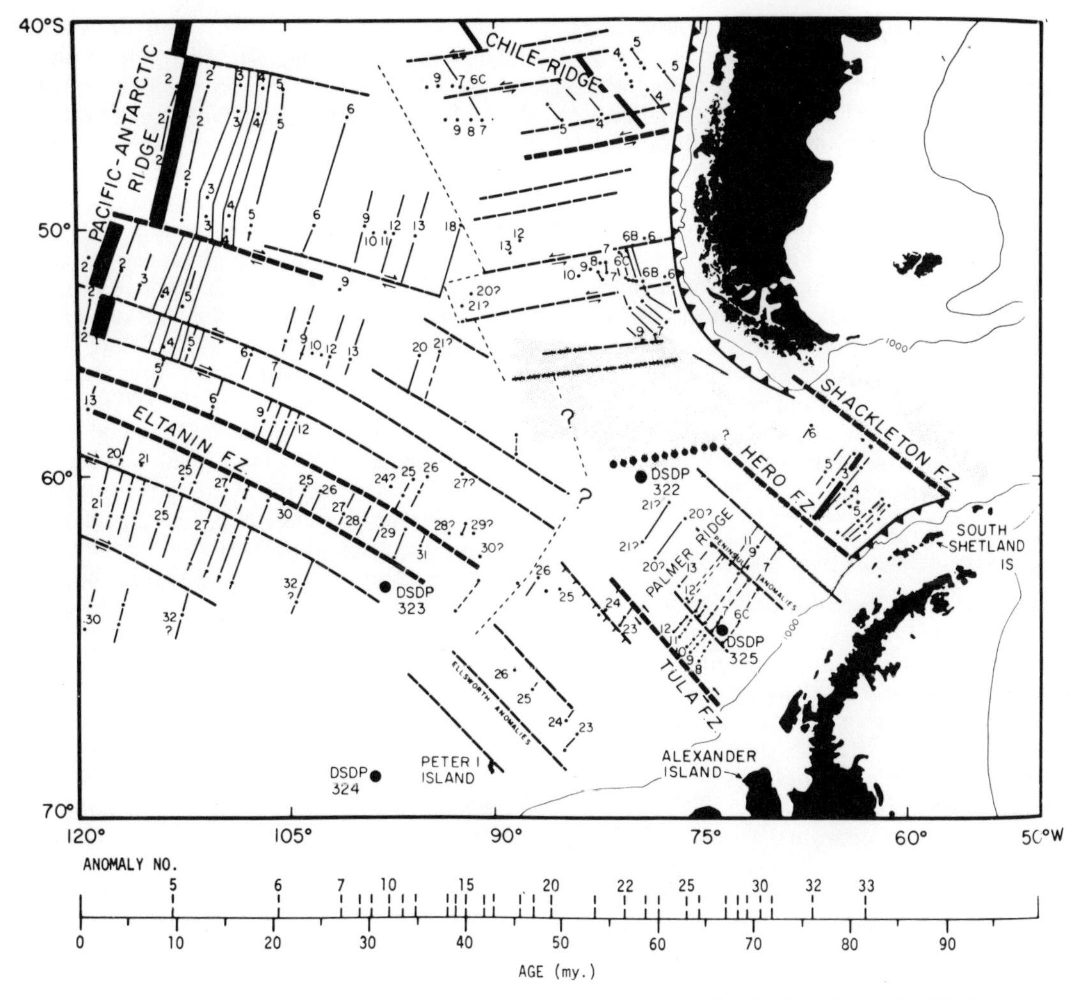

Fig. 3. Magnetic anomalies and principal structural features in the Southeast Pacific Basin. Scale at bottom shows ages of magnetic anomalies (after Herron & Tucholke, 1976)

sediment has been grouped into three facies: terrigenous silt and clay, siliceous ooze and calcareous ooze chalk. At several sites, these facies succeed each other in this order. At early stages of rifting and continental spreading when the continents were located close together deposition of terrigenous silt and clay facies took place. The age of this facies is diachronous throughout the region reflecting different ages for initiation of the basins; the Palaeocene and Eocene in the Tasman Sea Basin; and during the Eocene and Oligocene in the southeast Indian Ocean between Australia and Antarctica. As the oceanic realm became established with further spreading, increased basinal isolation from terrigenous sources occured and biogenic productivity increased in surface waters. The terrigenous facies was then succeeded by late Palaeogene biogenic ooze in the Tasmanian region. This information has been vital in the interpretation of history of circumpolar deep circulation south of the South Tasman Rise. At Site 280, immediately south of the South Tasman Rise, the earliest marine sediments overlying basement of Eocene age are of the terrigenous silt and clay facies, deposited at a basin margin relatively near a source area of detrital sediments rather than near the centre of the basin (Kennett, Houtz et. al., 1975). A lack of primary sedimentary structures and poor sediment sorting suggests sluggish bottom-water circulation. Additionally, high organic carbon content indicates that the basin was also poorly oxygenated during much of the time of deposition and very low biogenic content suggests low biogenic productivity although almost complete elimination of an original biogenic component cannot be ruled out. Biogenic sediments become increasingly important during the Oligocene replacing the early terrigenous sediments. Immediately to the south in the Antarctic continental margin area, the sequence at Site 274 records terrigenous sedimentation during the Early Oligocene, a hiatus during the Late

Oligocene and less terrigenous sediment input during the Neogene. The sites in the region indicate that although northward spreading of Australia commenced 53 MY ago in the Early Eocene, open-ocean conditions did not develop in the South Tasmanian sector until within the Oligocene. Furthermore, the sedimentological evidence does not support the presence of any major deep-water circum-polar circulation until during the Oligocene.

The Cainozoic sequences drilled in the Bellingshausen Basin and the Falkland Plateau regions record that open ocean processes have occurred in these regions during the entire Cainozoic although significant changes have occurred in the distribution of biogenic sedimentation as surface-water masses developed (Craddock & Hollister, 1976; Barker et al., 1977).

Unconformities

A further important criterion that has been employed in the study of Southern Ocean palaeoceanography is the temporal and geographic distribution patterns of deep-sea unconformities. Whether erosion, non-deposition or sediment accumulation occurs on the sea-floor is determined by the dynamic balance between the rate of supply of sediments and the rate at which they are removed. The rate of removal depends on processes at the ocean floor that either physically transport sediments to other regions or chemically dissolve them (Moore et al., in press). Several studies have demonstrated that well defined unconformities are widely distributed throughout various parts of the ocean basins including those in the southern high latitudes. In this region, the distribution of breaks in the Cainozoic sedimentary record is a record of the complex interplay between the development of ocean topography, the circum-Antarctic current, which although a wind-driven surface current, is attenuated to abyssal depths (Gordon, 1971) and the development of the climatic and glacial regime of Antarctica, especially as this has controlled the formation of bottom waters. In the vicinity of Australia and New Zealand, the distribution of unconformities has been summarized and discussed by Kennett et al. (1972), Edwards (1973), Hayes & Frakes (1975) and Kennett et al. (1975). Although local factors can complicate the record of unconformity distributions, in general the history of deep-sea erosion forms coherant patterns in Cainozoic sequences of the Tasmanian–Antarctic sector and further north in the Tasman and Coral Seas. In the more northern sites, the Palaeogene record tends to be disrupted by numerous unconformities some of which are of regional extent. In contrast, the Neogene sequences in this region are relatively continuous and have been little effected by deep-sea erosion (Kennett et al., 1972). In the Southern Ocean sector to the south, however, the situation is generally opposite with more continuous Eocene to Early Oligocene sections preserved, while the Middle to Late Oligocene and younger sediments have often been removed by deep-sea erosion. Kennett et al. (1975) have interpreted these changing distribution patterns as reflecting a major change in oceanic circulation during the Middle Oligocene (t = ~30 MY ago) as the circum-polar current developed south of the South Tasman Rise. The most strategically located DSDP sites to record this process are Sites 274, 280, 281 and 282 (Figure 1). All record continuous deposition from the time of formation of basement during the Middle to Late Eocene until about the Middle Oligocene at which time the sequences become highly disrupted by unconformities considered by Kennett et al. (1975) to have formed as a result of deep circum-polar current activity. Continuous deposition from the Late Eocene to present-day in site 269 far to the west of the Tasmanian–Victoria Land sector (Figure 1) took place because it was sufficiently removed from the highly active bottom currents that have occurred at times south of the South Tasman Rise.

Further to the north in the northern Tasman Sea and Coral Sea, widespread disconformities within the Palaeogene (Edwards, 1973) were created by northward flowing bottom waters generated in the Ross Sea sector of Antarctica. The changing importance of these bottom currents during the Palaeogene is considered by Kennett et al. (1972) to have resulted from climatic-glacial events in Antarctica which effected the production and intensity of bottom waters. After the circum-polar current developed during the Oligocene, northward flow through the Tasman Sea became greatly diminished and relatively continuous sedimentary sequences were deposited throughout the Neogene (t = 22 MY to present day).

Attempts have also been made to date the inception of circum-polar flow through the Drake Passage using the distribution of unconformities on the limited number of DSDP sequences in that region. In Bellingshausen Basin Site 323, Tucholke et al. (1976) record a possible important hiatus representing the Middle Palaeocene to Oligocene (t = ~60 to 22 MY BP). They also suggested that this unconformity was formed by the development of deep-water circum-polar circulation through the Drake Passage by the early Miocene. This interpretation however may not be correct because the site is located far to the west of the Drake Passage and the unconformity may merely represent an intensification of bottom-water activity in this region during the Palaeogene due to palaeoclimatic or other factors and in no way be related to the opening of the Drake Passage.

In the Falkland Plateau region, all of the DSDP

sites (Leg 36) contain disconformities representing the Middle and Late Eocene and Early Miocene. Factors controlling the intensification of bottom currents forming these breaks in sedimentation are not clearly known and insufficient deep-sea drilling in the region does not allow differentiation of bottom-current effects due to circum-polar current activity from those related to northward flow of bottom waters across the Falkland Plateau region from Antarctic sources, especially the Weddell Sea region.

Distribution of siliceous and calcareous sediments

In the present ocean, the Antarctic Convergence (Polar Front) sharply coincides with the deposition of calcareous biogenic oozes to the north from siliceous biogenic oozes to the south, resulting from the rapid temperature change in surface waters creating large differences in the planktonic biota. This observation has been used by several workers studying Antarctic Cainozoic sedimentary sequences as an index of prior positions of the Antarctic Convergence (Kennett, Houtz, et al., 1975; Hayes & Frakes, 1975; Tucholke et al., 1976). On the other hand, it is not known if this sedimentary boundary has reflected the same water-mass boundary or temperature changes during the Cainozoic as it does at the present-day Antarctic Convergence. For instance Kemp et al. (1975) note that although it is tempting to equate the factors which have in the past determined the siliceous-calcareous boundary with the present-day Antarctic Convergence, the parallel is not exact, since the modern oceanic boundary separates diatomaceous ooze from calcareous foraminiferal ooze, while throughout much of the Neogene, the ancient boundary separated diatomaceous from calcareous nannofossil ooze with less Foraminifera. Nevertheless a general association throughout the Cainozoic between the distribution of siliceous oozes and ice-rafted debris supports the contention that the siliceous oozes have approximated the position of surface waters with temperature characteristics like those of the present-day Antarctic water mass (Tucholke et al., 1976).

Studies of the temporal distribution of diatomaceous sediment have shown that the first sediments of this type were deposited closely adjacent to Antarctica and that the northern limits of this province have expanded northwards during the Neogene creating diachronous biogenic sedimentary facies (Kemp et al., 1975; Tucholke et al., 1976).

Another related approach which has been widely used, is the determination of siliceous biogenic sedimentation rates as a general measure of the productivity of surface-water masses in the region (assuming little dissolution of opaline silica) and thus the rate of upwelling of nutrient-rich intermediate waters. Changes in sedimentation rates may reflect either latitudinal movement of highly productive surface waters or changes in productivity with time at the same location. These two effects are difficult to differentiate although one DSDP site (Site 278) located at the present-day position of the Antarctic Convergence may have recorded changes in biogenic productivity through the middle and late Cainozoic. This is possible at this location because the Antarctic Convergence is presently diverted relatively southward and is fixed in position because of topographic control by the Macquarie Ridge which interferes with circum-polar flow (Gordon, 1971; Kennett, Houtz et al, 1975). As the Macquarie Ridge has been a prominant topographic feature throughout the Neogene (Kennett, Houtz et al., 1975), the Antarctic Convergence near Site 278 has probably been in a relatively stable position thus largely recording productivity changes through time. At this site, the Oligocene is represented by carbonate oozes. The first siliceous oozes are of Early Miocene age and at these latitudes probably reflect siliceous productivity associated with the early development of the convergence. Moderately low sedimentation rates indicate, however, that siliceous biogenic productivity was low and upwelling was sluggish. Alternations of siliceous-rich and calcareous-rich biogenic sediments in the Middle Miocene almost certainly record minor latitudinal fluctuations of the Antarctic Convergence. A considerable increase in siliceous biogenic productivity that began about the Middle to Late Pliocene and has continued to the present indicates an intensification of upwelling associated with the Antarctic Convergence. This is assumed to be related to higher rates of oceanic turnover in response to the Late Cainozoic glacial development (Kennett, Houtz et al., 1975).

SUMMARY OF CAINOZOIC PALAEOCEANOGRAPHIC HISTORY OF ANTARCTICA

Eocene ($t = 55–38$ MY)

Palaeomagnetic evidence shows that the Antarctic continent has essentially been in a polar position since the Late Cretaceous (McElhinny, 1973; Lowrie & Hayes, 1975). During the Palaeocene ($t = \sim 65$ to 55 MY ago) Australia and South America were joined to Antarctica (Weissel & Hayes, 1972; Herron & Tucholke, 1976). Insufficient deep-sea sequences of Palaeocene age have been obtained from Southern Ocean latitudes to provide information on palaeoceanographic conditions.

During the Early Eocene ($t = 53$ MY BP), Australia commenced drifting northwards from Antarctica forming an ocean between the two continents which has continued to increase in size (Weissel & Hayes, 1972; Edwards, 1975). The deep basin forming to the southwest of the South Tasman Rise during the Eocene as a result of spreading, received fine grained, poorly sorted detrital sediments with little biogenic material and high organic carbon content reflecting highly restricted deep-water circulation and a lack of any deep circum-polar current flow (Andrews et al., 1975; Hampton, 1975). Circum-polar current flow was blocked by continental masses associated with the present-day South Tasman Rise and the Drake Passage, which continued throughout the entire Eocene, although rather shallow-water marine connections of outer neritic-uppermost bathyal water depths (about 100 to 300 m) formed over the South Tasman Rise during the Late Eocene (t = ~40 MY ago; Kennett, Houtz et al., 1975). This shallow-water marine connection probably produced the first direct communications between shallow water and planktonic marine organisms of the southern Indian and Pacific Oceans.

The major avenues of interocean circulation during the Eocene were in the equatorial and low-latitude regions with unrestricted connections in the Indo-Pacific north of Australia and the Atlantic–Pacific through the middle America seaway. Middle latitude interocean circulation existed between the Indian and Atlantic Oceans south of Africa and in the Northern Hemisphere via the Tethyan seaway which was becoming highly restricted during this time (Frakes & Kemp, 1973).

The Early Cainozoic (Palaeogene) palaeotemperature record of the high southern latitudes

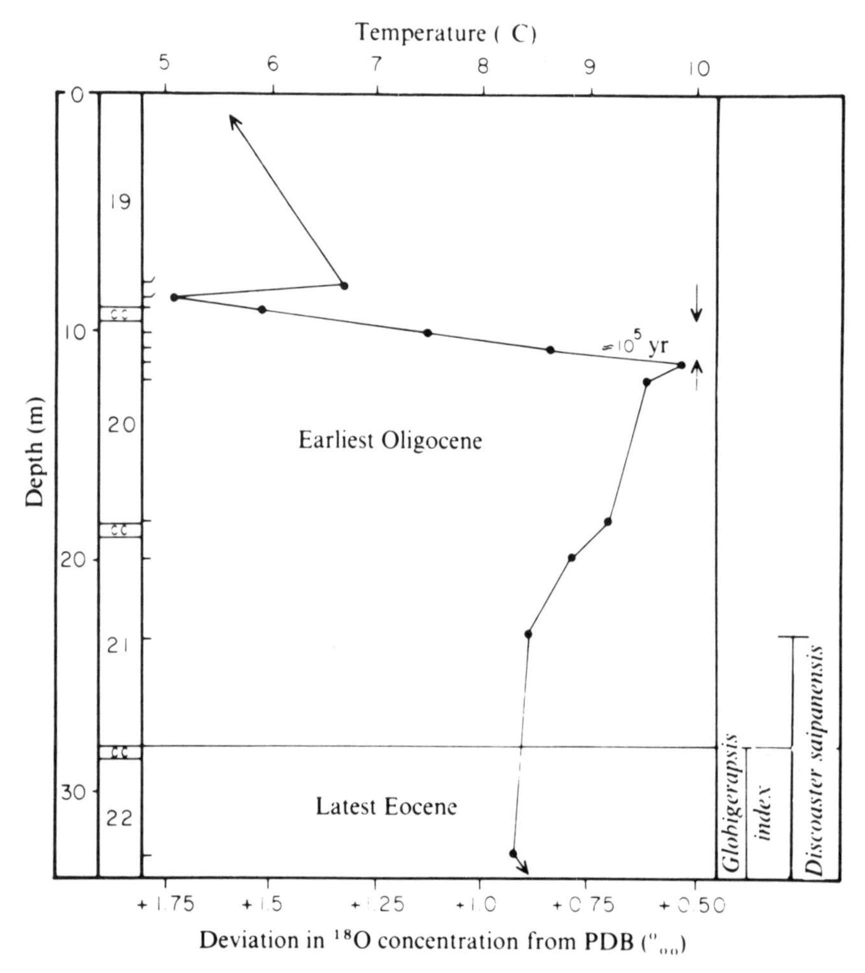

Figure 4. Palaeotemperature curve and oxygen isotope data for benthonic Foraminifera in latest Eocene–earliest Oligocene sediments of D.S.D.P. Site 277 (sub-Antarctic). Bottom temperature values were calculated using the procedure of Shackleton and Kennett (1975a). The duration of the temperature drop of 4° C within core 20 is calculated from sedimentation rates at between only 75 000 and 100 000 yr. Thickness is plotted from top of core 19 which is 168.5 m below the ocean floor (after Kennett & Shackleton, 1976)

(Antarctic and sub-Antarctic) is marked by rather rapid temperature decrease superimposed on a steady climatic cooling commencing in the Early Eocene (Shackleton & Kennett, 1975a). The Sub-Antarctic was marked by warm surface-water temperatures which decreased from 19°C in the Early Eocene to 11°C by the Late Eocene (Shackleton & Kennett, 1975a). Bottom waters were relatively warm reflecting high surface temperatures adjacent to the Antarctic continent. These also decreased through the Eocene similar to sub-Antarctic surface-water temperatures. Temperatures on at least parts of the continent were warm enough during the Eocene to support a cool temperate vegetation (*Nothofagus* flora) based on palynological evidence (Cranwell et al., 1960; McIntyre & Wilson, 1966). Palaeontological evidence from Eocene sequences

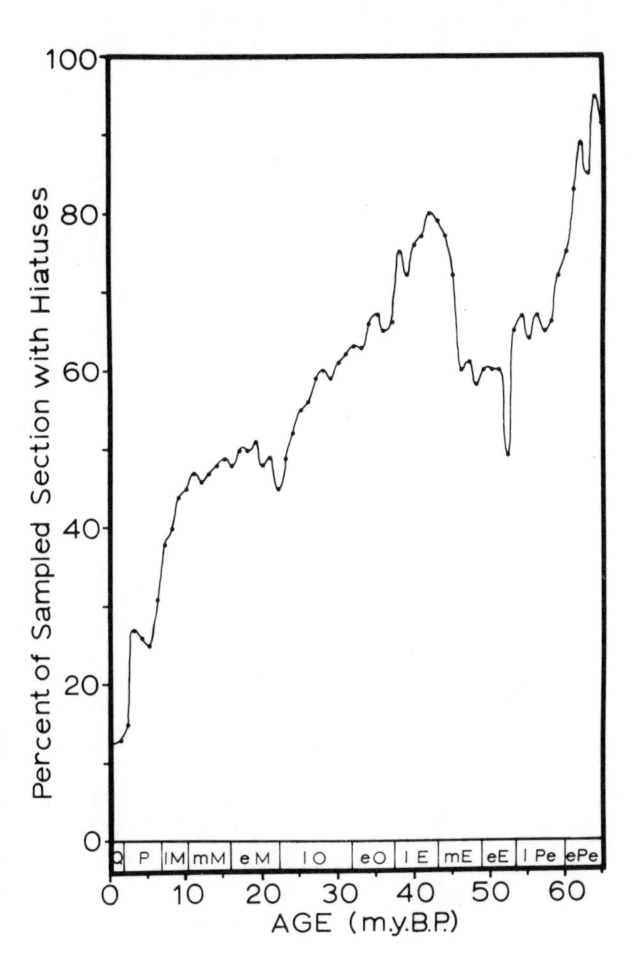

Fig. 5. Percentage of sampled D.S.D.P. sections containing hiatuses (Legs 1–32). Data from Moore et al. (in press) has been corrected for over-representation of areas which have been densely sampled. Note increase in hiatus distribution in Late Eocene to Early Oligocene (after Moore & Heath, in press).

elsewhere in the world show that warm, humid conditions were prevalent (Wolfe & Hopkins, 1967; Frakes & Kemp, 1973). During the late Eocene (t = ~38 MY ago) calcareous biogenic sediments were deposited adjacent to the continent and siliceous biogenic sedimentation is largely unknown from the region (Hayes & Frakes, 1975; Kennett et al., 1975).

The first major Antarctic climatic threshold was crossed about 38 MY ago near the Eocene–Oligocene boundary. Oxygen isotopic changes in deep-sea benthonic Foraminifera from the sub-Antarctic (Shackleton & Kennett, 1975a; Kennett & Shackleton, 1976) and the tropical Pacific region (Savin et al., 1975) indicate that bottom temperatures decreased rapidly by approximately 5°C (Figure 4) to approximate present-day levels at the respective water depths for each drilled sequence. This temperature reduction was calculated by Kennett & Shackleton (1976) to have occurred within 10^5 years which is remarkably abrupt for pre-glacial Tertiary times. and is considered to represent the time when large scale freezing conditions developed at sea-level around Antarctica forming the first significant sea-ice. It is inferred that at this time Antarctic bottom-water began to be produced, that bottom-water temperatures fell to approximate present-day values forming the psychrosphere (Benson, 1975a), and that thermohaline circulation developed much like in the present ocean. Savin et al. (1975) believe that bottom-water temperatures continued to drop later in the Cainozoic but Shackleton & Kennett (1975a) believe any such temperature change cannot be easily differentiated from ice-volume oxygen isotopic changes.

The drastic change in climate near the Eocene–Oligocene boundary increased bottom-water activity throughout much of the deep ocean basins particularly in western sectors. This is based on the widespread occurrence of deep-sea unconformities centred near the Eocene–Oligocene boundary (Figure 5; Rona, 1972; Kennett et al., 1972; Moore et al., in press; Moore & Heath, in press). In terms of deep-sea erosion, this event is one of the most dramatic for the entire Cainozoic. Also coinciding with this event was a major and apparently rapid deepening (about 2 000 m) of the Calcium Carbonate Compensation Depth (CCD) which created a sharp contrast between carbonate poor Late Eocene and carbonate rich Early Oligocene deep-sea sediments (Van Andel et al., 1975; Berger & Winterer, 1974). This conspicuous change in the CCD has been interpreted by Van Andel et al. (1975) as resulting from a major change in the nature of bottom waters effecting concentrations of CO_2 as proposed by Berger (1970, 1973). Highly oxygenated 'young' bottom waters are not as corrosive to the tests of Foraminifera and coccoliths as 'old' bottom waters that are rich in CO_2. Bottom waters during the Eocene are considered to have

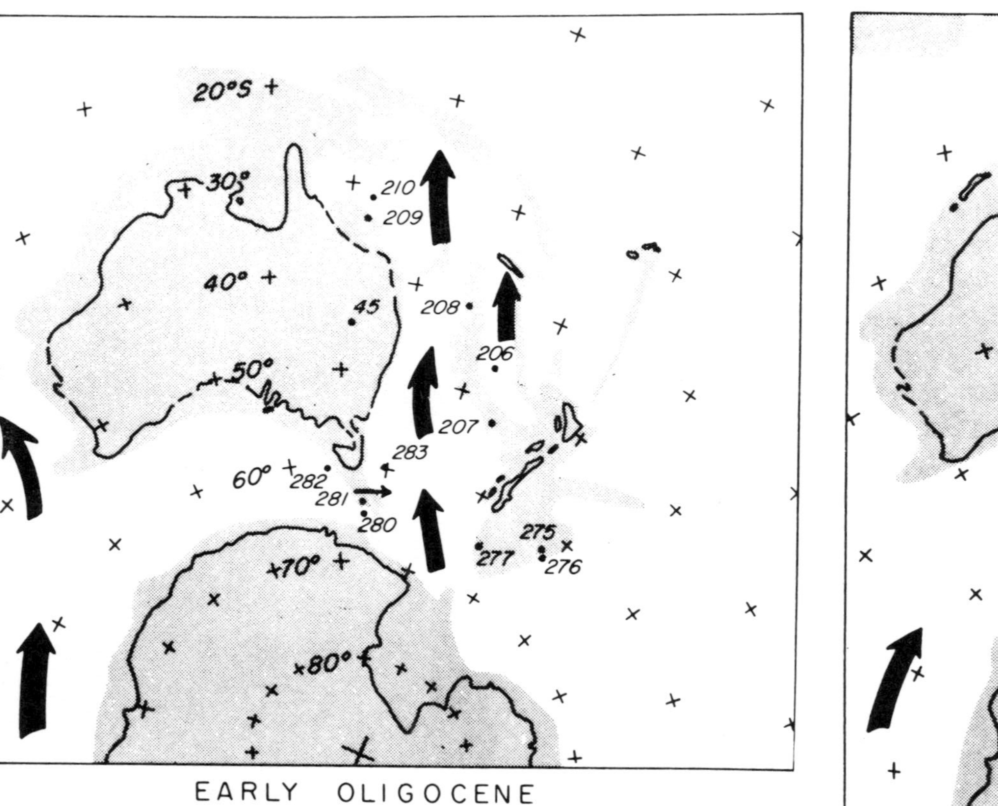

Fig. 6. Early Oligocene (37 MY BP) reconstruction of Australia, Antarctica and New Zealand, and associated ridges (after Weissel & Hayes, 1972) with suggested directions of bottom-water circulation (arrows) northward from Antarctic sources through the Tasman–Coral Sea area and west of Australia. No Circum-Antarctic Current had developed by this time, but currents flowed across the shallow south Tasman Rise (after Kennett et al., 1975).

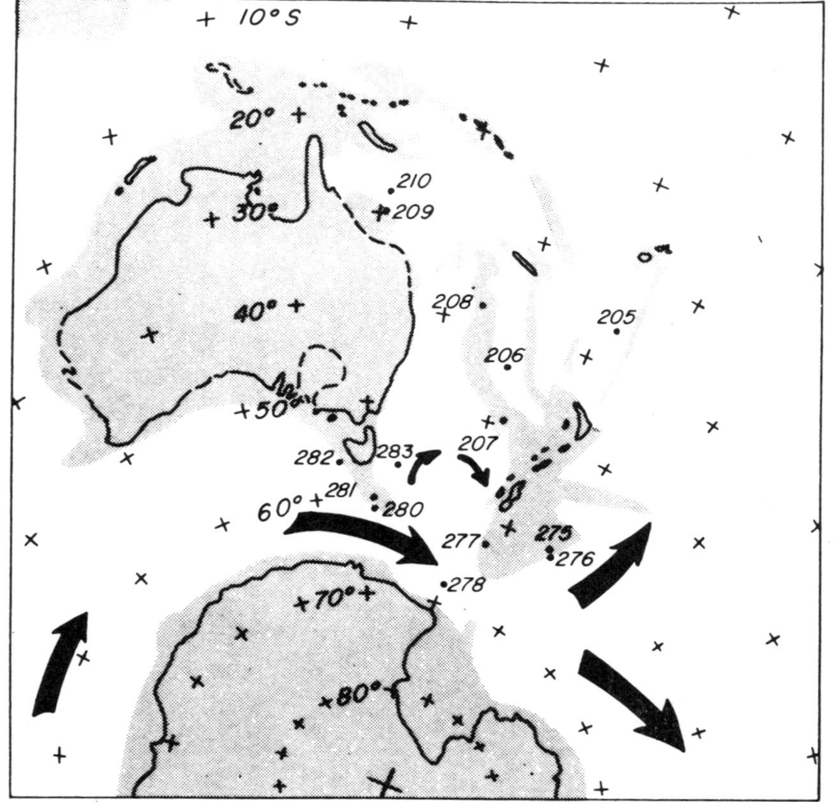

Fig. 7. Late Oligocene (30 MY BP) reconstructions of Australia, Antarctica and New Zealand, and associated ridges (after Weissel & Hayes, 1972) showing the direction of the Curcum-Antarctic Current south of Australia and New Zealand and a western-boundary current east of New Zealand. This direction has been retained until the present (after Kennett et al., 1975).

been richer in CO_2 because of lower production of oxygen rich polar waters, but the development of sea-ice in the Oligocene created a higher turnover of oxygen rich Antarctic bottom waters in the deep oceanic basins. As a result the CCD deepened.

The very sharp drop in bottom-water temperatures near the Eocene–Oligocene boundary also produced a major crisis in deep-sea benthonic faunas (Benson, 1975a; Douglas, 1974) effecting both ostracods and benthonic Foraminifera. Deep-sea ostracod faunas at this time show the most important changes for the Cainozoic. These changed from heavy, less ornamented Eocene forms to thinner, more highly ornamented Oligocene and younger forms. Benson (1975b) considers that the younger forms developed thinner walls to conserve calcium carbonate but added ornamentation to compensate for resulting loss of shell strength.

Decrease in surface-water temperature in the Antarctic also created major changes in high-latitude planktonic foraminiferal faunas which at the beginning of the Oligocene assumed characteristics typical of the present-day assemblage which includes very low diversity and relatively simple morphology (Kaneps, 1975).

Oligocene (t = 38–22 MY)

The conspicuous climatic cooling near the beginning of the Oligocene (Devereaux, 1967; Edwards, 1968; Jenkins, 1968; Shackleton & Kennett, 1975a) produced widespread glacial conditions throughout Antarctica. The extent of this glaciation is not known but the oxygen isotopic evidence indicates that no icecaps had yet developed on Antarctica (Shackleton & Kennett, 1975a; Savin et al., 1975). Sub-Antarctic surface water temperatures had fallen to about 7° C by the Early Oligocene which is similar to present-day values. Antarctic surface-water temperatures are also inferred to have reached values close to those of the present day (Shackleton & Kennett, 1975a). A diversity of palaeontological evidence throughout the world indicates relatively cool global climates (Hornibrook, 1971; Edwards, 1968; Wolfe & Hopkins, 1967; Frakes & Kemp, 1973). Calcareous biogenic sediments continued to be deposited close to the continent, while siliceous biogenic sediments, although expanding in importance, remained highly restricted and inconspicuous in distribution (Hayes & Frakes, 1975; Tucholke et al., 1976).

As a result of cool polar climates, vigorous bottom-water circulation was particularly pronounced through much of the Early Oligocene causing extensive deep-sea erosion. Widely distributed unconformities in the Tasman and Coral Seas resulted from the northward movement of actively eroding bottom waters derived from Antarctica (Figure 6; Kennett et al., 1972; Kennett et al., 1975).

By the Middle to Late Oligocene (t = ~ 30 to 25 MY ago), a substantial ocean had formed between Antarctica and Australia, while equatorial circulation north of Australia had become rather restricted. At this time major changes occurred in oceanic circulation in the Southern Hemisphere as deep circum-Antarctic flow developed south of the South Tasman Rise which by this time had sufficiently cleared Victoria Land, Antarctica (Figure 7). It is also possible that at this time the Drake Passage region began to open (Barker et al., 1976; Craddock & Hollister, 1976). Thus the circum-polar current became established (Kennett et al., 1975) and a major reorganization resulted in southeast Pacific and Southern Ocean sediment patterns. This was manifested by the development of Late Oligocene through Neogene disconformities in deep basins south of Tasmania and an almost reciprocal change to essentially uninterrupted Neogene sedimentation in northern Tasman Sea and Coral Sea regions (Kennett et al., 1972, 1975). The latter resulted from a reduction in northward flow of bottom waters through the Tasman-Coral Sea regions as the western boundary current became strongly established to the east of New Zealand (Figure 7). Sediment distribution patterns indicate that no subsequent major changes have occurred in the patterns of deep-sea circulation since the Late Oligocene, although conspicuous fluctuations in intensity have occurred in bottom-water activity.

During the Middle to Late Oligocene, the siliceous biogenic belt was narrow but expanding (Hayes & Frakes, 1975) and diatomaceous sedimentation commenced in the southwest Atlantic sector of the Southern Ocean (Barker et al., 1977). Cool-temperate vegetation persisted in the Ross Sea region until the end of the Oligocene at which time it disappeared (Kemp & Barrett, 1975).

Early Miocene (t = 22–14 MY)

By the early part of the Neogene, the basic patterns of water-mass distributions in the Southern Ocean had become established. During the Neogene, however, the relative importance of the various water masses and the intensity of oceanic processes did change substantially, thereby effecting sediment distribution- One of the more conspicuous changes is related to siliceous and calcareous biogenic sediment distributions as the former continued to expand at the expense of the latter, and sedimentation rates showed related increases (Kemp et al., 1975; Hayes & Frakes, 1975; Kennett et al., 1975; Craddock & Hollister, 1976). The Early Miocene also marks the establishment of a permanent steep temperature gradient between the polar and tropical regions related to the development of polar to tropical surface water-mass belts. These have largely retained their identity during the Middle and Late Cainozoic glacial changes but have oscillated latitudinally. This global change is also reflected by a parallel development of latitudinal provinciality in oceanic planktonic assemblages, which have also maintained their distinctive characteristics throughout the Neogene (Kennett et al., 1972). Of particular importance is the establishment by the Early Miocene of sharp differences between Antarctic and sub-Antarctic microfossil assemblages (Burns, 1975; Edwards & Perch-Nielsen, 1975) and a related sharp break between Antarctic siliceous and sub-Antarctic calcareous biogenic sedimentation. This boundary is strongly diachronous through the Neogene as the siliceous biogenic province expanded northwards (Kemp et al., 1975; Tucholke et al., 1976). Biogenic sedimentation rates, as monitored at Site 278 on the present-day Antarctic Convergence, show major changes during the Neogene (Kennett, Houtz et al., 1975). In the Early Miocene, low sedimentation rates suggest upwelling remained sluggish in northern Antarctic waters. Alternations of siliceous and carbonate, rich biogenic sediments in the Middle Miocene record minor fluctuations in the position of the Antarctic Convergence.

Oxygen isotopic data from the sub-Antarctic clearly records that in the Early Miocene, sub-Antarctic surface water temperatures increased by about 3°C to levels similar to those of the Late Eocene. Bottom-water temperatures also increased by about 2°C during the Early Miocene reflecting an increase in surface-water temperatures adjacent to the Antarctic continent (Shackleton & Kennett, 1975a). A related decrease in the flow of Antarctic Bottom Water in equatorial Pacific sequences is inferred by Van Andel et al. (1975) based on fewer hiatuses, a shoaling of the CCD, decreasing width of the equatorial carbonate-rich zone and other features resulting from CO_2 buildup and increased carbonate dissolution. The temperature increase recorded in Early Miocene Southern Ocean sequences is also reported in other regions of the world based on a variety of palaeontological evidence (Hornibrook, 1971).

Middle Miocene (t = 14–10 MY)

The next major global climatic threshold was reached near the beginning of the Middle Miocene (t = ~14 MY ago) at which time the East Antarctic ice cap developed (Shackleton & Kennett, 1975a; Savin et al., 1975). The oxygen isotopic evidence indicates that previous to this, no significant ice volumes had accumulated on the Antarctic continent. The cause of this event remains unexplained, and it is unknown why ice-cap development was delayed an additional 10 to 15 MY from the time of oceanic isolation of the continent and the development of the circum-polar current. Savin et al. (1975) have speculated that the ice buildup resulted from the initiation of circum-polar flow through the Drake Passage region in the Middle Miocene but the latest evidence suggests an earlier development. It may not be a coincidence that development of the ice-cap occurred at a time of warmer global climates, because increased precipitation on the Antarctic continent may have resulted from slightly warmer Antarctic surface waters. The buildup of the ice cap was essentially complete by the late Middle Miocene to early Late Miocene (t = ~ 10 to 12 MY ago; Shackleton & Kennett, 1975a).

By the Middle Miocene, the lysocline in equatorial Pacific regions became established close to its present position probably in response to full Antarctic glacial development (Van Andel et al., 1975). By the end of the Middle Miocene, the palaeontological and oxygen isotopic evidence indicates that polar surface water temperatures began to decrease again (Edwards & Perch-Nielsen, 1975; Burns, 1975; Shackleton & Kennett, 1975a).

Late Miocene (t = 10–5 MY)

Since the Middle Miocene, the East Antarctic ice cap has remained a semipermanent to permanent feature exhibiting some changes in volume. Global climates were cool during much of the Late Miocene (Ingle, 1967; Kennett, 1967; Wolfe & Hopkins, 1967). Towards the end of the Miocene and during the earliest Pliocene, a distinct and rapid northward movement (300 km) occurred in the siliceous biogenic sediment belt in both the SE Pacific and Indian Ocean regions (Kemp et al., 1975; Tucholke et al., 1976). This event is equated with a rapid northward movement of the Antarctic Convergence and Antarctic surface water mass (Kemp et al., 1975) and is considered to be related to a major expansion of the Antarctic ice-cap at this time (Shackleton & Kennett, 1975b), a northward extension of the Ross Ice Shelf much beyond present-day limits (Hayes & Frakes, 1975) and a possible increase in volume of grounded ice about 1.8 times the present volume (Queen Maud Glaciation: t = >4.2 MY ago, Mayewski, 1975). This is based on terrestrial geomorphological and sedimentological evidence. Rates of siliceous biogenic sedimentation also increased in Antarctic waters.

Outside of the Antarctic region this event is considered to be related to a major Late Miocene climatic cooling recorded in various regions (Kennett, 1967; Ingle, 1967; 1973; Bandy, 1967; Casey, 1972; Kennett & Vella, 1975; Barron, 1973) although this has been contested by Beu (1974) on the basis of

New Zealand molluscan fossils. Furthermore a distinct regression of the sea as recorded in shallower sequences of this age is probably the result of the related glacial-eustatic sea-level lowering (Kennett, 1967; Shackleton & Kennett, 1975b; Berggren & Haq, 1976; Ryan et al., 1974; Van Couvering et al., 1976). It has also been suggested that the latest Miocene (Messinian Stage) isolation of the Mediterranean Basin and resulting evaporitic sedimentation may have partly resulted from this glacial-eustatic lowering.

Pliocene and Quaternary

Although evidence exists for Late Miocene cooling in the Northern Hemisphere (Ingle, 1967; Wolfe & Hopkins, 1967) no evidence exists for ice-sheet development until the Late Pliocene about 2.5 to 3 MY ago (Shackleton & Kennett, 1975b; Berggren, 1972b) when a further global climatic threshold was passed. Since then, major oscillations have continued in Northern Hemisphere ice-sheets forming the classical Quaternary glacial and interglacial episodes (Emiliani, 1954; Shackleton & Opdyke, 1973). In the Southern Ocean, further increases occur in siliceous biogenic sedimentation rates which apparently reached a maximum during the Late Quaternary (Kennett et al., 1975; Tucholke et al., 1976; Craddock & Hollister, 1976) and are assumed to be the result of increased biological productivity reflecting higher rates of upwelling and oceanic turnover.

In the Southern Ocean, the Quaternary marks an apparent peak in activity of oceanic circulation as reflected by widespread deep-sea erosion. This is assumed to reflect major development of Antarctic Bottom Water, intensification of circum-polar flow and increased global water-mass turnover due to steep latitudinal temperature gradients in both hemispheres (Watkins & Kennett, 1972; Kennett & Watkins, 1975).

It is not yet known why Northern Hemisphere ice-sheet formation should take place so much later (~ 10 MY) than that on Antarctica but is probably related to late Cainozoic neotectonism (Hamilton, 1965; Kennett & Thunell, 1975; Vogt, 1972; Adams et al., 1976) and further changes in oceanic circulation patterns such as those resulting from the closure of the Middle American Seaway (Hamilton, 1965; Woodring, 1966; Crowell & Frakes, 1970; Keigwin, in press).

ACKNOWLEDGEMENTS

This research was supported by Grant Number OPP75-15511 (Division of Polar Programs) of the U.S. National Science Foundation.

REFERENCES

Adams, G.F., Fairbridge, R.W. & Rice, A.R. 1976. Contemporary acceleration of global instability, *Abstr. Am. Geophys. Union Ann. Meeting*, Washington, D.C.

Andrews, P.B., Gostin, V.A., Hampton, M.A., Margolis, S.V. & Ovenstone, A.T. 1975. Synthesis sediments of the southwest Pacific Ocean, southwest Indian Ocean and south Tasman Sea. In: J.P. Kennett, R.E. Houtz et al (eds.), *Initial reports of the DSDP* 29: 1147.

Bandy, O.L. 1967. Problems of Tertiary foraminiferal and radiolarian zonation, circum-Pacific area. In: K. Hatai (ed.), *Tertiary correlations and climatic changes in the Pacific*, 11th Pacific Science Congress, Tokyo, Papers of Symposium 25, Sendai, 95.

Barker, P.F. & Burrell, J. 1976. The opening of Drake Passage. *Proc. Joint Oceanog. Assembly*, Edinburgh: 103.

Barker, P., Dalziel, I. et al. 1977. *Initial reports of the DSDP* 36.

Barker, P. et al. 1977. Evolution of the southwestern Atlantic Ocean Basin: Results of Leg 36, Deep Sea Drilling Project. In: P. Barker, I. Dalziel et al. (eds.), *Initial reports of the DSDP* 36.

Barron J.A. 1973. Late Miocene–Early Pliocene paleotemperatures for California from marine diatom evidence. *Palaeogeography, Palaeoceanography, Palaeoecology* 14: 277.

Benson, R.A. 1975a. The origin of the psychrosphere as recorded in changes of deep-sea ostracode assemblages. *Lethaia* 8: 69.

Benson, R.A. 1975b. The Cenozoic ostracode faunas of the Sao Paulo Plateau and the Rio Grande Rise. In: P.R. Supko, K. Perch-Nielsen et al. (eds.), *Initial reports of the DSDP* 39.

Berger, W.H. 1970. Biogenous deep-sea sediments; fractionation by deep-sea circulation. *Geol. Soc. Am. Bull.* 81: 1385.

Berger, W.H. 1973. Deep-sea carbonates: Evidence for a coccolith lysocline. *Deep-Sea Res.* 20: 917.

Berger, W.H. & Winterer, E.L. 1974. Plate stratigraphy and the fluctuating carbonate line. *Spec. Publs. Inst. Ass. Sediment* 1: 11.

Berggren, W.A. 1972. A cenozoic time-scale – some implications for regional geology and paleobiography. *Lethaia* 5: 195.

Berggren, W.A. 1975b. Late Pliocene–Pleistocene glaciation. In: A.S. Laughton, W.A. Berggren et al. (eds.), *Initial reports of the DSDP* 12: 953.

Berggren, W.A. & Haq, B.U. 1976. The Andalusian Stage (Late Miocene): Biostratigraphy, biochronology and paleoecology. *Palaeogeography, Palaeoclimatology, Palaeoecology* 20: 67.

Beu, A.G. 1974. Molluscan evidence of warm sea

temperatures in New Zealand during Kapitean (Late Miocene) and Waipipian (Middle Pliocene) time. *N.Z. J. Geol. and Geophys.* 17: 465.

Burns, D.A. 1975. Distribution, abundance, and preservation of nannofossils in Eocene to Recent Antarctic sediments. *N.Z. J. Geol. and Geophys.* 18: 583.

Casey, R.E. 1972. Neogene radiolarian biostratigraphy and paleotemperatures, the experimental Mohole Antarctic core E14–8. *Palaeogeography, Palaeoclimatology, Palaeoecology* 12: 115.

Christoffel, D. & Falconer, R. 1972. Marine magnetic measurements in the southwest Pacific Ocean and the identification of new tectonic features. *Antarctic Research Series* 19, Am. Geophys. Union, Washington: 197.

Craddock, C. & Hollister, C.D. 1976. Geologic evolution of the southeast Pacific Basin. In: C.D. Hollister, C. Craddock et al. (eds.), *Initial reports of the DSDP* 35: 723.

Cranwell, L.M., Harrington, H.J. & Speden I.G. 1960. Lower Tertiary microfossils from McMurdo Sound, Antarctica. *Nature* 186: 700.

Crowell, J.C. & Frakes, L.A. 1970. Phanerozoic glaciation and the causes of ice ages. *Am. J. Sci.* 268: 193.

Deighton, I., Falvey, D.A. & Taylor, D.J. 1976. Depositional environments and geotectonic framework: Southern Australian continental margin. *Aust. Petroleum Exploration Ass. J.* 16: 25.

Devereux, I. 1967. Oxygen isotope paleotemperature measurements on New Zealand Tertiary fossils. *N.Z. J. Sci.* 10: 988.

Douglas, R.G. 1974. Biogeography and bathymetry of Late Cretaceous–Cenozoic abyssal benthic foraminifera. In: W. Berggren (ed.), *Woods Hole Oceanographic Inst. Symposium: Organisms and continents through time.*

Drewry, D.J. 1975. Initiation and growth of the East Antarctic ice sheet. *J. Geol. Soc.* London 131: 255.

Drewry, D.J. 1976. Deep-Sea drilling from Glomar Challenger in the Southern Ocean. *Polar Record* 18: 47.

Edwards, A.R. 1968. The calcareous nannoplankton evidence for New Zealand Tertiary marine climate. *Tuatara* 16: 26.

Edwards, A.R. 1973. Southwest Pacific regional unconformities encountered during Leg 21. In: R.E. Burns, J.E. Andrews et al. (eds.), *Initial reports of the DSDP* 21: 641.

Edwards, A.R. 1975. Further comments on the southwest Pacific Paleogene regional unconformities. In: J.E. Andrews, G. Packham et al. (eds.), *Initial reports of the DSDP* 30: 663.

Edwards, A.R. & Perch-Nielsen, K. 1975. Calcareous nannofossils from the southern southwest Pacific, Deep Sea Drilling Project, Leg 29. In: J.P. Kennett, R.E. Houtz et al. (eds.), *Initial reports of the DSDP* 29: 469.

Emiliani, C. 1954. Temperatures of Pacific bottom waters and polar superficial waters during the Tertiary. *Science* 119: 853.

Frakes, L.A. & Kemp, E.M. 1973. Paleogene continental positions and evolution of climate. In: D.H. Tarling & S.K. Runcorn (eds.), *Implications of continental drift to the earth sciences*: 539.

Gordon, A.L. 1971. Oceanography of Antarctic waters. *Antarctic Research Series* 15, Am. Geophys. Union, Washington: 169.

Hamilton, W. 1965. Cenozoic climatic change and its cause. In: J.M. Mitchell Jr. (ed.), *Meteorological Monographs* 8, Am. Meteorol. Soc., Boston, 159.

Hampton, M.A. 1975. Detrital and biogenic sediment trends at D.S.D.P. sites 280, and 281, and evolution of middle Cenozoic currents. In: J.P. Kennett, R.E. Houtz et al. (eds.), *Initial reports of the DSDP* 29: 1071.

Hayes, D.E., Frakes, L. et al. 1975. *Initial reports of the DSDP* 28, U.S. Gov. Printing Off., Washington, D.C.

Hayes, D.E. & Conolly, J.R. 1972. Morphology of the Southeast Indian Ocean. *Antarctic Research Series* 19, Am. Geophs. Union: 125.

Hayes, D.E. & Frakes, L.A. 1975. General synthesis Deep Sea Drilling Project Leg 28. In: D.E. Hayes, L.A. Frakes et al. (eds.), *Initial reports of the DSDP* 28: 869.

Hayes, D.E. & Ringis, I. 1973. Seafloor spreading in the Tasman Sea. *Nature* 243: 454.

Herron, E.M. & Tucholke, B.E. 1976. Sea-floor magnetic patterns and basement structure in the southeastern Pacific. In: C.D. Hollister, C. Craddock et al. (eds.), *Initial reports of the DSDP* 35: 263.

Hornibrook, N. de B. (in press). New Zealand Tertiary climate, New Zealand Geological Survey Report 47, 1971. (To be a chapter in: R.P. Suggate (ed.), *Geology of New Zealand*, Gov. Printer, Wellington).

Ingle Jr., J.C. 1967. Foraminiferal biofacies variation and the Miocene–Pliocene boundary in southern California. *Bull. Am. Paleontology* 52: 217.

Ingle Jr., J.C. 1973. Summary comments on Neogene biostratigraphy, physical stratigraphy and paleo-oceanography in the marginal northeastern Pacific Ocean. In: L.D. Kulm, R. Von Huene et al. (eds.), *Initial reports of the DSDP.*

Jenkins, D.G. 1968. Planktonic foraminifera as evidence of New Zealand Tertiary paleotemperatures. *Tuatara* 16: 32.

Kaneps, A.G. 1975. Cenozoic planktonic foraminifera from Antarctic Deep-Sea sediments, Leg 28 D.S.D.P. In: D.E. Hayes, L.A. Frakes et al.

(eds.), *Initial reports of the DSDP* 28: 573.
Keigwin Jr., L.D. (in press). Late Cenozoic planktic foraminiferal biostratigraphy and paleoceanography of the Panama Basin. *Micropaleontology.*
Kemp, E.M. & Barrett, P.J. 1975. Antarctic glaciation and early Tertiary vegetation. *Nature* 258: 507.
Kemp, E.M., Frakes, L.A. & Hayes, D.E. 1975. Paleoclimatic significance of diachronous biogenic facies, Leg 28, Deep Sea Drilling Project. In: D.E. Hayes, L.A. Frakes et al. (eds.), *Initial reports of the DSDP* 28: 909.
Kennett, J.P. 1967. Recognition and correlation of the Kapitean Stage (Upper Miocene, New Zealand). *N.Z. J. Geol. and Geophys.* 10: 1051.
Kennett, J.P., Burns, R.E., Andrews, J.E., Churkin, M. Davies, T.A., Dumitrica, P., Edwards, A.R., Galehouse, J.S., Packham, G.H. & van der Lingen, G.J. 1972. Australian–Antarctic continental drift, paleocirculation changes and Oligocene deepsea erosion. *Nature Physical Science* 239: 51.
Kennett, J.P., Houtz, R.E., Andrews, P.B., Edwards, A.R., Gostin, V.A., Hajos, M., Hampton, M.A., Jenkins, D.G., Margolis, S.V., Ovenshine, A.T. & Perch-Nielsen, K. 1974. Development of the circum-Antarctic current. *Science* 186: 144.
Kennett, J.P., Houtz, R.E. et al. (eds.). 1975. *Initial reports of the DSDP* 29, U.S. Gov. Printing Off., Washington D.C. 1197 pp.
Kennett, J.P., Houtz, R.E., Andrews, P.B., Edwards, A.R., Gostin, V.A., Hajos, M., Hampton, M.A., Jenkins, D.G., Margolis, S.V., Ovenshine, A.T. & Perch-Nielsen, K. 1975. Cenozoic paleoceanography in the south-west Pacific Ocean, Antarctic glaciation and the development of the circum-Antarctic current. In: J.P. Kennett, R.E. Houtz et al. (eds.), *Initial reports of the DSDP* 29: 1155.
Kennett, J.P. & Shackleton, N.J. 1976. Oxygen isotopic evidence for the development of the psychrosphere 38 MY ago. *Nature* 260: 513.
Kennett, J.P. & Thunell, R.C. 1975. Global increase in Quaternary explosive volcanism. *Science* 187: 497.
Kennett, J.P. & Vella, P. 1975. Late Cenozoic planktonic foraminifera and paleoceanography at D.S.D.P. Site 284 in the cool subtropical South Pacific. In: J.P. Kennett, R.E. Houtz et al. (eds.), *Initial reports of the DSDP* 29: 769.
Kennett, J.P. & Watkins, N.D. 1975. Deep-sea erosion and manganese nodule development in the southeast Indian Ocean. *Science* 188: 1011.
Lowrie, W. & Hayes, D.E. 1975. Magnetic properties of oceanic basalt samples. In: D.E. Hayes, L.A. Frakes et al. (eds.), *Initial reports of the DSDP* 28: 869.
Mayewski, P.A. 1975. *Glacial geology and Late Cenozoic history of the Transantarctic Mountains, Antarctica,* Inst. Polar Studies, Ohio State Univ. Report 56.
McElhinny, M.W. 1973. *Palaeomagnetism and place tectonics.* Cambridge University Press. 358 pp.
McIntyre, D.J. & Wilson, G.J. 1966. Preliminary palynology of some Antarctic Tertiary erratics, *N.Z. J. Botany* 4: 315.
Moore Jr., T.C. & Heath, G.R. (in press). Survival of deep-sea sedimentary sections,*Earth and Planetary Sci. Letters.*
Moore, T.C., van Andel, T.H., Sancetta, C. & Pisias, N. (in press). Cenozoic hiatuses in pelagic sediments. In: *Marine Plankton and Sediments*, Micropaleontology Press, American Museum of Natural History, New York.
Rona, P.A. 1972. Worldwide unconformities in marine sediments related to eustatic changes in sea level. *Nature Physical Science* 244: 25.
Ryan, W.B.F., Cita, M.B., Rawson, M.D., Burckle, L.H., Saito, T. 1974. A paleomagnetic assignment of Neogene stage boundaries and the development of isochronous datum planes between the Mediterranean, the Pacific and Indian Oceans in order to investigate the response of the world ocean to the Mediterranean salinity crisis. *Riv. Ital. Paleont.* 80: 631.
Savin, S.M., Douglas, R.G. & Stehli, F.G. 1975. Tertiary marine paleotemperatures. *Geol. Soc. Am. Bull.* 86: 1499.
Shackleton, N.J. & Kennett, J.P. 1975a. Paleotemperature history of the Cenozoic and the initiation of Antarctic glaciation: Oxygen and carbon isotope analyses in D.S.D.P. sites 277, 279 and 281. In: *Initial reports of the DSDP* 29: 743.
Shackleton, N.J. & Kennett, J.P.1975b. Late Cenozoic oxygen and carbon isotopic changes at D.S.D.P. Site 284: Implications for glacial history of the Northern Hemisphere and Antarctica. In: *Initial reports of the DSDP* 29: 801.
Shackleton, N.J. & Opdyke, N.D. 1973. Oxygen isotope and paleomagnetic stratigraphy of Equatorial Pacific core 28–238: Oxygen isotope temperatures on a 10^5 and 10^6 year time scale. *Quat. Res.* 3: 39.
Tucholke, B.E., Hollister, C.D., Weaver, F.M. & Vennum, W.R. 1976. Continental rise and abyssal plain sedimentation in the southeast Pacific basin. In: C.F. Hollister, C. Craddock et al. (eds.), *Leg 35 DSDP*, U.S. Printing Off., Washington, D.C.: 359.
Tucholke B.E. & Houtz, R.E. 1976. Sedimentary framework of the Bellinghausen Basin from seismic profiler data. *Ibid*: 197.
Van Andel, T.H., Heath, G.R. & Moore Jr, T.C. 1975. Cenozoic tectonics, sedimentation, and paleoceanography of the central equatorial Paci-

fic. *Geol. Soc. Am. Mem.* 143.
Van Couvering, J.A., Berggren, W.A., Drake, R.E., Aquirre, E. & Curtis, G.H. 1976. The terminal Miocene event. *Marine Micropaleontology* 3.
Vogt, P.R. 1972. Evidence for global synchronism in mantle plume convection and possible significance for geology. *Nature* 240: 338.
Watkins, N.D. & Kennett, J.P. 1972. Regional sedimentary disconformities and upper Cenozoic changes in bottom water velocities between Australia and Antarctica. *Antarctic Research Series* 19. Am. Geophys. Union, Washington: 273.
Weissel, J.K. & Hayes, D.E. 1972. Magnetic anomalies in the Southeast Indian Ocean. *Antarctic Research Series* 19, Am. Geophys. Union, Washington: 165.
Wolfe, J.A. & Hopkins, D.M. 1967. Climatic changes recorded by Tertiary land floras in north-western North America. In: *Tertiary correlations and climatic changes in the Pacific*, Pacific Sci. Congr. 11th Tokyo: 67.
Woodring, W.P. 1966. The Panama land bridge as a sea barrier. *Proc. Am. Philosophical Ass.* 110: 425.

* 6 *

A review of the Late Quaternary climatic history of Antarctic Seas

James D. Hays

CLIMAP, Lamont-Doherty Geological Observatory of Columbia University in the city of New York, Palisades, New York, U.S.A. and
Department of Geological Sciences, Columbia University, Palisades, New York, U.S.A.

Manuscript received 25th November 1977

CONTENTS

ABSTRACT

Detailed faunal and isotopic studies of Antarctic and sub-Antarctic deep-sea cores indicate that during most of the Late Quaternary (last 200 000 years) climatic changes in the area are in phase or nearly in phase with Northern Hemisphere glacial advances and retreats. However, changes in sub-Antarctic sea surface temperatures in much of the record precede slightly (≃3 000 years) changes in Northern Hemisphere glaciers. In the Holocene, for example, sub-Antarctic surface waters reached a temperature maximum ≃9 000 years BP, have been cooling since and are today half way between interglacial maximum and glacial minimum temperatures. This provides strong evidence that Southern Hemisphere climates are not being driven by changes in the volume of Northern Hemisphere ice sheets. Furthermore, changes in aerial extent of Antarctic sea ice, probably the most remarkable Pleistocene changes to occur in Antarctic Seas, also precede changes in Northern Hemisphere ice volume.

Power spectrum analysis of the climatic record in sub-Antarctic deep-sea cores shows that the dominant frequencies resolved represent periods of ~100 000 years, 41 000 years and 23 000 years. These are nearly identical to the dominant periods of volume change of Northern Hemisphere glaciers and changes in geometry of the Earth's orbit. This indicates that major climatic changes on the time scales of ice ages are induced by insolation changes in responses to the changing parameters of the Earth's orbit around the sun.

INTRODUCTION

The description of Tertiary events in the seas around Antarctica has benefited in the last ten years from the collection of over 2 000 piston cores, numerous dredges and the drilling activities of the *Glomar Challenger*. The rigorous climatic conditions of the Antarctic Seas discouraged early cruises and other parts of the world ocean have been more actively studied. However, in the last decade, the cruises of the USNS *Eltanin* and Lamont-Doherty research vessels *Vema* and *Conrad* have resulted in the collection of several thousand piston cores. These cores have now been used to map in some detail the distribution of sediments that blanket the sea floor (Goodell, 1973), describe numerous microfossils contained in these sediments, and map their distribution. Through the use of palaeomagnetic stratigraphy the gross Pleistocene stratigraphy of Antarctic sediments has been tied to the better known stratigraphy of lower latitudes (Hays & Berggren, 1971), and the patterns of regional erosional hiatuses mapped (Watkins & Kennett, 1971, 1972).

The high resolution deep-sea sediment stratigraphy provided by measurements of δO^{18} in the shells of benthic Foraminifera provides an opportunity to compare detailed Pleistocene climatic records from various parts of the world. Unfortunately, Antarctic sediments (south of the Antarctic Polar Front) are notoriously deficient in calcium carbonate. However, the abundance changes of a species of Radiolaria,

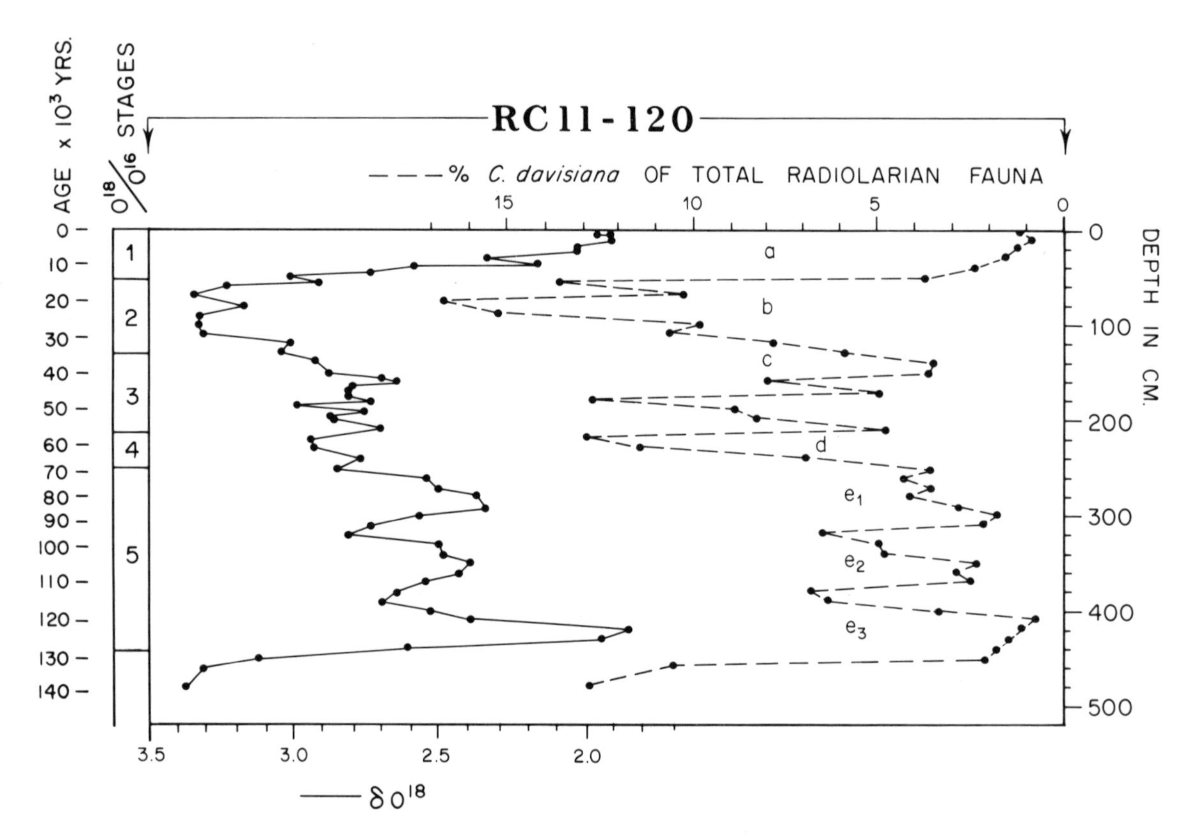

Fig. 1. Comparison of relative abundance changes of the radiolarian species *Cycladophora davisiana* with variations of δO^{18} in southern Indian Ocean core RC11–120 (Hays et al., 1976a).

Cycladophora davisiana, contain as much stratigraphic information as the δO^{18} record and are probably synchronous around the Antarctic continent (Hays et al., 1976a). It is through the use of this stratigraphy that Antarctic events can be put in a chronostratigraphic framework (Hays et al., 1976b) and then, through δO^{18} stratigraphy, related to Pleistocene events in other parts of the world.

The purpose of this paper is to review the current state of our knowledge of late Pleistocene Antarctic climatic events, as read from deep-sea sediments, and compare them with simultaneous events in the Northern Hemisphere.

STRATIGRAPHY

Shackleton & Opdyke (1973, 1976) have shown that late Pleistocene δO^{18} changes in the shells of Foraminifera preserved in deep-sea sediments are globally synchronous. In deep-sea cores raised from south of the Antarctic Polar Front, continuous records containing $CaCo_3$ are rare, thus useful oxygen isotope data is not available.

Hays et al. (1976a), have shown that one radiolarian species, *Cycladophora davisiana* undergoes substantial relative abundance fluctuations in Antarctic and sub-Antarctic sediments. They have shown by comparing the oxygen isotope record with the record of *C. davisiana*, that the faunal abundance changes are synchronous over a broad area of the sub-Antarctic. In Figure 1 (after Hays et al., 1976a), it can be seen that the variations in δO^{18} and relative *C. davisiana* abundance contains as much stratigraphic information as the δO^{18} curve over the last 140 000 years. This correlation has been extended further back in time by Hays et al. (1976b) (see Figs. 9 and 10).

Although the amplitude of the *C. davisiana* signal varies regionally, generally being maximal south of 50° S, the shape of the curves are similar throughout the Antarctic and sub-Antarctic providing a high resolution and probably synchronous stratigraphy for this region.

The *C. davisiana* stratigraphy permits correlation of discrete stratigraphic horizons around the Antarctic. The most recent maximum (b) in the *C. davisiana* curve closely approximates the last δO^{18} maximum which has been dated at ≃18 000 years BP.

SPECIES ASSEMBLAGE ANALYSIS

In order to determine how species assemblage patterns changed in Antarctic waters between the height of the last glaciation at 18 000 years BP and the Holocene, 19 species of Radiolaria were counted (total counts always exceed 500 specimens) in 208 core top samples and 98 samples at the 18K level. These data were factor analysed and the species divided into two dominant factors that between them account for more than 95% of the variance.

The two assemblages derived from the core top samples correspond to the polar water mass and the sub-Antarctic and southern subtropical water masses (Figs. 2 and 3). The 0.9 factor loading of the polar assemblage or the 0.25 factor loading of the sub-Antarctic assemblage closely follow the present position of the Antarctic Polar Front (Figs. 2 and 3).

The factor analysis of species counts from the 18 000 years BP level generates the same two assemblages, however, comparable factor loadings are shifted to the north (Figs. 4 and 5). The northward displacement of the polar assemblage as read from the position of the 0.9 factor loading, which is probably a reliable estimate of the position of the Antarctic Polar Front, is most pronounced in the Atlantic. However, there is a northward displacement from today's position of several degrees throughout the area.

SEA ICE COVER

In general, sediments south of the Polar Front are interbedded sequences of diatomaceous ooze and diatomaceous clay, the latter containing more silt and sand in the Atlantic and southwest Indian Oceans. These lithologic changes were first noted by Philippi (1912) and have been discussed by subsequent authors (Schott, 1939; Hough, 1956; Hays et al., 1976a; and Cooke & Hays, in press).

All authors since Philippi (1912) have suggested that these lithologic changes were in some way related to climatic change,either shifts in sediment belts around Antarctica due to glacial intervals (Philippi, 1912) or variations in concentrations of berg ice (Schott, 1939). With improved biostratigraphic control, Hays et al. (1976a) were able to show that the accumulation rates of the diatomaceous ooze was two to three times as fast as the accumulation rate of the diatomaceous clays. From this they argued that the diatomaceous clays could not be explained by increased influx of detrital material as had been suggested by Schott (1939) but rather that the diatomaceous clay was caused by a decrease of diatom input to the sediment relative to the diatom ooze. They further argued that the most likely inhibiting factor was increased sea ice during the critical summer months. Cooke & Hays (in press), through a much more extensive analysis of cores south of the Polar Front, showed, through comparison with the *C. davisiana* stratigraphy, that these lithologic changes were synchronous or nearly synchronous throughout the Atlantic and Indian Ocean sectors. They also showed that large amounts of coarse ($>63\mu$) volcanic material was disseminated through the diatomaceous clay between the South Sandwich Islands and Kerguelen, but was scarce in the diatomaceous ooze.

Discrete ash layers such as those studied by Ninkovich et al. (1964) and Federman (1977) east of the South Sandwich Islands occur only in the diatomaceous ooze layer. Due to the very wide distribution of the disseminated volcanic ash (more than 5 000 miles from its most probably source – the South Sandwich Islands), they suggested that its transport had been aided by sea ice which had carried it farther to the east before releasing it to fall to the bottom.

Since volcanic eruptions are independent of season, volcanic ejecta will fall on sea ice at any time of year. The northern limit of this sea ice transported material then provides a means of estimating winter (maximum) sea ice cover. Figures 6 and 7 show estimates of winter and summer sea ice in the Antarctic modified from Cooke & Hays (in press) 18 000 years ago, the maximum of the last Northern Hemisphere glaciation. The methods used to estimate the northern limits do not provide a means for estimating the percentage of the sea surface covered by ice south of these limits. The area south of the summer ice limit, Cooke & Hays (op. cit.) estimated, was $25 \times 10^6 km^2$ roughly ten times the area south of the present summer sea ice limit. Their estimate of the area south of the winter sea ice limit was $40 \times 10^6 km^2$ about twice the area south of today's winter sea ice limit.

ESTIMATES OF PAST SEA SURFACE TEMPERATURE

Estimates of past sea surface temperature can be made using the faunal counts of radiolarian species through techniques developed by Imbrie & Kipp (1971) and first applied to Antarctic Radiolaria by Lozano & Hays (1976). The transfer functions used to estimate the temperatures presented here were developed by Morley & Hays and will be reported on fully in the future (Hays et al., in prep.). Figure 8 shows estimates of winter sea surface temperature at 18K beyond the northern limit of our estimate position of sea ice.

These temperatures, as one might expect from the fossil assemblage patterns 18 000 years ago, show a

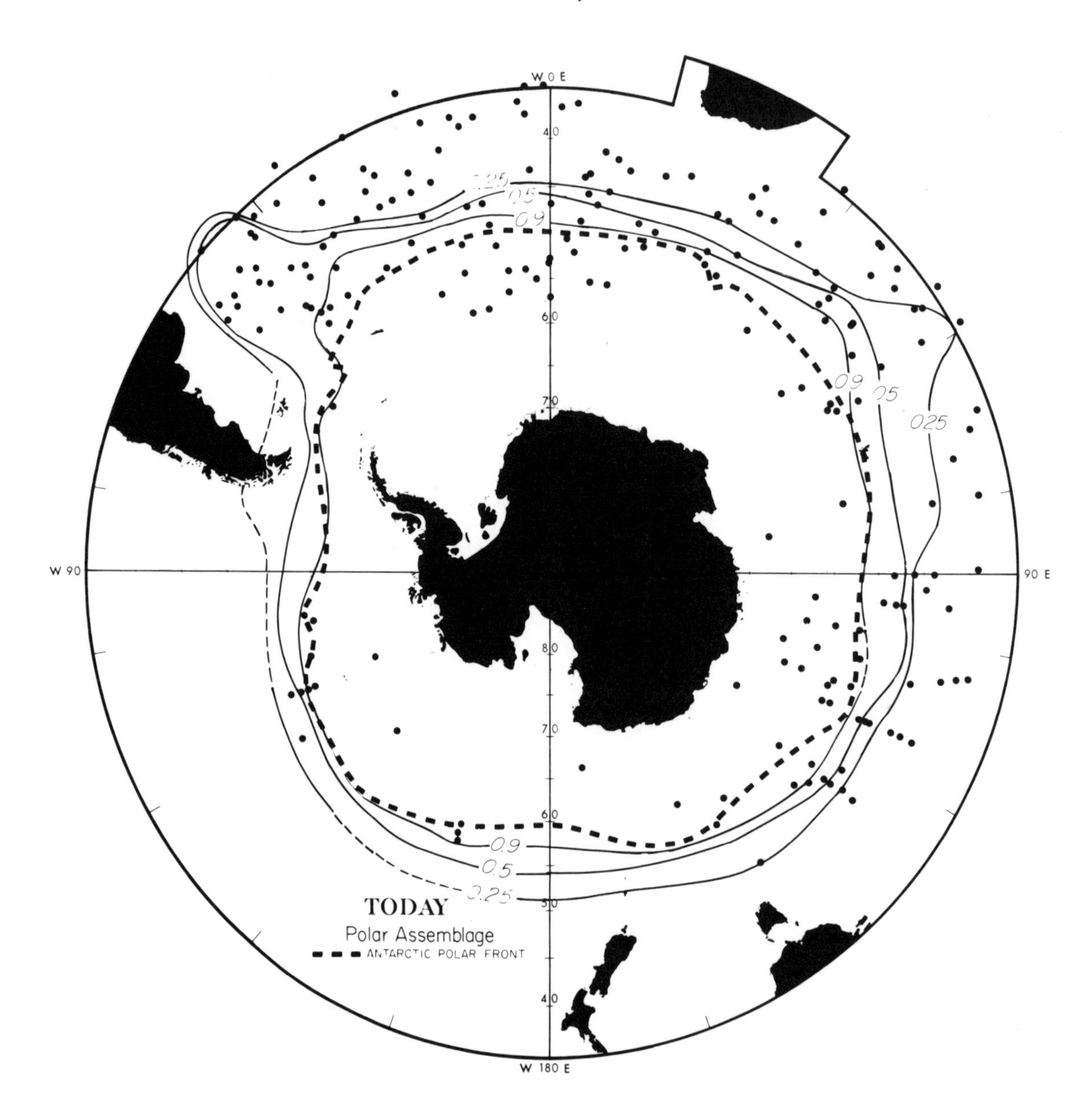

Fig. 2. Distribution of polar assemblage in 208 core top samples

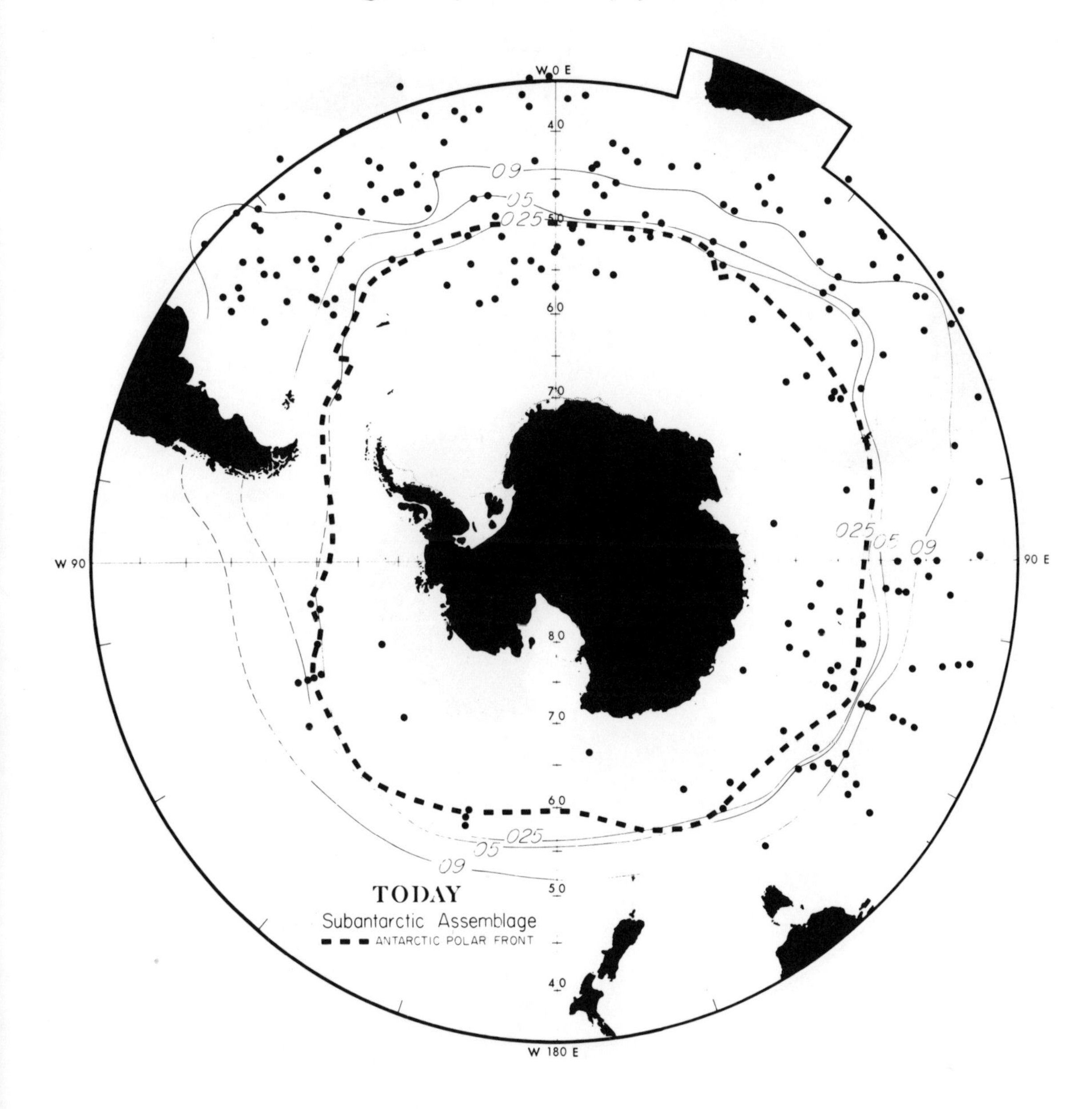

Fig. 3. Distribution of sub-Antarctic assemblage in 208 core top samples

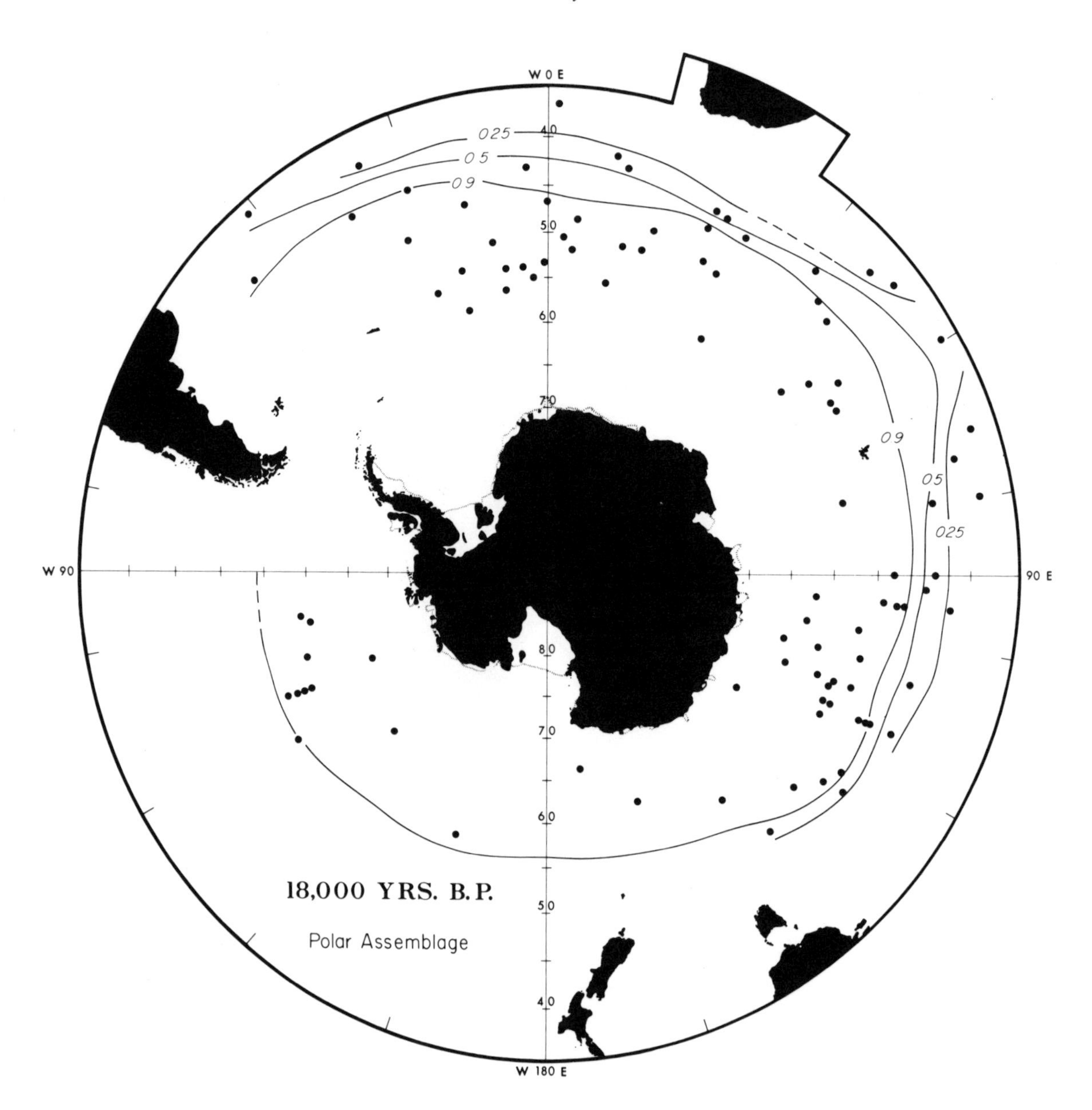

Fig. 4. Distribution of polar radiolarian assemblage at 18 000 years BP

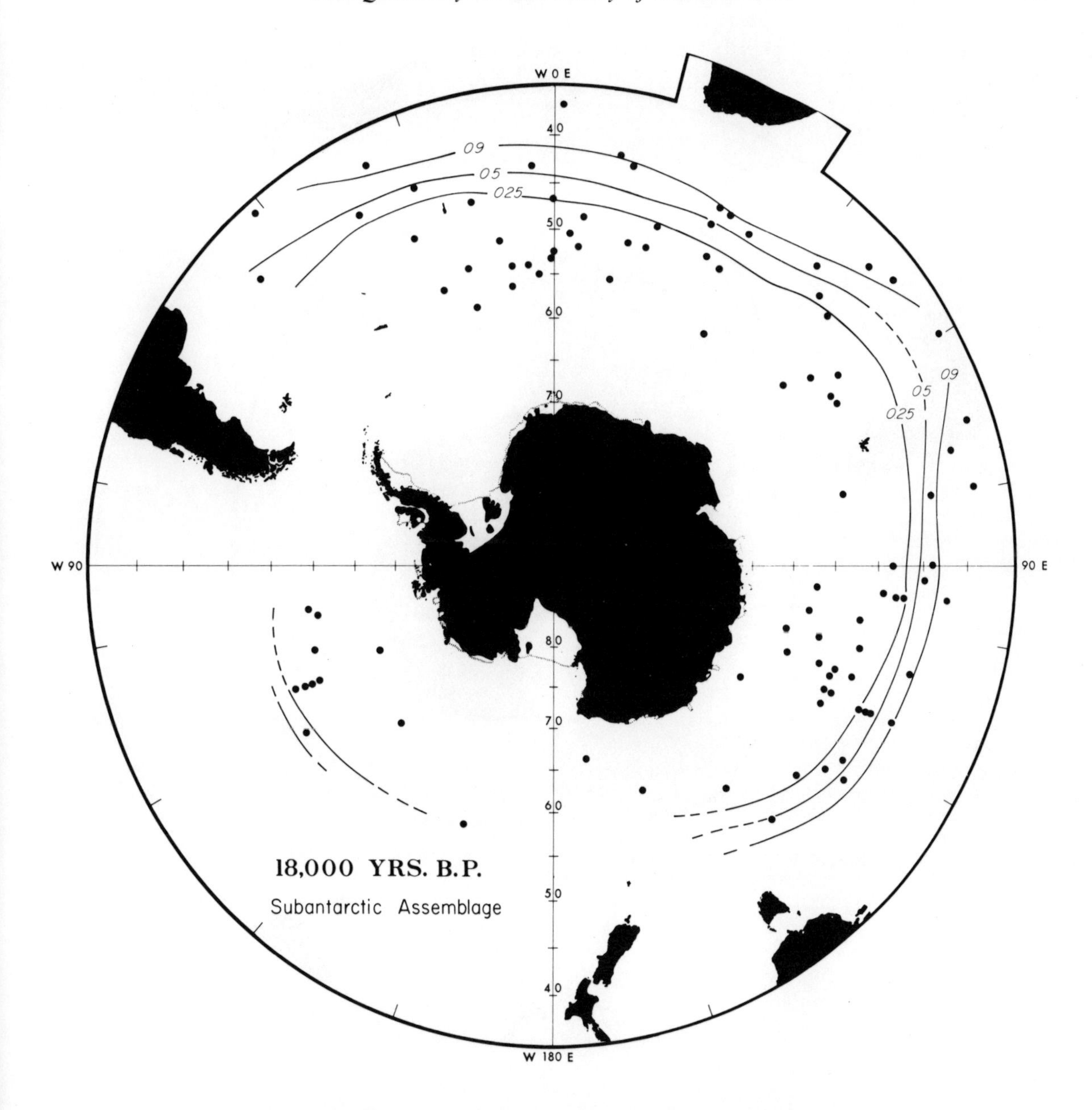

Fig. 5. Distribution of sub-Antarctic assemblage 18 000 years BP

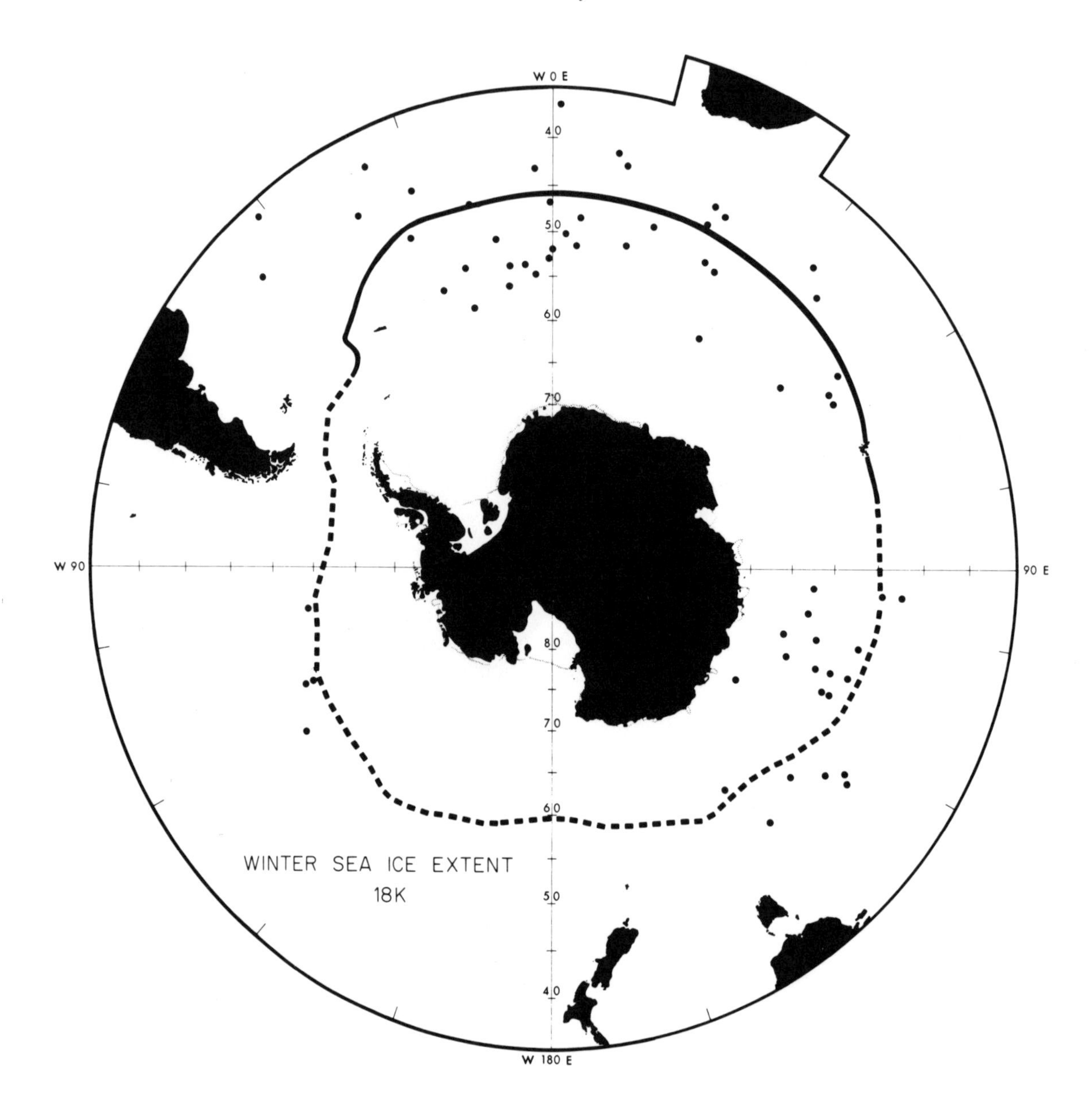

Fig. 6. Estimate of winter sea ice extent 18 000 years BP

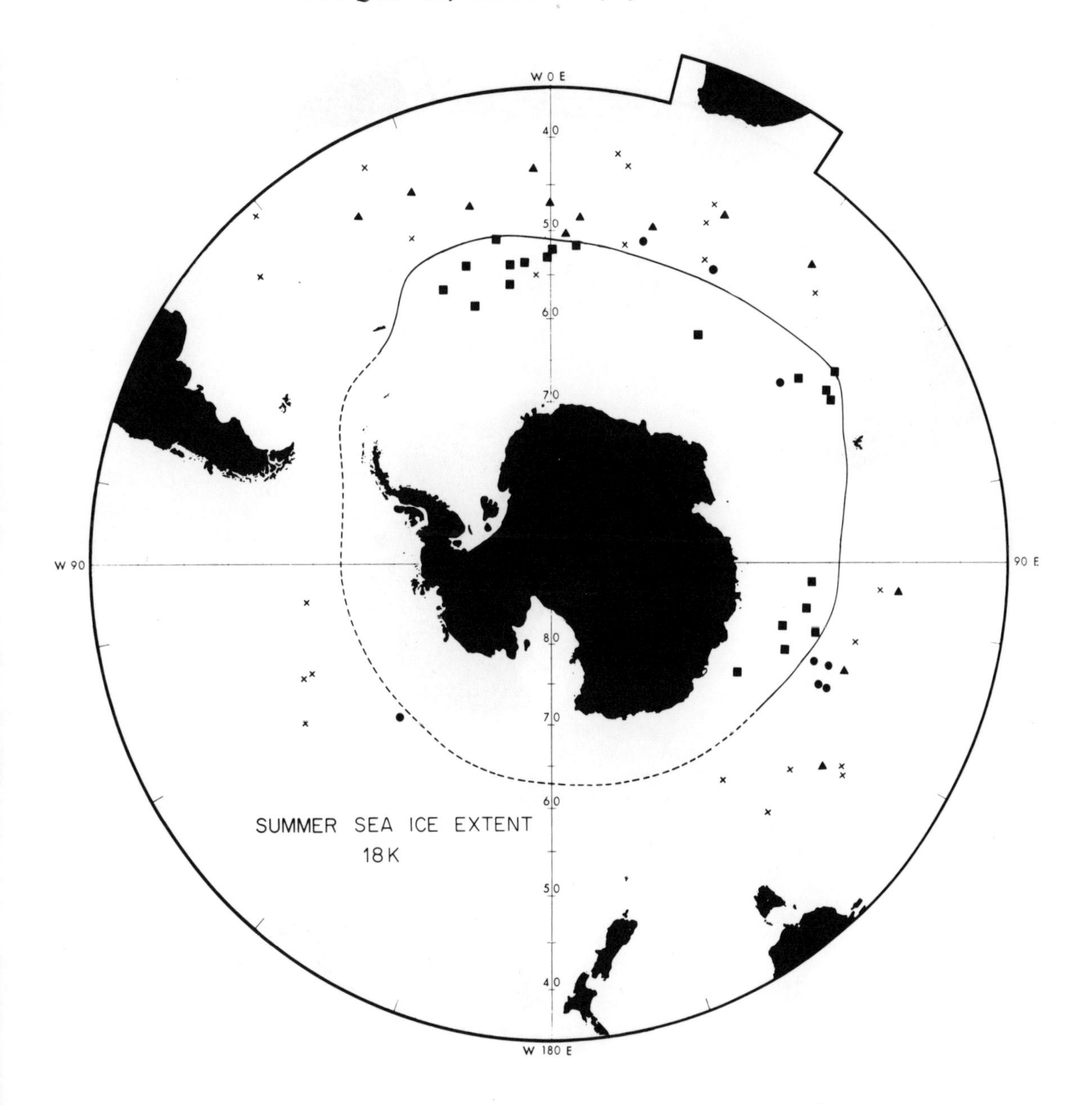

Fig. 7. Estimate of summer sea ice extent 18 000 years BP

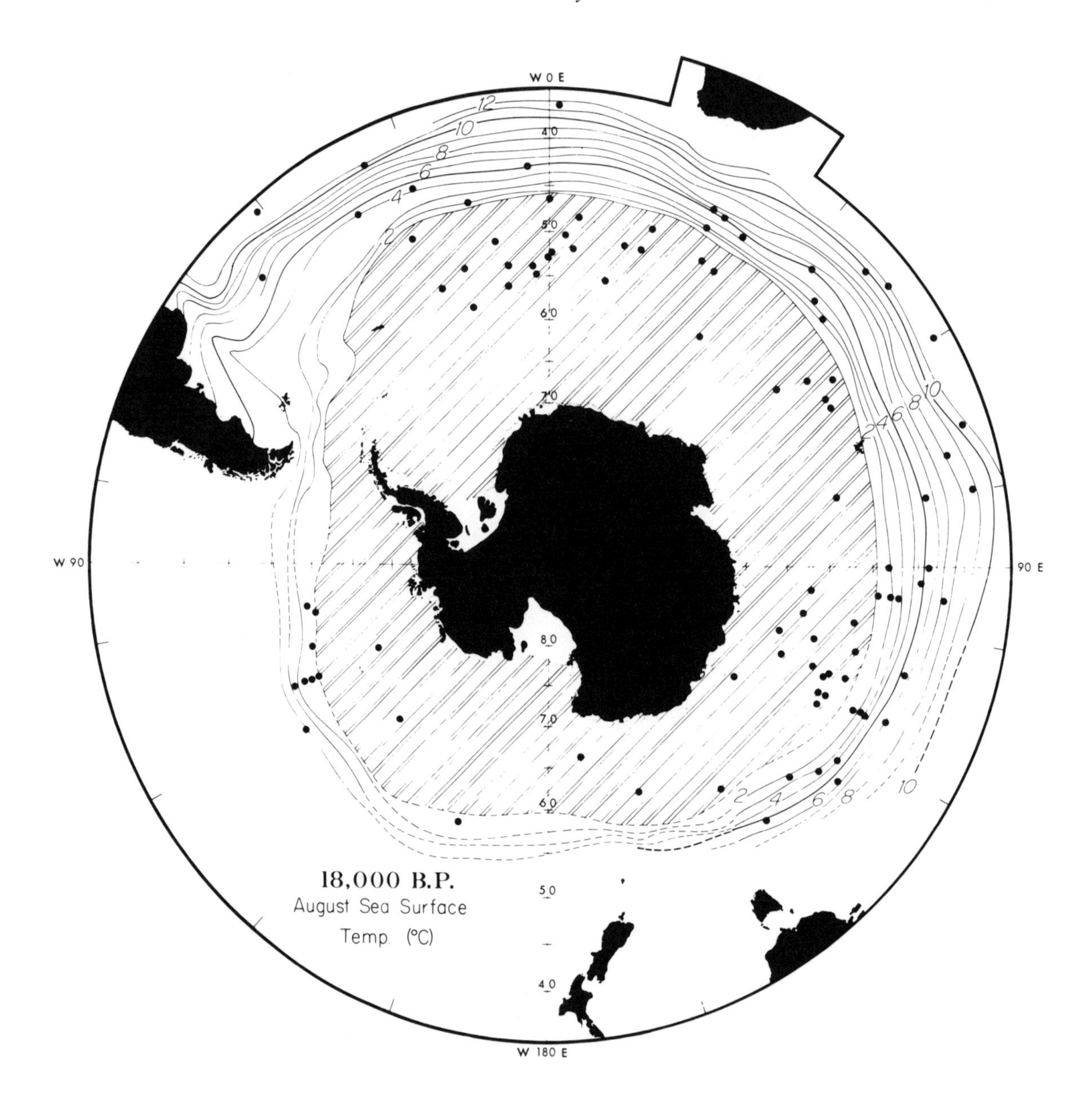

Fig. 8. Estimate of winter sea surface temperatures north of the estimated ice limit 18 000 years BP

marked cooling in sub-Antarctic waters relative to today, but little temperature change from today's values in subtropical waters.

In summary, the most important features that distinguish Antarctic conditions 18 000 years ago from today are 1) a northward shift of the Polar Front, not exceeding a few degrees from its present location; 2) cooler sub-Antarctic sea surface temperatures; 3) a large increase of both summer and winter sea ice cover.

DOWN CORE ANALYSES

Down core analyses of Antarctic and sub-Antarctic sediments reveal a sequence of constantly changing conditions that alternate between extremes represented by today's temperatures and sea ice cover and the conditions I have described for 18 000 years BP. To determine this past sequence, work has been done in both the sub-Antarctic (Hays et al., 1976b) and Antarctic (Hays et al., 1976a; Cooke & Hays, in press).

SUB-ANTARCTIC RECORD

Hays et al. (1976b) studied the record of two sub-Antarctic cores that together spanned the last 485 000 years (Figs. 9 and 10). These cores were located in the sub-Antarctic sector of the Indian Ocean, roughly midway between South Africa, Australia and Antarctica. They were raised from the flank of the mid-oceanic ridge system so their sedimentation rates were little affected by variations in detrital influx from Antarctica. Three parameters were measured in these cores, δO^{18} in the shells of benthonic Foraminifera, estimated sea surface temperature through statistical treatment of radiolarian assemblage data, and the relative percent of a single species, *Cycladophora davisiana*.

Comparing the δO^{18} curve (effectively, a measure

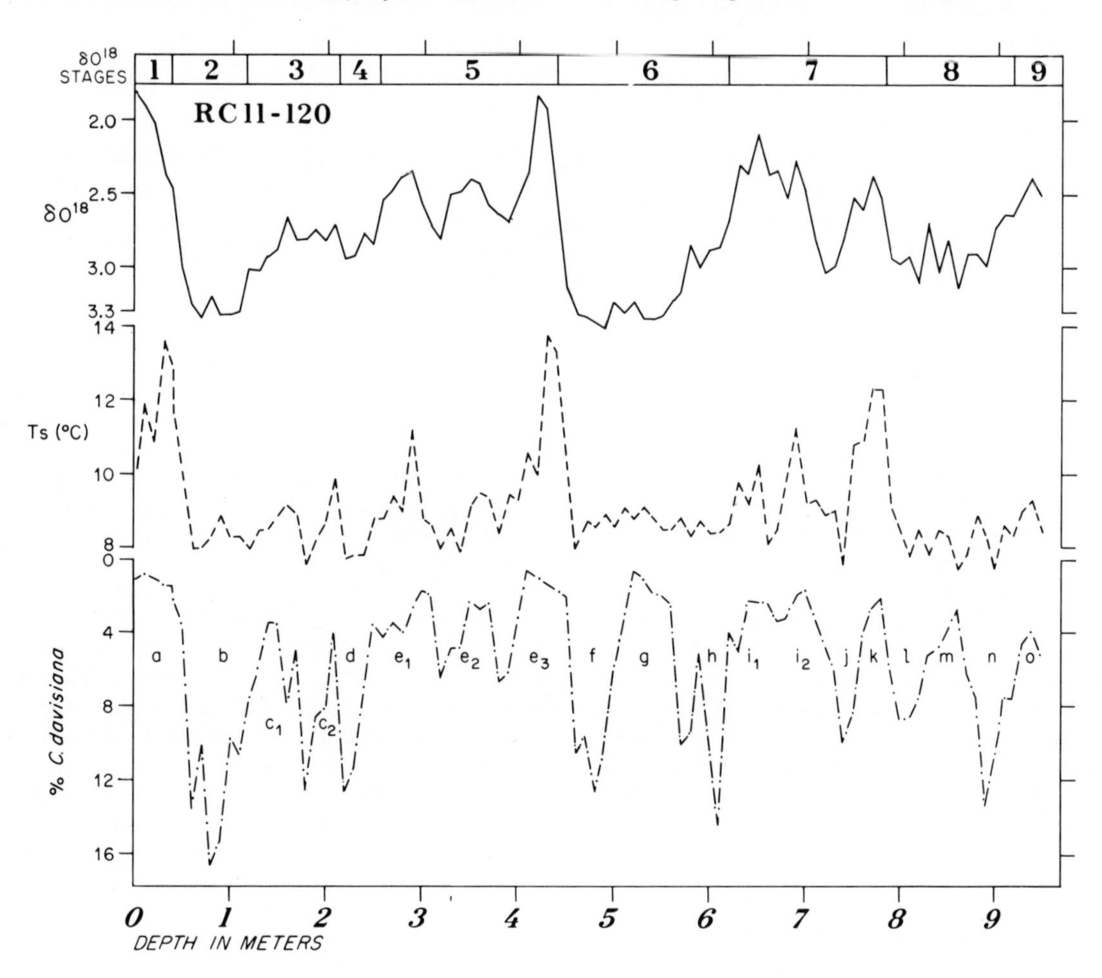

Fig. 9. Three parameters: Ts estimates of summer sea surface temperatures, δO^{18}, and percent *C. davisiana*, measured in sub-Antarctic Indian Ocean core RC11–120 after Hays et al. (1976b)

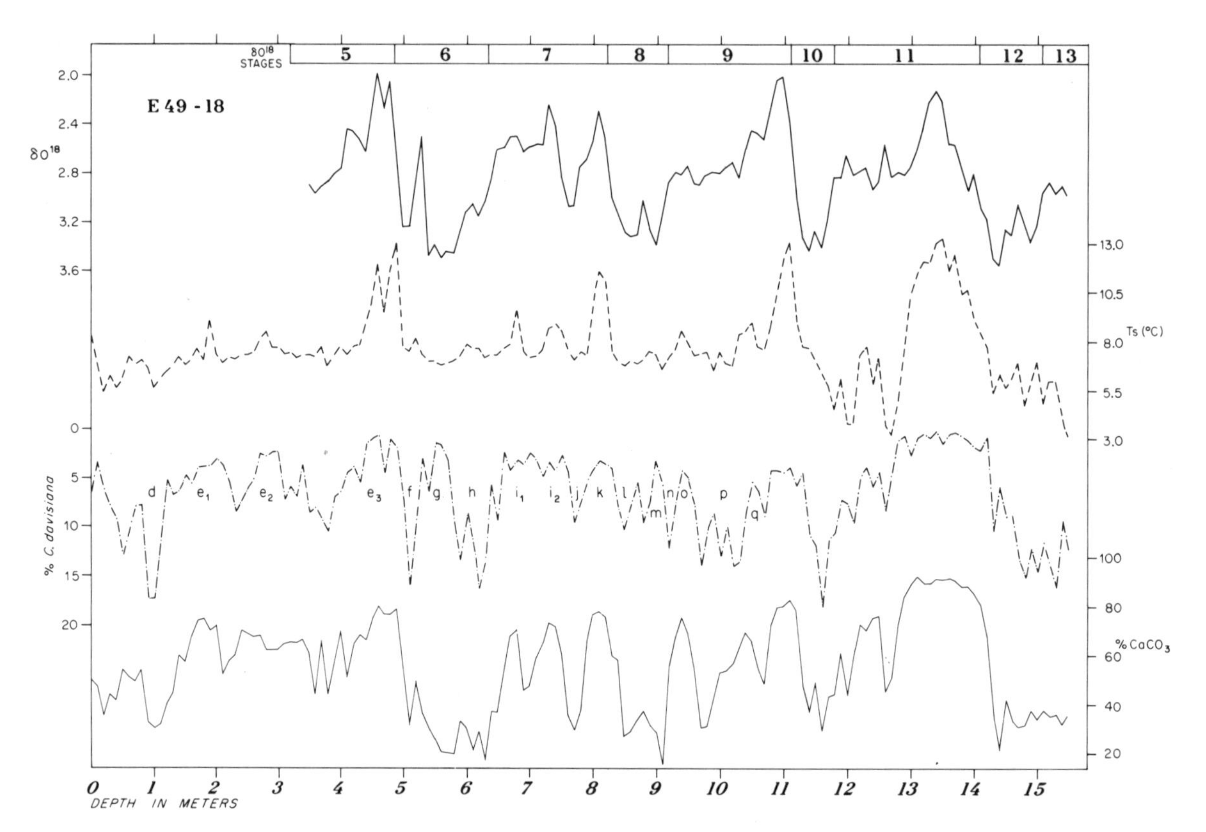

Fig. 10. Four parameters: Ts estimates of summer sea surface temperature, δO^{18}, percent *C. davisiana*, and percent calcium carbonate, measured in sub-Antarctic Indian Ocean core E49–18 after Hays et al. (1976b)

of Northern Hemisphere ice volume) with the estimates of local sea surface temperatures (Ts), we can determine the synchroneity of major climatic changes between the two hemispheres. An examination of Figures 9 and 10 shows that in general, major climatic changes in the two hemispheres are nearly synchronous. However, closer scrutiny of these records shows that changes in Ts (sub-Antarctic sea surface temperatures) lead δO^{18} (Northern Hemisphere ice volume) by a small amount. This is most easily seen in the top of RC11–120 (fig. 9), where sub-Antarctic sea surface temperatures have been declining since early Holocene, but ice has not yet begun to build up in the Northern Hemisphere. The importance of this small lead of the Southern Hemisphere over the Northern Hemisphere can not be over-emphasized. First, it rules out the possibility that changes in the size of Northern Hemisphere ice sheets control changes in Southern Hemisphere climate. Secondly, although the lag in the response of Northern Hemisphere ice may be due to longer time constants for ice sheets than sea surface waters, it leaves open the possibility that Southern Hemisphere changes may influence the Northern Hemisphere. We will look again at this possibility later.

To determine what, if any, periodicities are contained in these records, Hays et al. (1976b) performed power spectrum analyses on the three parameters measured. The results of this analysis showed that the dominant periods contained in the geological data closely approximate the dominant periods of the change of the geometry of the Earth's orbit around the sun, about 106 thousand years (eccentricity), 42 thousand years (obliquity), 23 and 19 thousand years (precession). From this, they concluded that these orbital changes acted as a timer or pacemaker of ice age climates during at least the last half million years. They also showed, through cross spectral analysis, that on average, changes in Northern Hemisphere ice volume lagged changes in Southern Hemisphere sea surface temperature by about 3 000 years.

CHANGES SOUTH OF THE POLAR FRONT

Temperature changes through at least the last several hundred thousand years south of the Polar Front are not detectable in Antarctic deep-sea cores (Hays et al., 1976a). The major faunal change that does occur is the gross change in abundance of *Cycladophora davisiana.*

The reasons for the changes in abundance of this species are not understood. The lithologic changes in cores south of the Polar Front, however, provide important clues to other environmental changes. Following Hays et al. (1976a) and Cooke & Hays (in press), these changes can be interpreted as changes of sea ice cover, so the sequence of lithologic changes may provide a record of changing sea ice cover through time. Using the stratigraphy provided by changing relative abundance of *C. davisiana*, Cooke & Hays (in press) were able to correlate cores from south of the Polar Front to RC11–120 north of the Polar Front. Using the chronology developed for this core by Hays et al. (1976b), sedimentation rates for cores south of the Polar Front can be calculated. Cooke & Hays (in press) were able to show that even in cores with accumulation rates as high as 30 cm/1 000 years, the lithologic changes occur within 5 cm. This indicates that the environmental changes responsible for these lithologic changes must have occurred in less than 200 years.

Perhaps even more important, they were able to determine the timing of these lithologic changes relative to changes in Northern Hemisphere ice volume.

Figure 11 shows the *C. davisiana* record for three cores: RC11–120, RC13–271 and RC13–259. The latter two of these were raised from south of the Antarctic Polar Front. RC11–120, raised from the sub-Antarctic Indian Ocean has a δO^{18} record by N. Shackleton (Hays et al., 1976a).

Assuming that the *C. davisiana* curves are synchronous north and south of the Polar Front, it was possible for them to relate the timing of lithologic changes to the δO^{18} curve.

In Figure 11, Cooke & Hays (in press), identified four widespread lithologic changes that could be correlated with the δO^{18} curve. The most recent change (1) consists of a change from diatomaceous

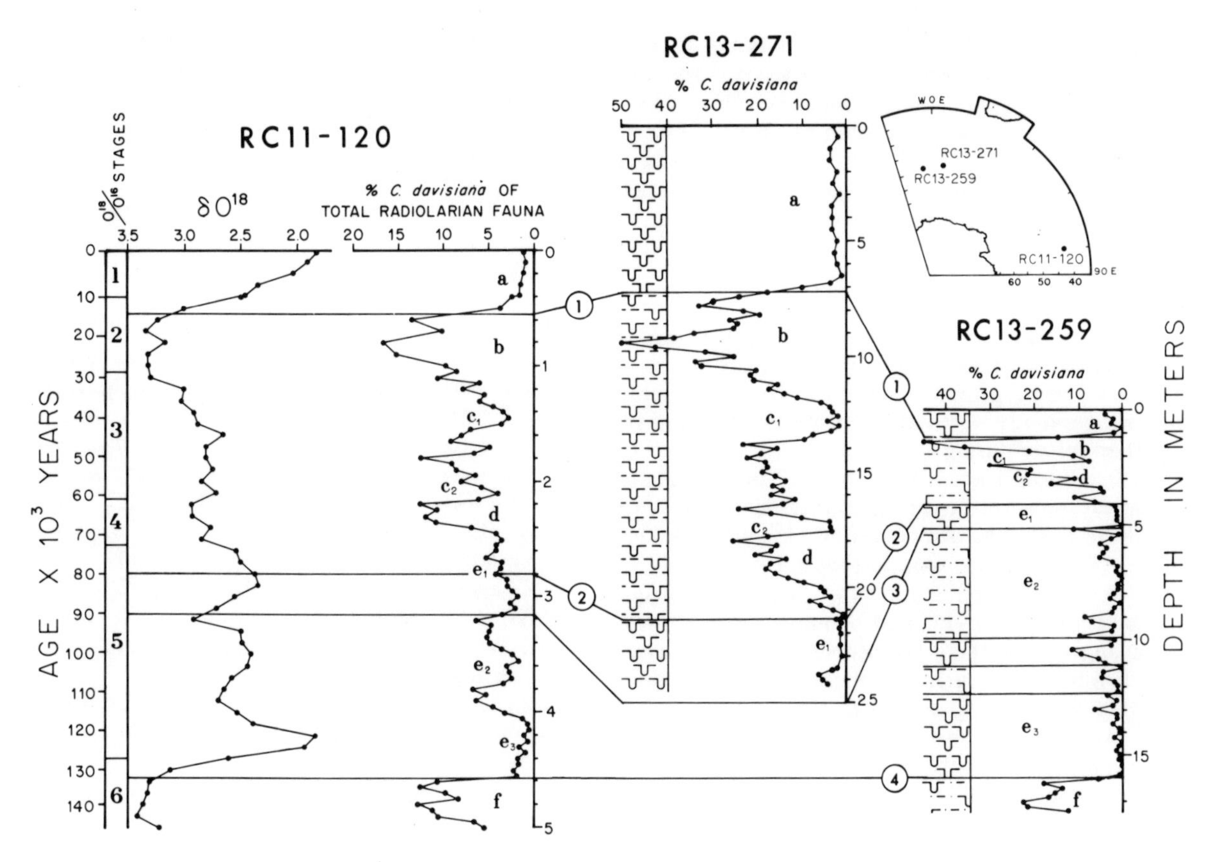

Fig. 11. Correlation of *C. davisiana* stratigraphy and lithostratigraphy of cores from south of the Antarctic Polar Front, RC13–271 and RC 13–259, with *C. davisiana* and δO^{18} stratigraphy of RC11–120 from north of the Polar Front (from Cooke & Hays, in press)

clay below to diatom ooze above. This may represent, as Cooke & Hays (op. cit.) suggest, a change from large amounts of summer ice present in the Antarctic Ocean to open, ice free, summer conditions like today. This lithologic change occurs in the upper half of the transition from *C. davisiana* zone (b) to *C. davisiana* zone (a) in cores RC13–271 and RC13–259. In the δO^{18} curve of RC11–120 this corresponds to the very base of the transition from δO^{18} stage 2 to stage 1. This suggests that the Antarctic became free of summer sea ice prior to, or just at the start of, Northern Hemisphere ice sheet melting.

Lithologic change (2) is the reverse of (1) marking a transition from diatom ooze below to diatomaceous clay above. Presumably, this represents an environmental change also opposite to (1) namely a change from open summer conditions to an Antarctic Ocean with large amounts of sea ice. This lithologic change (2) can be correlated to the δO^{18} curve of RC11–120 and it occurs in the middle of stage 5a near the ice minimum of that substage and before the build-up of ice that lead to the volume maximum of δO^{18} stage 4. Lithologic changes 3 and 4 are similar in the sequence of lithologies to lithologic change (1). Like lithologic change (1) they occur at the beginning of transitions from heavy to light oxygen isotopic ratios (more to less ice volume). From this we can conclude that like sub-Antarctic sea surface temperature changes from cooler to warmer, changes in sea ice from more to less ice in Antarctic Seas precede changes from more to less glacial ice in the Northern Hemisphere.

CONCLUSIONS

From detailed work on Antarctic deep-sea cores to date, we can draw the following conclusions:

1. There is general synchroneity between climatic changes in the Northern Hemisphere and Southern Hemisphere during the last 500 000 years.

2. During glacial times in the Northern Hemisphere, sub-Antarctic temperatures were cooler, the Polar Front, the northern limit of summer sea ice, and winter sea ice were further north.

3. In detail changes in sub-Antarctic sea surface temperatures an inferred Antarctic sea ice lead changes in Northern Hemisphere land ice. The lead of Antarctic sea surface temperature averages about 3 000 years during the last half million years.

4. These conclusions preclude the possibility that changes in Northern Hemisphere ice volume drive Southern Hemisphere climate on the ice age time scale.

They leave open the possibility that Southern Hemisphere changes, particularly changes in Antarctic sea ice, may influence Northern Hemisphere climate.

ACKNOWLEDGEMENTS

I am grateful to my CLIMAP colleagues for many stimulating discussions and thoughtful criticisms. In particular, Bill Ruddiman, John Imbrie, Nick Shackleton, David Cooke and Joseph Morley.

I wish to acknowledge technical held from the following: faunal counts – Grace Irving, Alice Pesanell, Karen Jare; drafting – Mary Perry, Barbara Walter; typing and proofreading – Rose Marie Cline and Erika Free.

CLIMAP studies are jointly funded by the National Science Foundation's Office of Climate Dynamics and International Decade of Ocean Exploration, Grant OCE77–22893. Core material was provided by ONR N00014–75–C–0210 and NSF–OCE76–18049. Lamont-Doherty Geological Observatory Contribution Number LDGO 2709.

REFERENCES

Cooke, D.W. & Hays, J.D. (in press). Estimates of Antarctic Ocean seasonal sea ice cover during glacial intervals. *Third Symposium on Antarctic Geology and Geophysics, Madison, August, 1977.*

Federman, A.M. 1977. The chemical characterization of abyssal tephras from the South Sandwich Islands. *Trans. Am. Geophys. Un.* 58(6): 528.

Goodell, H.G. 1973. The sediments. In: *Marine sediments of the southern oceans.* Am. Geog. Soc. Antarctic Map Folio Ser., folio 17.

Hays, J.D. & Berggren, W.A. 1971. Quaternary boundaries. In: B.M. Funnel & W.R. Reidel (eds.), *The micropaleontology of oceans*, Cambride University Press: 669–691.

Hays, J.D., Lozano, J.A., Shackleton, N.J. & Irving, G. 1976a. Reconstruction of the Atlantic and western Indian Ocean sectors of the 18 000 B.P. Antarctic Ocean. In: R.M. Cline & J.D. Hays (eds.), *Investigation of Late Quaternary paleoceanography and paleoclimatology,* Boulder, Geol. Soc. Am., Memoir 145: 337–372.

Hays, J.D., Imbrie, J. & Shackleton, N.J. 1976b. Variations in the Earth's orbit: Pacemaker of the ice ages. *Science* 194 (4270): 1121–1132.

Hays, J.D., Morley, J.J., Burckle, L. & Cooke, D.W. (in prep.). Late Quaternary Antarctic paleoceanography.

Hough, J.L. 1956. Sediment distribution in the southern oceans around Antarctica. *J. Sed. Petrol.* 26(4): 301–306.

Imbrie, J. & Kipp, N.G. 1971. A new micropaleontological method for quantitative paleoclimatology: Application to a late Pleistocene Caribbean core. In: K.K. Turekian (ed.), *The late Cenozoic glacial ages*, Yale Univ. Press, New Haven, Conn.: 71–79.

Lozano, J.A. & Hays, J.D. 1976. Relationship of radiolarian assemblages to sediment types and physical oceanography in the Atlantic and western Indian Sectors of the Antarctic Ocean. In: R.M. Cline & J.D. Hays (eds.), *Investigation of Late Quaternary paleoceanography and paleoclimatology*, Boulder, Geol. Soc. Am., Memoir 145: 303–336.

Ninkovich, D., Heezen, B.C., Connolly, J.R. & Burckle, L.H. 1964. South Sandwich tephra in deep-sea sediments. *Deep-Sea Res.*11(4):605–619.

Philippi, E. 1912. Die Grundproben der Deutschen Südpolar-Expedition 1901–1903. In: *Deutsche Südpolar-Expedition 1901–1903*, 2, Berlin, Geographie und Geologie: 415–616.

Schott, W. 1939. Deep-sea sediments of the Indian Ocean. In: P. Trask (ed.), *Recent marine sediments*, Tulsa, Am. Assoc. of Petroleum Geologists: 396–408.

Shackleton, N.J. & Opdyke, N.D. 1973. Oxygen isotope and paleomagnetic stratigraphy of equatorial Pacific core V28–238: Oxygen isotope temperatures and ice volume on a 10^5 and 10^6 year scale. *Quat. Res.* 3(1): 39–55.

Shackleton, N.J. & Opdyke, N.D. 1976. Oxygen-isotope and paleomagnetic stratigraphy of Pacific core V28–239: Late Pliocene to Latest Pleistocene. In: R.M. Cline & J.D. Hays (eds.), *Investigation of Late Quaternary paleoceanography and paleoclimatology*, Boulder, Geol. Soc. Am., Memoir 145: 449–464.

Watkins, N.D. & Kennett, J.P. 1971. Antarctic bottom water: Major changes in velocity during the late Cenozoic between Australia and Antarctica. *Science*, 173: 813–818.

Watkins, N.D. & Kennett, J.P. 1972. Regional sedimentary disconformities and upper Cenozoic changes in bottom water velocities between Australia and Antarctica. *Am. Geophys. Union Antarctic Research Ser.*, 19: 273–293.

* 7 *

Glacial development and temperature trends in the Antarctic and in South America

J.H. Mercer

Institute of Polar Studies, Ohio State University, Columbus, Ohio 43210, U.S.A.

Manuscript received 17th August 1977

CONTENTS

ABSTRACT

Variations in surface temperature of the Pacific sector of the Southern Ocean result in similar changes downwind in South America. Northward in South America the influence of the Southern Ocean decreases and that of the Atlantic Ocean increases. Northward retreat of tropical vegetation in southern South America during the Late Eocene and Early Oligocene reflects contemporaneous cooling of the Southern Ocean. The glacial history of southern South America after the onset of major glaciation 3.5 MY ago likewise must reflect Early Pliocene cooling of the Southern Ocean.

Oxygen isotopic studies of deep-sea cores suggest that conditions in coastal Antarctica first became frigid at the beginning of the Oligocene ca 37 MY ago, when the sea froze and glaciers extended to sea level. However, other deep-sea core studies suggest that conditions did not become severe until the end of the Oligocene ca 26 MY ago, when ice rafting began and tree species were exterminated. The later date is considered the more reliable because it is based on more direct evidence.

According to oxygen isotopic studies, the East Antarctic ice sheet accumulated from 14 to 10 MY ago. If so, the West Antarctic ice sheet, whose situation largely below sea level precludes it ever having been composed of temperate ice, must have formed after 10 MY ago. Its emplacement would have caused a rather abrupt northward expansion of cold Antarctic water; such an event has been noted in Early Pliocene time soon after 5 MY ago. Evidence for an earlier comparably cold episode in Late Miocene time ca 6 MY ago is equivocal. This suggests that the West Antarctic ice sheet first formed during the Early Pliocene. If so, its formation may have been triggered by the ocean freshening caused by massive evaporite deposition in the Mediterranean basis 5.5–5 MY ago. Some geological evidence, however, and some recent reassessments of oceanographic evidence, suggest that the West Antarctic ice sheet was present by about 7 MY ago. Initiation of glaciation in southern South America

3.5 MY ago does not disprove a 7 MY ago for the West Antarctic ice sheet, but it favours a younger age.

During the Gauss Epoch 3.3–2.4 MY ago oceanographic studies suggest an unlikely interhemispheric contrast: a return to comparative warmth in high southern latitudes, and the start of major mid-latitude glaciation in the Northern Hemisphere. Unfortunately, little is known about glacial events in South America during this interval. After ~ 2.1 MY ago repeated glaciations in southern South America undoubtedly correspond to repeated severe coolings of the Southern Ocean, but the individual episodes cannot yet be correlated.

During the last interglacial the oxygen isotopic content of ocean water implies less global ice cover than today's. In southern Chile, deep chemical weathering at that time suggests exceptional warmth. These observations support the hypothesis that the ca + 6 m sea level ca 125 000 years ago resulted from deglaciation of West Antarctica when temperatures rose above the critical level for ice shelves. During the last glaciation world ice volume was greatest ca 18 000 BP, implying that the Northern Hemisphere mid-latitude ice sheets were then largest. In south-central Chile, however, glaciers were largest – and by inference, temperatures were lowest – before 56 000 BP. After a smaller readvance culminating ca 19 500 BP the Chilean glaciers shrank during a major interstade and later readvanced, probably until about 13 000 BP. This oscillation is not evident in the sub-Antarctic Indian Ocean cores, but may be shown by the poorly-dated Byrd Station (Antarctica) core. After 13 000 BP, southern South America warmed rapidly, and by 11 000 BP glaciers were within their present borders, where they remained during the Younger *Dryas* Stade, ca 11 000–10 000 BP, the final European interval of severe cold. This early recession (compared to the North Atlantic area) is compatible with maximum postglacial warmth ca 9 400 ± 600 BP in the sub-Antarctic Indian Ocean, and strongly suggests that the North Atlantic was a highly atypical part of the world ocean during deglaciation of North America and Eurasia. Former large lakes in South Victoria Land, Antarctica, dammed by grounded Ross Sea ice, imply temperatures at least as high as today's. A Late Wisconsin age has been suggested, but this seems unlikely, because southern South America was then much colder. The lakes may have formed early in the Southern Hemisphere hypsithermal, starting ca 11 000 BP, when eustatic sea level, which probably controlled the extent of the grounded Ross Sea ice, was still low. In Peru at lat 14° S, at a site dominated by air from the equatorial North Atlantic during the accumulation season, initial results suggested that a minor readvance occurred during Younger *Dryas* time, but further studies consistently indicate that the advance culminated at least 500 years earlier, ca 11 000 BP. By 10 000 BP the Peruvian glaciers were little, if any, larger than they are today.

During Neoglacial time glaciers in southern South America were largest about 4 500 BP, whereas in the Northern Hemisphere a contemporaneous advance was relatively smaller. This suggests that the inferred cooling was caused by an event in high southern latitudes, perhaps greatly increased calving from West Antarctica. A later Neoglacial advance in southern South America culminated ca 2 700–2 200 BP, as in many other parts of the temperate zones of both hemispheres. Throughout this interval in Peru, however, ice was less extensive than it is today.

ANTARCTIC AND SOUTH AMERICAN TEMPERATURE TRENDS AND GLACIAL EVENTS FROM THE END OF THE PALAEOCENE TO THE LATE MIOCENE (t=55–10 MY)

The latitudinal positions of both South America and Antarctica have remained nearly constant throughout the Cainozoic (Heirtzler, 1971: 675), so that palaeotemperature trends there do not reflect drift into other climatic zones. During the early Cainozoic Australia was still attached to Antarctica, the present Drake Passage between South America and the Antarctic Peninsula may have been closed, and no Antarctic ice sheet existed. Thus ocean currents in the southern South Pacific Ocean, and the zonal temperature gradient between South America and Antarctica, must have been very different to what they are today. By the Early Miocene (t = ca 22 MY) Australia and Antarctica were well-separated, and a deep channel probably had opened between South America and West Antarctica. The result was the initiation of the circum-Antarctic current and enhanced thermal isolation of Antarctica, leading to increasing land and sea ice cover. These glacial developments in and aroung Antarctica, especially West Antarctica, strongly influenced the climate of southern South America.

South America

Early and Middle Cainozoic temperature trends in South America have been inferred mainly from floral and faunal changes in the southern part of the continent. The tropical South American flora extended furthest south in the Late Palaeocene and Early Eocene; in the Late Eocene and Early Oligocene it began to retreat northwards, and was replaced by a cool-temperate flora dominated by *Nothofagus* species (Menéndez, 1971: 361). Menéndez & Caccavari de Felice (1975: 182) have examined a core from northern Tierra del Fuego whose age ranges from Late Cretaceous to Oligocene, according to the microplankton present. The major dominance and variety of *Nothofagus*, indicating a temperate climate, began at the end of the Eocene, replacing a more warmth-loving Early Eocene assemblage. Fasola (1969) has studied the Late Eocene to Middle Oligocene vegetation in the Strait of Magellan area. He concludes that average annual temperature was then still about 6° C higher than it is today. As will be seen, these trends broadly agree with the palaeotemperature curve for the Southern Ocean, deduced from oxygen isotopic changes (Shackleton & Kennett, 1975a) and with the microfaunally-deduced temperature trends on the Falkland Plateau (Ciesielski & Wise, in press).

Southern Ocean and Antarctica

Temperature trends in the Southern Ocean and the timing of the initiation and growth of Antarctic ice cover before the establishment of the ice sheet have been estimated from 1) the oxygen isotopic composition of ocean water, 2) the distribution of ice-rafted detritus, 3) the occurrence of subglacial vulcanism, 4) marine microfaunal assemblages, and 5) terrestrial floral assemblages. Previous discussions have been by Denton et al. (1971), Mercer (1973) and Drewry (1975).

Oxygen isotopic measurements. Interpretations of the changing oxygen isotopic ratios in benthic and planktic Foraminifera in the equatorial Pacific Ocean (Douglas & Savin, 1973) and in middle latitudes of the southeast Indian Ocean (Shackleton & Kennett, 1975a) have provided continuous and detailed palaeotemperature curves for surface and deep water since the Late Palaeocene. The thermal climate of coastal Antarctica has been estimated from the calculated temperatures of surface and deep water in the southeast Indian Ocean. Shackleton & Kennett (1975a) conclude that conditions were warmest in the Late Palaeocene and Early Eocene; at the DSDP (Deep Sea Drilling Project) Site 277 on the Campbell Plateau at lat. 52° S (Fig. 1), water temperatures then were 20° C at the surface and 17° C on the bottom at present 1 000 m depth. A slow, sporadic cooling began in the mid-Early Eocene (t = 51 MY) and by the end of the Eocene (t = 38 MY) temperatures had dropped by about 10° C. An abrupt and dramatic cooling followed at the Eocene–Oligocene transition, with temperatures falling to 7° C at the surface and to 5° C on the bottom. Shackleton & Kennett (1975a: 752) believe that deep water as cold as this must have formed at about freezing point around Antarctica, in association with sea ice, and that this earliest Oligocene cooling therefore must mark a critical stage in the development of Antarctic glaciation, with glaciers extending to sea level.

After this sharp cooling, temperatures changed little during the rest of the Oligocene; however, during the Early Miocene (t = 19 MY) and again during the earliest Middle Miocene (t = 14 MY), temperatures rose almost to Late Eocene levels at DSDP Site 279 at lat. 51° S on the MacQuarie Ridge (Shackleton & Kennett, 1975a: 752–753).

After the early Middle Miocene warming about 14 MY ago, the trend of the oxygen isotopic curve would imply that water temperatures off Antarctica were well below freezing point – an impossible condition – unless corrected by assuming another important stage in the development of Antarctic glaciation: the rather rapid accumulation between 14 MY and 10 MY ago of the isotopically light East Antarctic ice sheet (Shackleton & Kennett, 1975a: 752). When the estimated correction is made, calculated temperatures at DSDP Site 281 (lat. 48° S) fell while the ice sheet was forming, but later rose by about 5° C in the middle Late Miocene (t = ca 8 MY). Temperatures were then as high as during the Early Miocene warm episode, and higher than at any time during the Oligocene, but neither the East Antarctic ice sheet, nor the deep-water temperatures, were significantly affected, demonstrating the stability of the ice sheet once it had formed. After this warm episode, temperatures fell during the remainder of the Miocene (Shackleton & Kennett, 1975a: 751–752).

Ice-rafting. Different workers give a wide range of dates for the start of ice-rafting. Geitzenauer, et al. (1968) and Margolis & Kennett (1971) conclude, from the presence of quartz grains identified as of glacial origin in *Eltanin* piston cores from the South Pacific, that icebergs were reaching as far from the Antarctic continent during the earliest Eocene (t = 53 MY) and the Middle Eocene (t = 46 MY) as they do today. Supporting evidence is provided by low foraminiferal diversity, which implies low temperatures, at the same horizons as the glacial quartz grains. In later publications, however, Kennett's and Margolis' views about Eocene ice-rafting are unclear. For example, Shackleton & Kennett (1975a: 754), after analyzing material obtained on DSDP Leg 29, conclude that Antarctic glaciers did not reach sea level until the beginning of the Oligocene, and Kennett et al. (1975: 1165) reach the contradictory conclusions that during the Eocene, Antarctic glaciation was confined to high elevations and that the evidence for ice-rafting indicates some glaciation at or near sea level. Margolis (1975) notes the absence of ice-rafted material below the Lower Miocene at any DSDP Leg 29 site; he emphasizes that this does not necessarily disprove the Oligocene and Eocene ice-rafting noted further east by Margolis & Kennett (1971), but he points out that sanidine and volcanic glass grains can closely resemble glacial quartz, and were in fact so identified on board ship.

The scientists of DSDP Leg 28 conclude that ice-rafting in the southern Ross Sea began in the Late Oligocene (t = 25 MY) from a source somewhere in West Antarctica (Allis et al., 1975: 884; Barrett, 1975: 758; Hayes & Frakes, 1975: 929). They have studied the stratigraphy at Site 270 at lat. 77° 30′ S, where a preglacial glauconitic sandstone unit dated by K-Ar at 26 ± 0.4 MY is overlain by a thick glacial marine sequence identified as such by its content of dropstones, many of them striated. The basal glacial unit contains sparse ice-rafted material, and Middle to Late Oligocene dinoflagellates. Its estimated age of about 25 MY, obtained by downward extrapolation of the magnetic stratigraphy, is consistent with the 26 MY age of the underlying sandstone. At Site 268, near the coast of Wilkes Land at lat. 64° S, ice-rafted material originating in East Antarctica has been identified down to the middle Lower Miocene (= ca 18 MY) and less certainly down to the uppermost Oligocene (Piper & Brisco, 1975: 740). During DSDP Leg 35 in the Bellingshausen Sea area (Site 325), the oldest ice-rafted material noted was Lower Miocene (Hollister, Craddock et al., 1976: 171).

After its initiation, the range of ice-rafting (north of western Wilkes Land) gradually and steadily extended further north throughout the Miocene, following close behind the northward-moving boundary between earlier calcareous and later siliceous sedimentation; from this evidence, Hayes & Frakes (1975: 922 and 927) conclude that Antarctic temperatures fell slowly and continuously throughout the Miocene.

Subglacial vulcanism. Le Masurier (1972) notes that palagonite breccia, which is diagnostic of subaqueous vulcanism, occurs widely in West Antarctica at high elevations. A submarine origin, he believes, would require an improbable amount of subsequent uplift, and he therefore concludes that the palagonite was erupted beneath an ice sheet. Outcrops of palagonite have been dated by K-Ar as about 31 MY, 22 MY, and 20 MY old, and from 10 MY old to recent; a widely quoted age of 42 ± 9 MY that implied the presence of a Late Eocene ice sheet in West Antarctica (LeMasurier, 1972: 65; Rutford et al., 1972) has recently been re-dated at about 15 MY (LeMasurier, personal communication).

Marine microfauna. Ciesielski & Wise (in press), from observations made on the Falkland Plateau in the South Atlantic Ocean during DSDP Leg 36, note the presence of the warm water coccolith *Discoaster* during the Palaeocene and Eocene, and its absence during the Oligocene. This, they conclude, shows significant mid-Tertiary cooling. Further cooling occurred at the beginning of the Miocene, when the opening of a deep-water channel between South America and the Antarctic Peninsula resulted in the initiation of the circum-Antarctic current (Barker & Burrell, in press, quoted by Ciesielski & Wise, in press).

Ciesielski (1975: 651 and 653) has calculated water temperatures in general terms at DSDP Leg 29 sites, using changes in silicoflagellate populations. At Site 274 off north Victoria Land, he finds that the Early Oligocene (t = 37–32 MY) was rather warm compared to the Miocene; he has no data for the Middle and Late Oligocene. At Site 266 near MacQuarie Island at lat. 56° S, he concluded that the Early Miocene was rather warm, compared to Middle and Late Miocene times.

Terrestrial flora. Near McMurdo Sound, Antarctica, siltstone erratics on Minna Bluff and Black Island contain plant microfossils dominated by *Nothofagus* species. On the evidence of associated dinoflagellates, the siltstone is thought to be of Late Eocene age (McIntyre & Wilson, 1966). Kemp (1975) has examined core material from Site 270 in the southern Ross Sea for plant remains. The lowest glacial marine unit (Unit 2J) immediately above the glauconitic sandstone dated at 26 MY old contains abundant plant remains, predominantly *Nothofagus* species. The floral assemblage is almost identical to that in the Late Eocene Black Island and Minna Bluff erratics, and Kemp admits the possibility that the plant remains in unit 2J have been recycled from older deposits. However, she believes they have not because neither Palaeozoic plant microfossils from the Transantarctic Mountains nor Eocene dinoflagellates of the type in the Minna Bluff and Black Island Eocene erratics are present as recycled elements in unit 2J. The flora remained virtually unchanged throughout the Oligocene, Kemp believes, because of the isolation of post-Eocene Antarctica; new cold-tolerant species were unable to immigrate as temperatures fell, and instead the pre-existing assemblage persisted, but the vegetation became stunted and sparse, eventually dying out sometime during the Miocene.

Fig. 1. Antarctica and part of the Southern Ocean, showing Deep Sea Drilling Program sites. *1* McMurdo Sound; *2* Black Island; *3* Minna Bluff; *4* South Shetland Islands; *5* Reedy Glacier; *6* Shackleton Glacier; *7* Beardmore Glacier; *8* Taylor Valley; *9* Wright Valley

SEYMOUR ISLAND
4
WEDDELL
SEA
ANTARCTIC
325
PENINSULA
JONES MTNS.
WEST
ANTARCTICA
70°S
EAST
TRANSANTARCTIC
ANTARCTICA
ROSS ICE SHELF
WEST ICE SHELF
5
6
7
3
2
1
8
9
ROSS SEA
270
VICTORIA LAND
MTNS.
WILKES LAND
268
274
266
60°S
50°S
ICE
ICE SHELF
LAND
277
279
281
40°S
NEW ZEALAND
AUSTRALIA
284

Near West Ice Shelf (at about lat. 67° S, long. 77° E), glacial marine sediments contain recycled pollen and spores of different ages. The youngest pollen is identified as Middle Eocene by the associated dinoflagellates, and implies the presence of *Nothofagus* forests on shore (Kemp, 1972: 153).

In the northern Antarctic Peninsula, Tertiary plant beds, apparently spanning a considerable time interval, are known on Seymour Island and in the South Shetland Islands. On Seymour Island the oldest plant beds are of Palaeocene age (Cranwell, 1969), but the age of the youngest is uncertain. In the South Shetland Islands none of the plant beds have been reliably dated. Thus how long *Nothofagus* forests survived in the Antarctic Peninsula and offshore islands after the Palaeocene is at present not known.

Discussion. Different lines of investigation suggest very different times for the inception, at sea level in Antarctica, of severe conditions defined by the start of ice-rafting and extensive freezing of the sea, and the imminent extinction of tree species on land: before the earliest Eocene (before t = 53 MY) (Geitzenaer et al., 1968; Margolis & Kennett, 1971); earliest Oligocene (t = 37 MY) (Shackleton & Kennett, 1975a); before 31 MY ago (LeMasurier, 1972); or Late Oligocene (t = 25 MY) (Allis et al., 1975; Barrett, 1975; Hayes & Frakes, 1975; Kemp, 1975).

As has been pointed out before (Mercer, 1973), calving outlet glaciers from mountain icefields (e.g., present-day Alaska or Patagonia) produce bergs that are much too small to travel far from their source. Only ice sheets can produce bergs that are large enough to survive a journey across several degrees of latitude; such bergs may be either of Greenland type, calved from large, exceptionally fast-flowing temperate outlet glaciers, or of Antarctic type, calved from ice shelves as tabular bergs. Thus Early Eocene ice-rafting to middle latitudes, as claimed by Geitzenauer et al. (1968) and Margolis & Kennett (1971) would require that an ice sheet then covered at least part of Antarctica and extended to the coast. This would imply temperatures low enough for the formation of extensive sea ice offshore and, therefore, the formation of bottom water near freezing point in the Early Eocene, and push the first appearance of calving mountain glaciers back into the Palaeocene. Furthermore, one of the authors claiming Early Eocene ice-rafting (Kennett) also reaches the incompatible conclusion that the Early Eocene was warm and that Antarctic glaciers did not reach sea level until the Early Oligocene (Shackleton & Kennett, 1975a: 754), and another (Margolis) notes the absence of any evidence for ice-rafting until the Early Miocene at the sub-Antarctic sites drilled during DSDP Leg 29. LeMasurier's (1972) date of 42 ± 9 MY for subglacial vulcanism in West Antarctica formerly was the sole supporting evidence for ice-rafting during the Eocene; but this date has now been rejected. Although neither Kennett nor Margolis has retracted his conclusion that bergs were reaching middle latitudes by the Early Eocene, independent confirming evidence is needed before this can be accepted.

Shackleton & Kennett (1975a: 753) conclude, from their calculated bottom water temperatures at sites 277 and 279 (DSDP Leg 29), that extensive sea ice was forming in Antarctic coastal waters at the beginning of the Oligocene and they infer from this that Antarctic glaciers must have reached sea level at the same time. This presumably implies that ice-rafting began in Antarctic coastal waters at the beginning of the Oligocene. Hayes & Frakes (1975: 927), however, identify the oldest glacial marine unit in the southern Ross Sea (Site 270, DSDP Leg 28) as Late Oligocene in age; this is supported by Ciesielski's conclusion, from the silicoflagellate content of marine sediments, that the water temperature off north Victoria Land (Site 274) was quite high during the Early Oligocene. According to evidence obtained on Leg 28 from the Ross Sea and the Indian Ocean between the Wilkes Land coast of Antarctica and Australia, ice-rafting was confined to coastal waters for several million years after its inception and did not extent to palaeolatitudes 59° S until 12–13 MY ago, and 53° S until 4 MY ago (Hayes & Frakes, 1975: 928). This picture of iceberg-free Antarctic seas until the start of ice-rafting in the Late Oligocene, followed by slow but persistent cooling and northward extension of ice-rafting throughout the Miocene, is incompatible with Shackleton's and Kennett's sequence of freezing offshore waters throughout the Oligocene followed by mainly higher temperatures during the Early and Middle Miocene, when temperatures in coastal Antarctic waters on occasion rose significantly above 0° C.

LeMasurier & Wade (1976) conclude that an ice sheet was present in West Antarctica by the end of the Early Oligocene (t = 31 MY). This ice sheet, whose major portion is grounded far below sea level, depends for its existence on the presence of fringing ice shelves (Mercer, 1968a; Hughes, 1977). Ice shelves are always cold glaciers (Robin & Adie, 1964: 105), so that the West Antarctic ice sheet must always have been composed of cold ice; its climatic limit appears to be the summer sea level isotherm of 0° C (Mercer, 1968a). At first sight, Shackleton's and Kennett's conclusion that sea ice was forming round Antarctica by the Early Oligocene (t = 37 MY) might seem to constitute supporting or at least permissive evidence for the presence of cold glaciers at sea level 6 MY later. However, Shackleton and Kennett also conclude that the East Antarctic ice sheet did not accumulate before 14 MY ago. Now, the East Antarctic ice sheet is land-based, and thus would have been a temperate glacier in its initial states. A cold West Antarctic ice sheet could not have formed before a temperate East Antarctic ice sheet, and therefore, according to Shackleton's and Kennett's timetable, it must be less than 14 MY old. Thus, the concept of a cold ice sheet in West Antarctica during the Oligocene is not supported by the oxygen isotopic composition of ocean water. Furthermore, it is not supported by either the character of the molluscan content of the Oligocene to Miocene glacial marine sediments in the Ross Sea. These sediments were derived largely from West Antarctica; Barrett (1975: 756) concludes from their physical properties that they were deposited by bergs from temperate glaciers, and Hayes & Frakes (1975: 220) conclude that the molluscan content of the sediments in Unit 2, Site 270 (t = 25–16 MY) show relatively ice-free conditions, perhaps similar to the waters round Tierra del Fuego today. If the West Antarctic palagonite breccias were shown unequivocally to be subglacial and not submarine in origin, and if the age determinations were known to be reliable, LeMasurier's conclusion would have to be seriously considered, however puzzling its implications; but because neither situation holds, his concept of an ice sheet in West Antarctica during the Late Oligocene and Early Miocene must be rejected.

Thus, if we reject Eocene and Oligocene ice-rafting to middle latitudes, and the presence of an ice sheet in West Antarctica during the Oligocene, two incompatible timetables for the first appearance of calving Antarctic glaciers remain: according to Shackleton & Kennett (1975a), frigid offshore waters and thus, by inference, calving mountain glaciers were first present in the earliest Oligocene (t = 37 MY), whereas according to Hayes & Frakes (1975), ice-rafting did not begin until the end of the Oligocene 12 MY later. The DSDP Leg 28 scientists' conclusion, based on direct, on-site observation of ice-rafted material must be preferred to Shackleton's and Kennett's inference, from palaeotemperature reconstructions at a distant site, that ice-rafting began earlier.

The question then arises: could Shackleton's and Kennett's conclusions be made compatible with those of Hayes and Frakes by rejecting only their inference about ice-rafting while accepting their palaeotemperature reconstruction? In other words, could calving glaciers have been absent in coastal Antarctica for 10 MY or so despite a frigid climate at sea level? If coastal mountains were present the answer must be no; but if both East Antarctica and the West Antarctic archipelago were low-lying during the Early and Middle Oligocene, they perhaps did not become glaciated despite low temperatures. This state of affairs would probably require anti-cyclonic conditions with predominantly clear skies in summer, in order to remove the winter snow cover; the likelihood of this is hard to estimate, in the absence of a modern analogous situation. If Antarctica was frigid but unglaciated during the Early and Middle Oligocene, the start of ice-rafting may mark, not a threshold of climatic cooling, but a tectonic development: the uplift of the Transantarctic Mountains *and* the West Antarctic archipelago to the point that they intercepted sufficient moisture to support actively calving glaciers.

A key role in the reconstruction of Antarctic palaeotemperatures during the Oligocene is held by the vegetational history: specifically, the extinction date of tree species. Trees could not have survived in the Ross Sea area if offshore water temperatures were near freezing, with extensive formation of sea ice. In southern Tierra del Fuego today, where numerous small, short-lived bergs are calved from very active tide-water glaciers, average midsummer air and offshore water temperatures are both about 8° C and trees grow up to about 500 m above sea level. Assuming a lapse rate of 0.7° C per 100 m in this maritime climate, the average midsummer temperature at the treeline is about 4.5° C. A reasonable inference is that *Nothofagus* species will not survive where summer temperatures are below this level. By the same token, however, stunted trees would survive at sea level in Tierra del Fuego if temperature fell by 4° C; under these conditions ice-rafting, which is already in progress, would increase greatly in scale, and the cooler ocean would enable the bergs to travel moderate distances offshore.

Thus Hayes' and Frakes' observations are permissive for the survival of trees in the Ross Sea area during the early stages of ice-rafting in the Late Oligocene, whereas Shackleton's and Kennett's palaeotemperature reconstruction requires that Antarctic forests were extinguished by sharp cooling at the Eocene–Oligocene transition. Unfortunately the evidence that trees survived in the Ross Sea region until the end of the Oligocene is equivocal, but the likelihood that they did suggests that Shackleton and Kennett may have inferred excessively low Antarctic temperatures for the Oligocene from their oxygen isotope curve.

ANTARCTIC AND SOUTH AMERICAN TEMPERATURE TRENDS AND GLACIAL DEVELOPMENT FROM THE LATE MIOCENE (t=10 MY) TO THE EARLY PLIOCENE (END OF THE GILBERT EPOCH, T=3.3 MY)

Southern South America lies downwind of that part of the Southern Ocean that abuts West Antarctica; thus its Late Cainozoic climate has been very sensitive to environmental changes in West Antarctica. The West Antarctic ice sheet is present during the current interglacial, and therefore its prior formation was a minimal requirement for the repeated major glaciations that have affected southern South America since the Middle Pliocene. During these glaciations the ice, which is now confined to the Andean Cordillera, extended eastward over the Patagonian plains. Repeated major South American glaciations probably began when variations in the Earth's orbit (Hays, Imbrie & Shackleton, 1976) were superimposed on the regional cooling induced by the presence of the West Antarctic ice sheet and its associated belt of cold water. Emplacement of this ice sheet must also have had a considerable effect on circulation patterns in the southern South Pacific Ocean as former direct connections between the Ross and Weddell seas were closed.

Southern South America: Miocene–Pliocene changes in molluscan fauna

According to Zinsmeister (in preparation), significant changes in the shallow-water molluscan faunas occurred along the west coast of southern South America during the latest Miocene and earliest Pliocene. The presence of distinctively warm-water genera of molluscs indicates that during Middle and Late Miocene time the coastal waters of southern Chile were warm subtropical. These warm water faunas are known as far south as the Taitoa Peninsula (lat. 46° 50′ S), and similar warm water faunas occur in southernmost Argentine Patagonia and Tierra del Fuego, showing that waters were subtropical around the coasts of southern South America. These warm-water faunas were replaced by distinctively cool temperate faunas in latest Miocene or earliest Pliocene time: the absence of microfossil control prevents a more precise age determination. Zinsmeister believes that this sudden replacement of warm water by cool-temperate water was associated with the development of the West Antarctic ice sheet and the resultant closing of the seaways through the West Antarctic archipelago and intensification of the northward flowing Humboldt Current.

Southern South America: the start of glaciation

A detailed description of mid-Pliocene glacial deposits in southern South America has been given elsewhere (Mercer, 1975 and 1976) and an outline only will be given here. The oldest known glacial deposit is exposed in the escarpment face of a basaltic plateau, the Meseta Desocupada, on the north side of Lago Viedma (Fig. 3) (lat. 49° 28′ S, long. 72° 25′ W, elevation 1 430 m), where a bed of till is interbedded between reversely magnetized lava flows 3.55 ± 0.07 MY old (above) and 3.48 ± 0.09 MY old (below). Near by, on the Meseta Chica, till covers lava 3.68 ± 0.03 MY old and is covered by lava 3.55 ± 0.19 MY old at exposures 1 km apart; basaltic clasts in the till are 3.49 ± 0.08 MY and 3.53 ± 0.05 MY old. In the Río Santa Cruz valley 120 km to the southeast, coarse outwash gravel underlies lava 2.95 ± 0,07 MY old, confirming that major glaciation occurred before 3 MY ago.

On the Meseta Desocupada no till or outwash gravel is interbedded with the 14 underlying reversely magnetized lava flows, the oldest of which is about 4.5 MY old, suggesting that the 3.5 MY-old glaciation was the first at this site. However, this does not necessarily imply that this glaciation marked the first cold episode of Pleistocene severity, because the history of uplift of the Patagonian Andes is not well known; possibly the mountains were not high enough to support large glaciers before 3.5 MY ago.

Age of the West Antarctic Ice Sheet

As was pointed out earlier, the West Antarctic ice sheet must always have been composed of cold ice (except at great depth) and could not have been present while summer temperatures in West Antartica remained above 0° C at sea level; the glaciers would then have been temperate, and could not have extended beyond the coastlines of the archipelago, however active they were. A threshold was crossed when summer temperatures reached the critical level of about 0° C that allows ice shelves to form. As Bentley & Ostenso (1961: 981) suggest, once ice shelves had begun to form, they spread and gradually extended between the islands of the West Antarctic archipelago and East Antarctica, thickening and eventually grounding to form the West Antarctic ice sheet.

According to Shackleton's & Kennett's (1975a: 752) isotopic measurements, the East Antarctic ice sheet accumulated between 14 and 10 MY ago. This 4 million year interval is at least two orders of magnitude longer than the response time of a large ice sheet to climatic change, so that the East Antarctic ice sheet is likely to have been more or less in equilibrium with the climate at all times during its buildup, and would have reached full size by the time summer temperatures had dropped to 0° C at sea level. If so, the West Antarctic ice sheet formed after 10 MY ago.

Geological evidence

Field evidence has been interpreted as showing that an ice sheet was present in West Antarctica during the Late Oligocene (LeMasurier, 1972) and during the Late Miocene (Rutford et al., 1972). Reasons have been given earlier for rejecting an Oligocene

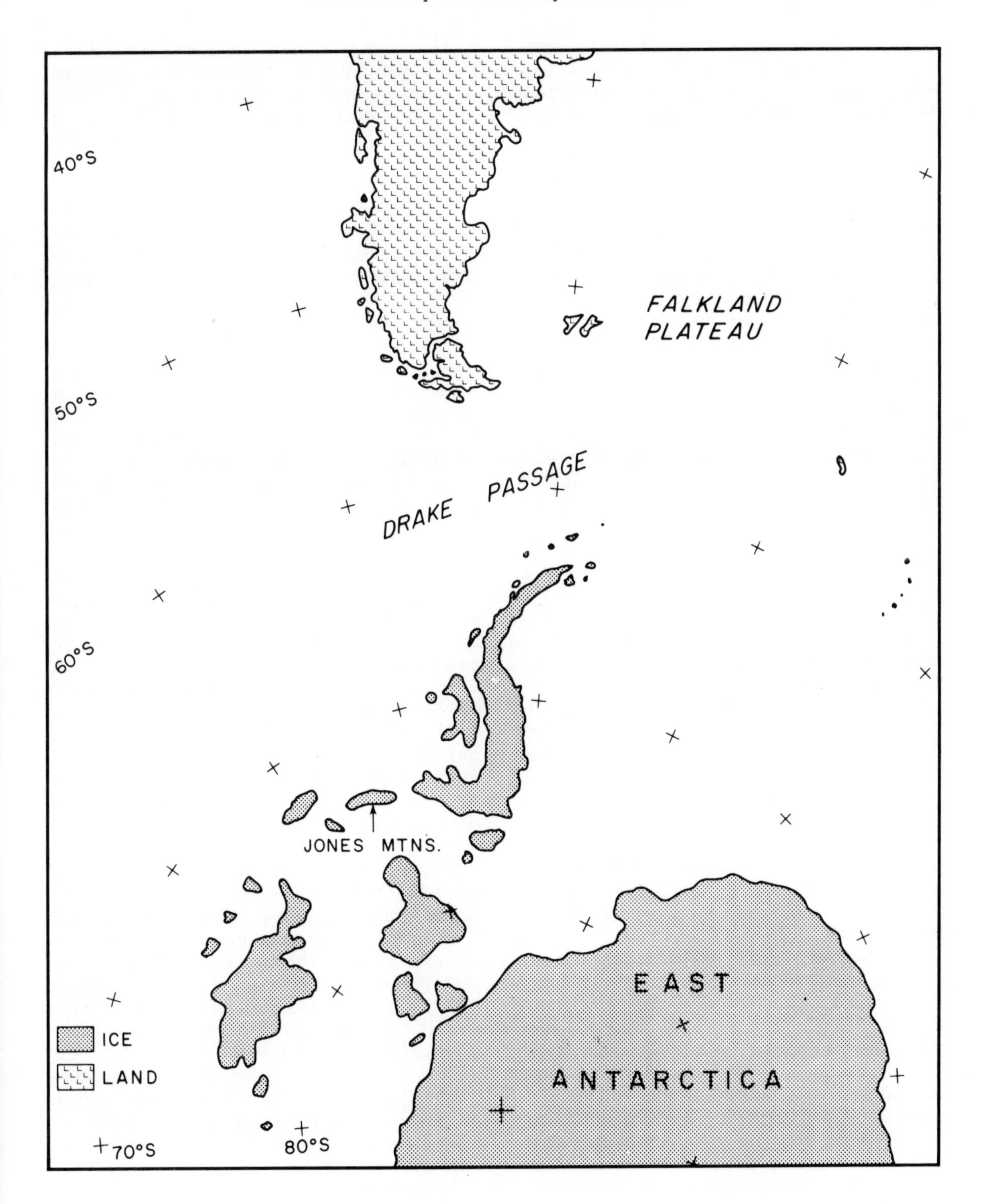

Fig. 2. Southern South America, West Antarctica and part of East Antarctica, in Late Miocene time immediately before the formation of the West Antarctic ice sheet

age for the West Antarctic ice sheet; but the evidence presented by Rutford et al. (1972) for a Late Miocene age is much stronger. In coastal West Antarctica, 500 m of basaltic volcanic rocks in the Jones Mountains unconformably cover a striated and polished basement complex on which pockets and lenses of tillite rest; the topography of the surrounding area strongly suggests the glaciation was by an ice sheet. Within 3 m above the unconformity, K-Ar ages range from 9 to 300 MY. More than 100 m above, however, ages are more consistent, ranging from 6.1 MY to 24 ± 12 MY, with several clustering between 7 and 10 MY. Rutford et al. (1972) conclude that the West Antarctic ice sheet was present at least 7 MY ago.

Mayewski (1975) presents evidence for a major glaciation of Antarctica before 4 MY ago, during which the West Antarctic ice sheet was much larger than it is today, and the East Antarctic ice sheet submerged the Transantarctic Mountains. He has examined the compact, semi-lithified till of the so-called Sirius Formation (Mercer, 1970a: 427) that occurs widely at high elevations in the Transantarctic Mountains, and has properties diagnostic of deposition by temperate ice. Mayewski concludes that this till was deposited beneath a cold ice sheet that was thick enough over the Transantarctic Mountains to be at the pressure melting point at the ice-rock interface. He displutes a contrary view of Mercer (1968b) that the till was deposited by local temperate glaciers. Mayewski (1975: 143 and 115) calculates that the ice sheet that submerged the Transantarctic Mountains during what he calls the Queen Maud Glaciation had nearly twice the volume of the present ice sheet, by assuming that the mountains had already been fully uplifted. He believes that Wright Valley, Victoria Land, must then have been ice-filled; therefore, an unmodified 4.2 MY-old volcanic cone in this valley gives a minimal age for the glaciation.

These inferences from the presence of the compact, semi-lithified till are unacceptable for the following reasons: First, neither the time of uplift of the Transantarctic Mountains nor the time of deposition of the till are known, except within wide limits; therefore, the elevations at which the till was deposited are not known. Consequently, the volume of a supposed all-submerging ice sheet cannot be calculated. Second, both Mayewski and Mercer agree that a phase of temperate glaciation in the Transantarctic Mountains must have preceded the present cold ice cover. If so, the principle of Occam's Razor favors interpreting the compact till as havig been produced by temperate glaciers, and not on a later occasion beneath a hypothetical cold ice sheet of monstrous dimensions, unless strong evidence is presented that temperate glaciers could not have emplaced the material.

While Mayewski's interpretation of the so-called Sirius Formation is considered unacceptable, a reappraisal of the contrasting settings in which the till occurs suggests that the views of Mercer (1968, 1970b) also require considerable modification. Probably, the till was deposited during a long time interval, possibly of the order of millions of years, by different types of glaciers, while the Transantarctic Mountains were being uplifted and while the East Antarctic ice cover was increasing from mountain glaciers to temperate ice sheet. If so, Mercer's (1970a: 427) original suggestion that this till constitutes a formation seems neither useful nor appropriate, and is best avoided. The oldest till was deposited by plateau ice caps before the East Antarctic ice sheet had developed, and is preserved on plateau remnants. It predates great dissection of the landscape; an example is exposed at 2 500 m elevation on the Wisconsin Plateau, Reedy Glacier area (Mercer, 1968b: 479). Till deposited later by outlet glaciers of a temperate East Antarctic ice sheet is preserved on the sides of valleys at present occupied by ice sheet outlet glaciers; for example, at 2 100 m elevation on Bennett Platform above the Shackleton Glacier, and at 1 900–2 200 m near Plunket Point about the Beardmore Glacier. Deposition of till by temperate East Antarctic glaciers implies that the West Antarctic ice sheet had not yet formed.

Reconstruction of Antarctic ice cover during deposition of the compact till is made difficult by lack of firm evidence about the time of uplift of the Transantarctic Mountains; in other words, there is no way of knowing what were the original elevations of the sites now covered by till, or how thick the East Antarctic ice sheet was when it first began to drain through the mountains to the Ross Sea. Grindley (1967: 584) suggested that the East Antarctic ice sheet formed before the Transantarctic Mountains were uplifted, but Drewry (1975: 269) points out that the existence of subglacial valleys on the inland side of the mountains, revealed by radio-echo soundings, disproves this.

An ice sheet that elsewhere extended to sea level is

Fig. 3. Part of South America, West Antarctica and part of East Antarctica, showing present-day ice cover. Inset, southern Argentina immediately east of the south Patagonian icefield, showing sites where glacial drift is associated with Pliocene to Early Pleistocene lava flows

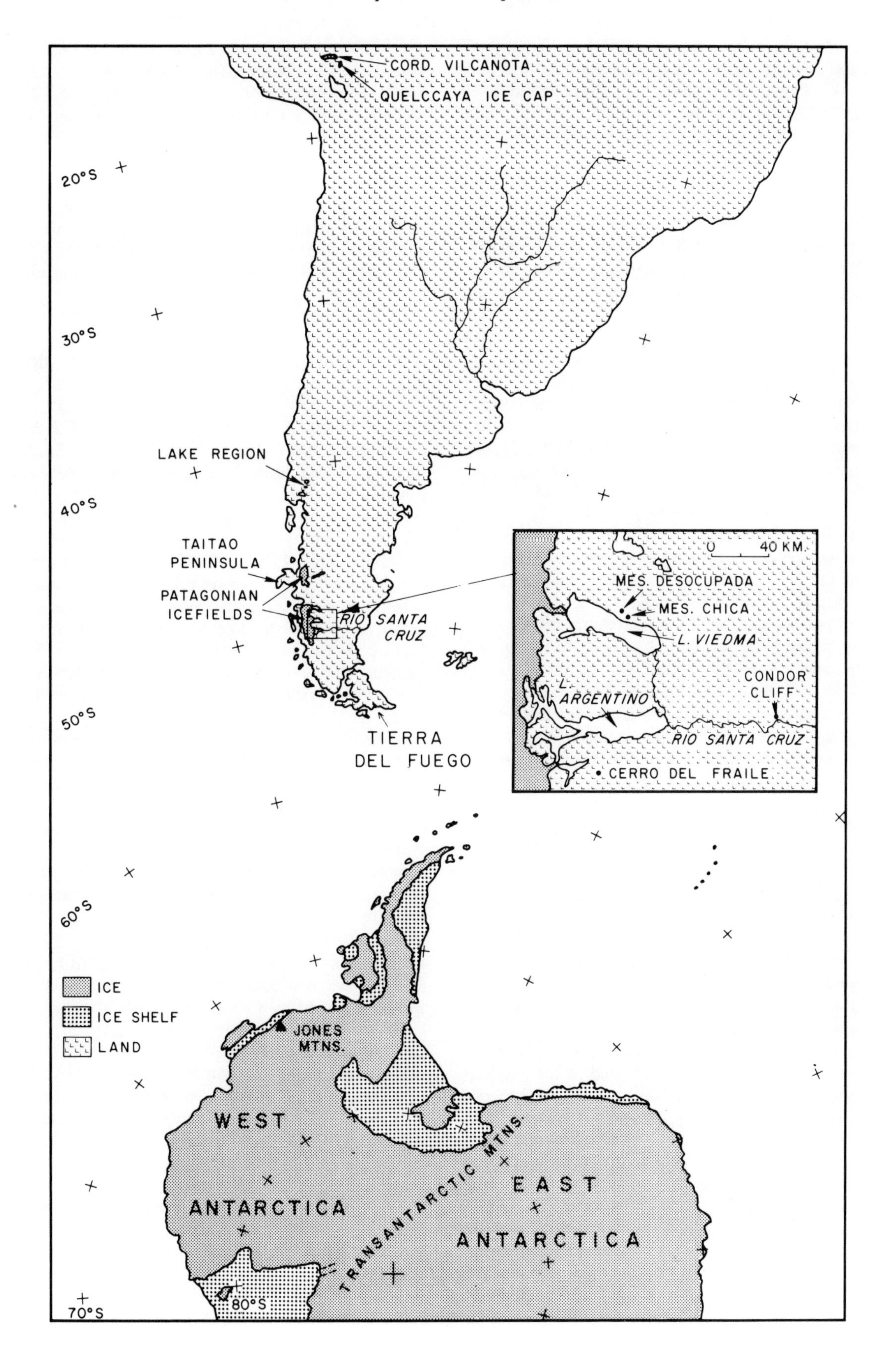
CORD. VILCANOTA
QUELCCAYA ICE CAP
20°S
30°S
LAKE REGION
40°S
TAITAO PENINSULA
PATAGONIAN ICEFIELDS
RIO SANTA CRUZ
50°S
TIERRA DEL FUEGO
0 40 KM.
MES. DESOCUPADA
MES. CHICA
L. VIEDMA
CONDOR CLIFF
L. ARGENTINO
RIO SANTA CRUZ
CERRO DEL FRAILE
60°S
ICE
ICE SHELF
LAND
JONES MTNS.
WEST ANTARCTICA
EAST ANTARCTICA
TRANSANTARCTIC MTNS.
80°S
70°S

unlikely to have been temperate at the present 2 000 m elevation of such sites as Bennett Platform and Plunket Point. This implies that if the compact till at these sites was deposited by outlets of the East Antarctic ice sheet, considerable uplift, probably of the order of at least 1 000 m, has occurred there since the ice sheet reached full size, about 10 MY ago according to Shackleton & Kennett (1975a: 753). Rather similar estimates of uplift have been made by Drewry (1975) and Webb (1972). Drewry (1975: 270) concludes that uplift probably began in the Late Eocene, about the same time as the Antarctic and Australian plates separated; his diagram (Drewry, 1975: 267) very plausibly illustrates the coeval growth of the East Antarctic ice sheet and uplift of the Transantarctic Mountains, and shows how till may be stranded far above present glacier surfaces. He believes that since the Late Oligocene, uplift has amounted to 1 500–2 000 m. Webb (1972: 232) concludes that uplift in south Victoria Land may have amounted to ca 80 m/MY in the last 3½ MY, which is a rate of about 1 000 m per 12 MY.

Probably, therefore, the so-called Sirius Formation was deposited by temperate glaciers over a long time interval; the oldest till is older than the start of ice-rafting in the Late Oligocene, and the youngest till was deposited while the East Antarctic ice sheet was building up between 14 and 10 MY ago, and most likely towards the end of that interval. The compact till is all older than the West Antarctic ice sheet and the Late Miocene–Early Pliocene unconformity in the Ross Sea (see below). There is no reason to believe that the Transantarctic Mountains have ever been submerged beneath a cold ice sheet.

Evidence from latest Miocene–Early Pliocene oceanographic change

The transformation of West Antarctica from a heavily glaciated archipelago in an iceberg-infested sea to a cold marine ice sheet (Fig. 2 and 3) may have occurred quite rapidly – during tens or hundreds of thousands of years – because of positive feedback effects from the replacement of ocean surfaces by ice shelves. Thus a rather sudden northward extension of ice-rafting, and of cold Antarctic surface water, might be expected to have occurred as the ice sheet formed. The range of ice-rafting would have increased dramatically with the appearance of tabular bergs calved from the first ice shelves. Significantly, there is a great deal of evidence from ocean cores and from seismic profiling for just such an event some time during the interval 5–3 MY ago.

Ice-rafting, after its initiation in the Late Oligocene, continued to extend its range gradually northward until the very end of the Miocene, when there was a rather sudden expansion of cold, ice-laden water during a short interval. Kemp et al. (1975: 916–917) note that in the southeast Indian Ocean the diatom ooze – nannoplankton ooze boundary moved ca 300 km northward sometime during the interval 5.5–3.5 MY ago, after a northward movement of only 450 km during the previus 17 MY. According to Hayes & Frakes (1975: 935 and 937), this probably represents a sudden intensified chilling of surface waters, and the birth of the modern Antarctic Convergence or a close analog. As they point out, this abrupt cooling must have strongly influenced the course of the Late Cainozoic glaciation in lower latitudes, such as southern South America.

In the southern Ross Sea, seismic profiler data revealed a pronounced unconformity in the glacial marine sediments (Houtz & Davey, 1973), which was later recognized at several sites during DSDP Leg 28. The limiting ages for the unconformity are given as 5.5 and 3.7 MY (Hayes & Frakes, 1975: 936) or 5.0 and 3.0 MY (Hayes, Frakes et al., 1975: 233) (see Fig. 4). A minimal age of 3 MY appears better founded because according to Hayes, Frakes et al (1975: 232) 'the oldest confidently dated strata above the unconformity are (?) early to middle Gauss in age (Site 270)' [The Gilbert-Gauss boundary is about 3.3 MY]. Weaver (in preparation), however, without giving his palaeontological evidence in detail, estimates the age of the unconformity at 7.0–4.2 MY ago. Hayes & Frakes (1975: 936) interpret the unconformity as being the result of erosion by a greatly expanded Antarctic ice sheet, at a time when the grounding line of the Ross Ice Shelf lay north of its present front. They conclude that these two events – a sudden increase in the extent of cold ocean water, and an advance of grounded ice in the Ross Sea – are different aspects of the same glacial-climatic episode: a great expansion of the Antarctic ice sheet. Hayes & Frakes (1975: 936) conclude that although neither the ice sheet expansion nor the oceanic cooling can be dated palaeontologically more closely that 5.5–3.7 MY ago, these associated events occurred about 4.5 MY ago, by assuming first that they preceded the interval 4.3–3.9 MY ago when water temperatures in the southeast Indian and southwest Pacific oceans between latitudes 56° S and 67° S were 5–10° C higher than they are today (Ciesielski & Weaver, 1974: 514), and second that they were contemporaneous with the maximum Late Cainozoic volume of the Antarctic ice sheet 4.7–4.3 MY ago deduced by Shackleton & Kennett (1975b) from oxygen ixotopic measurements.

According to Shackleton & Kennett (1975b: 805), the oxygen isotope record obtained during DSDP Leg 29 at Site 284 in the Tasman Sea shows that the global volume of land ice increased during the Kapitean Age. They assign the Kapitean to the interval 4.7–4.3 MY ago, following the work of Kennett & Watkins (1974) in New Zealand. They

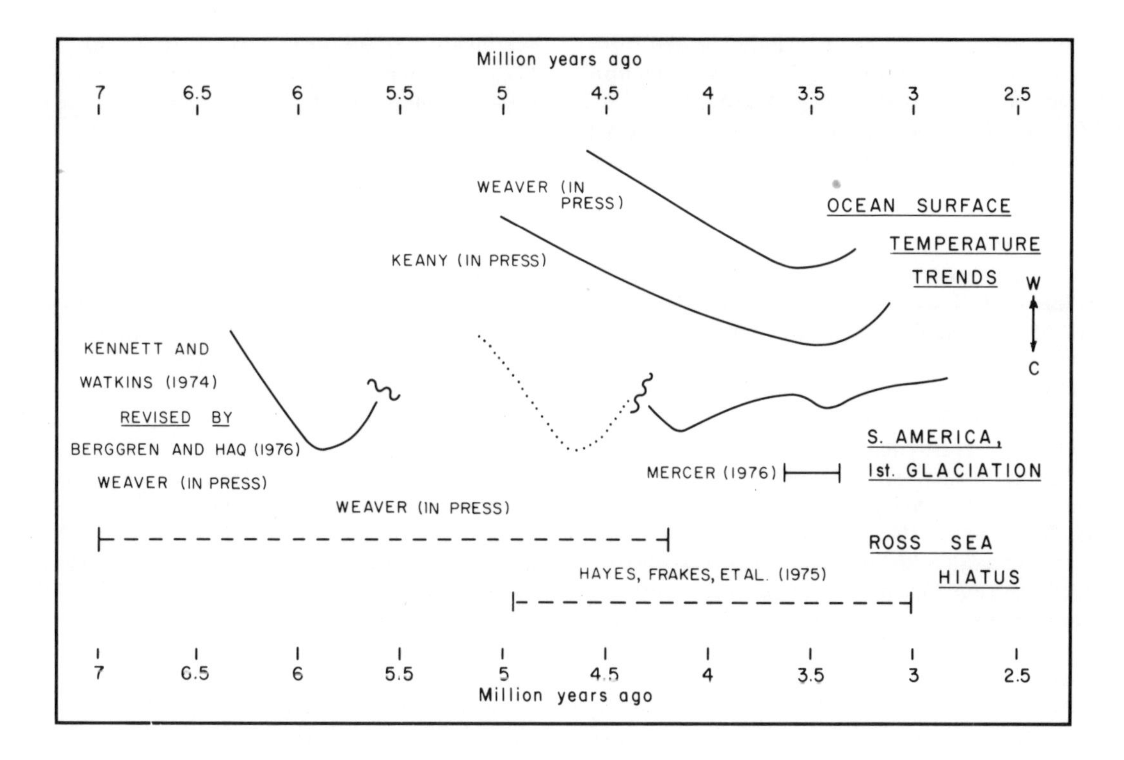

Fig. 4. Reconstructions of ocean surface temperature trends in high southern latitudes; initiation of South American glaciation; and estimated date of formation of Ross Sea unconformity.

assume that the ice whose volume they are monitoring was in Antarctica because there is no evidence for any extensive buildup of Northern Hemisphere ice until about 2.6 MY ago (Shackleton & Kennett, 1975b: 804). They note that their inferred increase in Antarctic ice volume occurred at the same time as a severe cooling deduced by Kennett & Vella (1975: 779) from the microfaunal content of the same core. The isotopic curve can be interpreted in two ways, depending on different assumptions about deep water temperature trends at Site 284: either the Antarctic ice sheet reached its present size during the Kapitean and then lost half its volume, or it reached a size 50% greater than at present and then shrank to its present size. They choose the second interpretation because of Hayes' and Frakes' evidence for an expanded West Antarctic ice sheet at that time. As Mercer (1976: 135) points out, this contains an element of circular reasoning.

Ryan et al (1974: 644) and Berggren & Haq (1976: 97), however, believe that the base of the New Zealand Kapitean Age was 6.5 or 6 MY ago, not 4.7 MY as was deduced by Kennett & Watkins (1974), and Weaver (in preparation) agrees with this assessment. If so, the inferred Southern Ocean cooling and Antarctic ice sheet buildup occurred about 1.5 MY earlier than Shackleton & Kennett (1975b) believed, and about 1.5 MY earlier than the Ross Ice Shelf expansion according to Hayes, Frakes et al (1975: 233). According to Weaver's date for the unconformity (7–4.2 MY), however, the two events could still be contemporaneous. Evidently, the situation is still very confused, and much more work needs to be done to determine the ages of the Ross Sea unconformity, the inferred oceanic cooling, and the buildup of the Antarctic ice sheet. Until this has been done the temporal relation between these events will remain in doubt.

Blank & Margolis (1975) examined *Eltanin* cores that they believed were of Pliocene age. Their palaeotemperature curve for the interval 5.1–2.7 MY ago is rather similar to that of Kennett & Vella (1975), except that the cooling peaked at ca 4 MY ago instead of 4.5 MY ago. (Mercer, 1976: 135, fig. 7). However, Weaver & Dinkelman (1976) have since shown that the age determinations were much too young, probably by at least 10 MY.

Ciesielski & Wise (in press) studied Late Cainozoic sedimentation on the Falkland Plateau to the east of southern South America during DSDP Leg 36. They note the presence of a pronounced and widespread disconformity, bracketed by Miocene beds at least 8.5 MY old and by Pliocene beds between 4.6 and 3.7 MY old. They ascribe this to vigorous bottom current activity, which they conclude resulted from the Late Miocene increase in Antarctic glaciation inferred by other workers (e.g. Hayes & Frakes, 1975; Shackleton & Kennett, 1975b; Mayewski, 1975). This severe Late Miocene cooling, they believe, was followed by much warmer conditions beginning at about 4.3 MY ago, as deduced in the southeast Indian Ocean by Ciesielski & Weaver (1974). To the south of this area Anderson (1972: 136) has studied sediments in the Weddell Sea; he infers the presence of dry-based ice shelves, indicating severe cold from the beginning of the record about 5.7 MY ago, lasting until the end of the Gilbert epoch, ca 3.3 MY ago. However, many other workers have concluded that the Southern Ocean was warm in the early Gilbert epoch, that is immediately after 5.1 MY ago (Hays & Opdyke, 1967; Bandy et al., 1971; and Keany, in press). The most recent work is that of Keany (in press), who has studied the radiolarian content of cores from the Southern Ocean south of Australia and New Zealand, with records from the Gilbert-Epoch 5 boundary to the middle Gauss (t = 5.1–3 MY). He concludes that Antarctic temperatures were highest at the beginning of this interval; they then fell, particularly after ca 4.5 MY ago, reaching a prolonged minimum between 3.7 and 3.2 MY ago.

Discussion

Much evidence from different lines of investigation points to drastic cooling in high southern latitudes (Antarctica and southern South America) during the interval 6–3 MY ago. An unsolved problem is whether two cold episodes occurred (later Miocene, ca 5.5 MY ago and Early Pliocene ca 3.5 MY ago), separated by a pronounced earliest Pliocene warming (ca 5–4 MY ago); or only one (Early Pliocene, ca 3.5 MY ago). If the New Zealand Kapitean Age ca 6–5 MY ago was severely cold, why did a dramatic northward expansion of cold Antarctic water not occur until 1–2 MY later (Kemp, et al., 1975: 916–917)? Evidence for Early Pliocene severe cooling ca. 3.7–3.5 MY ago has been inferred from totally independent lines of evidence (Ciesielski & Weaver, 1974; Mercer, 1976; Keany, in press), yet a severe cooling at that time is not shown by those oceanographic investigations (e.g. Kennett & Vella, 1975: 779) that infer a severe cooling during the latest Miocene; why not? Although Ciesielski & Wise (in press) and Weaver (in preparation) infer latest Miocene and Early Pliocene cold episodes, separated by a marked warming, in fact the faunal content of no single core shows such an oscillation. A single dramatic cooling is suggested by the sudden 300 km northward expansion of cold water (Kemp et al., 1975: 916) and by the Ross Sea unconformity. On present evidence the first severely cold episode (that is, as cold as a Pleistocene glacial age) in high southern latitudes probably culminated about 3.5 MY ago. Evidence for an equally cold, or colder, episode at the end of the Miocene is confusing and equivocal. More data are needed to settle the question.

The Early Pliocene cooling was probably the effect, rather than the cause, of the initial, rather rapid formation of the West Antarctic ice sheet and its fringing ice shelves, associated with the northward extension of sea ice and cold Antarctic surface water; if not, one is forced to the unlikely conclusion that formation of this marine ice sheet had no marked effect on the Southern Ocean. The change in marine environment would have been greatest in those areas adjacent to West Antarctica; that is, in the Pacific sector of the Southern Ocean that most affects the climate of southern South America. Once the West Antarctic ice sheet was present, the astronomical factors discussed by Hays, Imbrie & Shackleton (1976) would have produced repeated glaciations in southern South America and other south temperate areas. Probably, therefore, the West Antarctic ice sheet began to form ~ 5 MY ago and was fully formed by ca 3.8 MY ago. Early in its existence, it reached its greatest Late Cainozoic dimensions, although the reasons that it then exceeded its Pleistocene dimensions may have been topographic (a shallower Ross Sea) rather than climatic. The strongest evidence for the earlier presence – by 7 MY ago – of the West Antarctic ice sheet is that of Rutford et al. (1972). However, in view of the problems encountered in dating many of the rocks, this conclusion needs confirmation.

An intriguing possibility is that formation of the West Antarctic ice sheet was triggered by a rather sudden decrease in salinity of the world ocean – the Messinian salinity crisis in the Mediterranean. Ryan et al. (1974: 660) calculate that salinity decreased by at least 6 percent in less than 1.1 MY when a million cubic km of evaporite was deposited in the Mediterranean basin, and Berggren & Haq (1976: 97) believe that most of the salts were deposited between 5.5 and 5 MY ago. Ryan et al. (1974) conclude that this freshening of the world ocean would have slightly raised the temperature at which sea ice forms and would also have had a profound effect on enhancing the density stratification of surface waters at high latitudes. The result would have been increased formation and expansion of polar sea ice cover, and a pronounced drop in the air temperatures over the polar oceans.

ANTARCTIC AND SOUTH AMERICAN TEMPERATURE TRENDS AND GLACIAL DEVELOPMENT FROM THE BEGINNING OF THE GAUSS MAGNETIC EPOCH TO THE PRESENT (t = 3.3–0 MY)

The Gauss magnetic epoch (t = 3.3–2.4 MY)

As has been pointed out earlier (Mercer 1976: 137) the evidence for global temperature trends during the Gauss epoch is rather enigmatic, suggesting warming in the Southern Hemisphere and cooling in the Northern Hemisphere. Most workers believe that, in the Southern Ocean, the early and middle Gauss was a comparatively warm interlude between late Gilbert and late Gauss–early Matuyama cold episodes (e.g. Bandy et al., 1971: 18; Anderson, 1972: 138; Fillon, 1973: 179; 1975: 787; Barrett, 1975: 165; Kennett & Vella, 1975: 780). Weaver (1973: 122) and Ciesielski & Wise (in press), however, believe that cold conditions prevailed in high southern latitudes during the Gauss epoch. In the Northern Hemisphere, temperatures apparently fell throughout the Gauss: the evidence that ice-rafting in the North Atlantic began about 3 MY ago and increased greatly about 2.5 MY ago (Berggren, 1972) is consistent with the trend in oxygen isotopic ratios in ocean water, which implies that the Northern Hemisphere ice sheets first accumulated about 2.6 MY ago (Shackleton & Kennett, 1975b: 804).

Opposite temperature trends in the two hemispheres during the Gauss epoch seem inherently unlikely. A knowledge of Gauss-age climate trends in South America would help greatly in resolving this problem, but unfortunatey virtually no information is yet available. In only one place – near Condor Cliff in the Santa Cruz valley (lat. 50° S long. 71° W) – are lava flows of known Gauss age associated with glacial drift. At this site much more outwash covers a flow 2.7 MY old than underlies it, suggesting that glaciations were infrequent after the initial 3.5 MY old glaciation until the late Gauss or early Matuyama. This evidence is far from compelling, and the character of the Gauss climate in southern South America remains in doubt.

The Matuyama magnetic epoch (t = 2.4–0.7 MY)

In southern South America six units of till are interbedded with lava flows of Matuyama age on Cerro del Fraile, Lago Argentino (lat. 50° 30′ S, long. 72° 40′ W) (Fleck et al., 1972; Mercer, 1976: 138). The ages of the tills in millions of years are: 2.08, ca 2.05, 2.05–1.86, 1.86–1.67 (two tills), and 1.47–1.03. Much the thickest till is the one 2.05–1.86 MY old. At a nearby site till covers, underlies and is interbedded between two lava flows, both of which are about 2 MY old (Mercer, 1976: 131). The till between 1.47 and 1.03 MY old is thought to have been deposited during the greatest glaciation; further east, till of this glaciation covers basalt 1.24 ± 0.01 MY and 1.17 ± 0.05 MY old. If so, the greatest glaciation occurred between 1.5 and 1.2 MY ago.

This record of repeated South American glaciations during the Matuyama epoch is consistent with the deep sea core evidence from sub-Antarctic waters south of Australia and New Zealand; Keany & Kennett (1972) note about seven colder–warmer oscillations between 2 and 1 My ago, and conclude that the Matuyama epoch was on the whole slightly cooler than the following Brunhes. The resolution of the South American glacial chronology is for the most part too poor to attempt to match units of till with these oceanic coolings; however, the thick till 2.05–1.86 MY old may correspond to the pronounced oceanic cooling that peaked about 1.9 MY ago, and the greatest glaciation, though to have occurred 1.5–1.2 MY ago, may correspond to the prolonged cold interval between 1.35 and 1.1 MY ago.

The Brunhes magnetic epoch (t = 0.7–0 MY)

During the Brunhes epoch the oxygen isotopic record of the oceans shows repeated buildup and decay of ice sheets (Shackleton & Opdyke, 1973) accompanied by temperature fluctuations in the Southern Ocean (Hayes, Imbrie & Shackleton, 1976: 1123). Many glacial advances took place during this interval in southern South America (Mercer, 1976: 143) but none except the last have been dated. There is thus little point in attempting to compare specific late Pleistocene glacial or climatic events in the Antarctic area and South America before the last interglacial.

The Last Interglacial (t = ca 128–75 × 10^3 Y)

Evidence suggesting unusually high temperatures in southern South America during the last interglacial has come from the Chilean lake region near lat 40° S (Mercer, 1976: 151). During the last glaciation, glaciers were most extensive at a time beyond the reach of C-14 dating (more than 56 000 years BP). Little-weathered till of this glaciation covers deeply weathered till inferred to date from the previous glaciation. This deep weathering must have taken place during the last interglacial because the weathering horizon has been sealed in and protected by the overlying impervious till since early in the last glaciation. The implication is that very much more weathering occurred during the last interglacial than has occurred during the last 56 000 years, which includes the present interglacial so far. The difference in weathering is so great that higher temperatures are implied.

If the last interglacial in southern Chile was exceptionally warm, the same should apply to the Southern Ocean and Antarctica, and there is some evidence that this is so. According to the generally-used stratigraphic scheme of Emiliani (1955) and the chronology of Shackleton & Opdyke (1973: 49), Stage 5 (128–75 $\times$ 10^3 BP) comprises the last interglacial and substage 5 e near the beginning of the interglacial is the warmest part. J. Hays (personal communication) finds that Southern Ocean cores consistently show substage 5 e (ca 128–118 000 BP) 1° -2° C or, exceptionally, 4° C warmer than the Holocene peak. In Antarctica Hendy et al (in press) have dated algal limestones deposited in lakes in Taylor Valley, Victoria Land, which show that considerably more meltwater was present for much of the interval 75–130 000 years BP (Stage 5) than is present today.

In addition, there is intriguing but inconclusive evidence that Antarctic temperatures may have been high enough during the last interglacial to destroy the West Antarctic ice sheet. As was explained above in discussing its original formation, this ice sheet depends for its existence of the buttressing effect of its fringing ice shelves. If average midsummer temperature rises above 0° C they recede, the grounding line between them and the grounded ice sheet also recedes and the ice sheet thins. A rise in Antarctic temperature of about 5° C would start shrinkage of the West Antarctic ice sheet, and a further warming of a few degrees would destroy it (Mercer, 1968a). The West Antarctic ice sheet would distintegrate very rapidly by catastrophic calving under unfavorable climatic conditions; unlike grounded ice sheets, such as the East Antarctic and Greenland ice sheets and the former Laurentide and Scandinavian ice sheets, it has virtually no climatic inertia. Mercer (1968a) pointed out that the probable ca + 6 m sea level of the last interglacial was close to the ca 4–5 m rise that would result from disintegration of the West Antarctic ice sheet, suggesting cause and effect; Shackleton & Opdye (1973) have since shown that the ocean did indeed contain more water during substage 5 e, implying absence at that time of a large body of ice that exists today.

Absence of the West Antarctic ice sheet could perhaps partly explain the remarkably high temperatures that appear to have prevailed in southern Chile during the last interglacial. Conditions would once more have resembled those in the Late Miocene, before the initial formation of the West Antarctic ice sheet (Fig. 2). As was pointed out earlier during discussion of the probably Early Pliocene formation of the West Antarctic ice sheet, the circum-Antarctic current today is forced to flow through the Drake Passage between the Antarctic Peninsula and South America, and a branch of this current mixed with subtropical water is diverted up the west coast of South America as the Humboldt Current. When the West Antarctic ice sheet is absent some of the cold current in this sector is able to flow further south between the Ross and Weddell Seas; at that time the Humboldt Current would be weaker and would contain more subtropical and less Antarctic water. The shift in the circum-Antarctic current would have been confined to the South American sector, so that in other Southern Hemisphere temperate land areas, such as South Africa, Australia and New Zealand, the interglacial warming would have been much less marked. A similar but smaller effect would have resulted from a narrowing of the belt of pack ice surrounding Antarctica during the last interglacial, even if the West Antarctic ice sheet remained intact. Gardner & Hays (1976: 263) discussing glacial-interglacial changes in the South Atlantic, point out that atmospheric circulation would weaken because of decreased latitudinal temperature gradient if the pack ice belt receded south, causing the circum-Antarctic current and the north-flowing Benguela Current on the west coast of Africa to weaken. Presumably the same would apply to the Humboldt Current on the west coast of South America.

Summing up the evidence for exceptional warmth in Antarctica and southern South America during the last interglacial: 1) meltwater was more abundant than it is today in South Victoria land; 2) both sea level and the isotopic composition of the oceans were compatible with the absence at that time of the West Antarctic ice sheet, which is peculiarly vulnerable to climatic warming of more than 5° C; 3) Subantarctic surface water temperatures were 1–4° C warmer; and 4) chemical weathering in southern Chile was much more intense than during the present interglacial. This evidence is not compelling; more decisive evidence will come from coring of sediments beneath the Ross Ice Shelf, from more refined isotopic measurements of ocean cores,and from a reconstruction of the thermal environment of southern Chile from faunal and floral analyses.

The Last Glaciation (t = ca 75–13 $\times$ 10^3 Y)

The oxygen isotopic composition of the oceans shows that the last glaciation comprised two global build-ups of ice: a minor one culminating about 75 000 years BP, and a major one culminating about 20 000 years BP (Shackleton & Opdye, 1973: 49). In southern Chile, however,an early advance (Early Llanquihue Glaciation), beyond the range of C-14 dating and, therefore, probably dating from about 75 000 years BP, was greater than the Late Llanquihue Glaciation about 20 000 years BP. Maximum severity of climate about 75 000 years BP has also been noted in the equatorial Atlantic (Gardner & Hays, 1976: 237), and in the southern South Atlantic ice-rafting was most intense about that time (Cooke, 1976). In Peru at lat. 14° S (Cordillera Vilcanota) the extent of the glaciers 75 000 years BP is not known, but the Late Wisconsin-age maximum that culminated about 25 800 years BP was a comparatively minor affair, with glaciers only abut half as large as during several earlier glaciations (Mercer, 1977).

After the Late Llanquihue glacial maximum about 20 000 years BP the Chilean glaciers halved in length during the Varas Interstade, before readvancing shortly before 13 000 years BP (Mercer, 1976: 154). The Varas Interstade was at least partly coeval with the Erie Interstade in the Northern Hemisphere which was centred about 15 600 years BP (Mörner & Dreimanis, 1973: 120). A slight warming centred about 16 000 years BP has been inferred from a study of an ocean core obtained 160 km east of Tierra del Fuego (Burckle, 1972: 324).

Johnsen et al (1972) showed four possible calibrations of the ice core from Byrd Station (lat. 80° S), with different assumptions about accumulation and divergence of ice flow. Their preferred time scale (Fig. 5c in Johnsen et al., 1972: 433), which was chosen because it gave the best fit with events in the Northern Hemisphere (Greenland and Scandinavia), is incompatible with climatic trends in southern South America. However, another time scale (Fig. 5d in Johnsen et al., 1972: 433) gives a much better fit with South American events; furthermore, according to Johnsen et al (1972: 432) it is based on a more likely value of the ice flow divergence. It makes more sense to compare events at Byrd Station with those in southern South America, 30° of latitude away, rather than with Scandinavia and Greenland, 140°–160° of latitude away. Using Fig. 5d of Johnsen et al. (1972), the oxygen isotope curve qualitatively reflects the South American glacial maximum at 19–20 000 years BP, the Varas Interstade 16–14 000 years BP, and the readvance after 14 000 years BP (Fig. 5).

'Postglacial' time (t = ~ 13 000–0 BP)

During the 3 000 or so years after the final readvance, climatic trends in southern South America differed markedly from those in much of the Northern Hemisphere, but were similar to those over the sub-Antarctic ocean, clearly demonstrating the dominant influence of the Antarctic area on the climate of southern South America during the termination of glacial conditions.

Glaciers in both southernmost (lat. 54° S) and south-central (lat. 40°) Chile had receded into the mountains before 12 000 years BP (Mercer, 1970b: 19; and 1976: 155) and by 11 000 years BP were smaller than they are today, suggesting that hypsithermal conditions were reached much earlier than round the North Atlantic Ocean (Mercer, 1970b: 21). The South American geological evidence that the glacial-interglacial transition was more rapid in the Southern Hemisphere than in the Northern Hemisphere has recenty been supported by oceanographic investigations; in the southern Indian Ocean, summer temperature reached a peak about 9 400 ± 600 years BP and thereafter slowly declined (Hays, Imbrie & Shackleton, 1976: 1124).

In Peru at lat. 14° S, long. 71° W, a minor readvance of the Quelccaya Ice Cap culminated at about 11 000 years BP. This was initially believed to be an age equivalent of the European Younger *Dryas* Stade, but several age determinations have now been obtained that are consistent in dating this readvance at about 500 years earliers; its relation to the Younger *Dryas* event is not clear, but it is important to note that the main moisture-bearing winds in this part of Peru originate over the equatorial North Atlantic Ocean, and that the influence of the Antarctic region is minimal (Mercer, 1977.)

The reason for the interhemispheric contrast in climatic trends between 13 000 and 10 000 years BP may well lie in the different development of land and sea ice in the two hemispheres. As Hays, Lozano et al. (1976: 369) points out, increased Antarctic sea ice was apparently the glacial-age counterpart of Northern Hemisphere ice sheet development. However, the climatic inertia of the glacial-age Antarctic pack ice belt was much less than that of the Laurentide and Scandinavian ice sheets, so that the Southern Hemisphere was able to respond much more quickly to the orbital variations that are thought to control the major global glacial–interglacial cycles.

One consequence of the rapid 'postglacial'-warming of high southern latitudes was that high temperatures there coincided with the low glacio-eustatic sea level that resulted from the survival of large Northern Hemisphere ice sheets: at the Southern Hemisphere hypsithermal peak ca 9 400 years BP the southern margin of the Laurentide Ice Sheet still lay across Lake Superior. In Antarctica the Ross Ice Shelf may have been more extensive than it is now about 11 000 years BP because of low sea level at the same time as meltwater was more abundant because of higher temperatures.

This could perhaps explain the presence in south Victoria Land of extensive lakes, thought by Denton et al. (1971: 287) to have been impounded by a thickened Ross Ice Shelf, and assigned by them to full-glacial time, although this would imply that temperatures in south Victoria Land in full glacial time were then at least as high as they are today.

During Neoglacial time that followed the Hypsi-

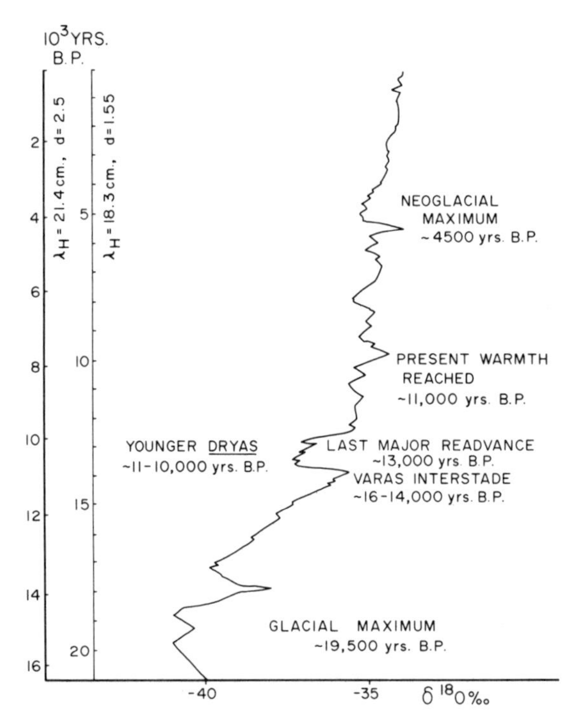

Fig. 5. Oxygen isotopic curve from Byrd Station, Antarctica, on two time scales, depending on different assumptions about net accumulation, λ_H, and ice flow divergence, d, (from Johnsen et al., 1972). To the left the Younger *Dryas* Stade is placed, assuming that $\lambda_H = 21.4$ cm, d = 2.5, as favoured by Johnsen et al. (1972). To the right are shown glacial events in Chile, in standard radiocarbon years, assuming that $\lambda_H = 18.3$ cm and d = 1.55

thermal Interval, times of glacial advance in the two hemispheres more or less coincided (ca 4 500, ca 2 500, and 300–100 years BP), but the relative importance of the episodes differed. In particular, the ca 4 500 years BP advance in southern South America was much the greatest of the three, whereas in the Northern Hemisphere it was relatively minor. There is intriguing but inconclusive evidence that the event may have been connected with greatly increased output of bergs from West Antarctica into the Southern Ocean. In the Byrd ice core, Johnsen et al. (1972: 434) note a change in oxygen isotopic composition of about 1 percent, starting ca 4 000 years BP according to their preferred time scale. They believe that this reflects, not a warming trend, but a lowering of the ice sheet surface. On the time scale that seems preferable for reasons discussed earlier (Fig. 5d of Johnsen et al., 1972: 433) the change in isotopic values starts at about 5 500 years BP, which is equivalent to ca 4 500 years BP on the C-14 time scale (Fig. 5).

REFERENCES

Allis, R.G., Barrett, P.J. & Christoffel, D.A. 1975. A paleomagnetic stratigraphy for Oligocene and Early Miocene marine glacial sediments at Site 270, Ross Sea, Antarctica. In: D.E. Hayes, L.A. Frakes et al. (eds.), *Initial reports of the DSDP* 28, U.S. Gov. Printing Office, Washington, D.C.: 879–884.

Anderson, J.B. 1972. *The marine geology of the Weddell Sea*, Contribution 35, Florida State University, Tallahassee.

Bandy, O.L., Casey, R.E. & Wright, R.C. 1971. Late Neogene planktonic zonation, magnetic reversals and radiometric dates, Antarctic to the tropics. In: J.L. Reid (ed.), Antarctic oceanology I, *Antarctic Research Series* 15, Am. Geophys. Union: 1–26.

Barker, P.F. & Burrell, J. (in press). The influence upon Southern Ocean circulation, sedimentation and climate of the opening of Drake Passage. *Third Symposium on Antarctic Geology and Geophysics*, Madison, Wisconsin, 1977.

Barrett, P.J. 1975. Textural characteristics of Cenozoic preglacial and glacial sediments at Site 270, Ross Sea, Antarctica. In: D.E. Hayes, L.A. Frakes et al. (eds.), *Initial reports of the DSDP* 28: 757–767.

Barton, C.M. 1964. Significance of the Tertiary fossil floras of King George Island, South Shetland Islands. In: R.J. Adie (ed.), *Antarctic Geology*, Proc. First Internat. Symposium on Antarctic Geology, Cape Town, 1963, North Holland, Amsterdam: 603–608.

Bentley, C.R. & Ostenso, N.A. 1961. Glacial and subglacial topography of West Antarctica. *J. Glaciology* 3: 882–911.

Berggren, W.A. 1972. A Cenozoic time-scale – some implications for regional geology and paleobiogeography. *Lethaia* 5: 195–215.

Berggren, W.A. & Haq, B.U. 1976. The Andalusian Stage (Late Miocene): biostratigraphy, biochronology, and paleoecology. *Palaeogeography, Palaeoclimatology, Palaeoecology*, 20: 67–129.

Blank, R.G. & Margolis, S.V. 1975. Pliocene climatic and glacial history of Antarctica as revealed by southeast Indian Ocean deep-sea cores. *Geol. Soc. Am. Bull.* 86: 1058–1066.

Burckle, L.H. 1972. Diatom evidence bearing on the Holocene in the South Atlantic. *Quat. Res.* 2: 323–326.

Ciesielski, P.F. 1975. Biostratigraphy and paleoecology of Neogene and Oligocene silicoflagellates from cores recovered during Antarctic Leg 28, Deep Sea Drilling Project. In: D.E. Hayes, L.A. Frakes et al. (eds.), *Initial reports of the DSDP* 28: 625–664.

Ciesielski, P.F. & Weaver, F.M. 1974. Early Pliocene temperature changes in the Antarctic seas. *Geology* 2: 511–515.

Ciesielski, P.F. & Wise, S.W. (in press). Geologic history of the Maurice Ewing Bank of the Falkland Plateau (Southwest Atlantic sector of the Southern Ocean) based on piston and drill cores. *Marine Geology.*

Cooke, D.W. 1976. Glacial-interglacial sedimentation changes in the Antarctic Ocean. *Abstracts with Programs* 8: 820–821. *Geol. Soc. Am. Annual Meeting,* Denver.

Cranwell, L.M. 1969. Antarctic and Circum-Antarctic palynological contributions. *Antarctic J. U.S.* 4: 197–198.

Denton, G.H., Armstrong, R.L. & Stuiver, M. 1971. The Late Cenozoic glacial history of Antarctica. In: K.K. Turekian (ed.), *The Late Cenozoic glacial ages,* Yale University Press, New Haven: 267–306.

Douglas, R.G. & Savin, S.M. 1973. Oxygen and carbon isotope analyses of Cretaceous and Tertiary foraminifera from the central North Pacific. In: P.H. Roth & J.R. Herring (eds.), *Initial reports of the DSDP* 17: 591–605.

Drewry, D.J. 1975. Initiation and growth of the East Antarctic ice sheet. *J. Geol. Soc.* 131: 255–273.

Dusén, P. 1911. Über die Tertiäre Flora der Seymour Insel. In: *Wissenschaftlichen Ergebnisse der Schwedischen Südpolar-Expedition 1901–1903,* 3: 1–27. Lithogr. Inst. des Generalstabs, Stockholm.

Emiliani, C. 1955. Pleistocene temperatures. *J. Geology* 63: 538–578.

Fasola, A. 1969. Estudio palinológico de la formación Loreto (Terciário medio), Província de Magallanes, Chile. *Ameghiniana* 6: 3–49.

Fillon, R.H. 1973. Radiolarian evidence of late Cenozoic oceanic paleotemperatures, Ross Sea, Antarctica. *Palaeogeography, Palaeoclimatology, Palaeoecology* 14: 171–185.

Fleck, R.J., Mercer, J.H., Nairs, A.E.M. & Peterson, D.N. 1972. Chronology of Late Pliocene and Early Pleistocene glacial and magnetic events in southern Argentina. *Earth and Planetary Sci. Letters* 16: 15–22.

Gardner, J.V. & Hays, J.D. 1976. Responses of sea-surface temperature and circulation to global climatic change during the past 200,000 years in the eastern Equatorial Atlantic Ocean. *Geol. Soc. Am. Mem.* 145: 221–246.

Geitzenauer, K.R., Margolis, S.V. & Edwards, D.S. 1968. Evidence consistent with Eocene glaciation in a South Pacific deep sea sedimentary core. *Earth and Planetary Sci. Letters* 4: 173–177.

Grindley, G.W. 1967. The geomorphology of the Miller Range, Transantarctic Mountains, with notes on the glacial history and neotectonics of Antarctica. *N.Z. J. Geol. and Geophys.* 10: 557–598.

Hayes, D.E. & Frakes, L.A. 1975. General synthesis, Deep-Sea Drilling Project, Leg 28. In: D.E. Hayes, L.A. Frakes et al. (eds.), *Initial reports of the DSDP* 28: 919–942.

Hayes, D.E., Frakes, L.A. et al. 1975. Sites 270, 271, 272. In: D.E. Hayes, L.A. Frakes et al. (eds.), *Initial reports of the DSDP* 28: 211–334.

Hays, J.D. 1969. Climatic record of Late Cenozoic Antarctic Ocean sediments related to the record of world climate. *Palaeoecology of Africa* 5, Balkema, Cape Town: 139–163.

Hays, J.D., Imbrie, J. & Shackleton, N.J. 1976. Variations in the Earth's orbit: pacemaker of the Ice Ages. *Science* 194: 1121–1132.

Hays, J.D., Lozano, J.A., Shackleton, N. & Irving, G. 1976. Reconstruction of the Atlantic and western Indian Ocean sectors of the 18,000 B.P. Antarctic Ocean. In: R.M. Cline & J.D. Hays (eds.), *Investigations of Late Quaternary Paleoceanography and Paleoclimatology*: 337–372.

Hays, J.D. & Opdyke, N.D. 1967. Antarctic radiolaria, magnetic reversals and climate change. *Science* 158: 1001–1011.

Heirtzler, J.R. 1971. The evolution of the southern oceans. In: L.Q. Quam (ed.), *Research in the Antarctic,* Am. Ass. Advancement Sci., Washington, D.C., Publ. 93: 667–684.

Hendy, C.H., Healy, T.R., Rayner, E.M., Shaw, J. & Wilson, A.T. (in press). Global climate and the Late Pleistocene glacial chronology of the Taylor Valley, Antarctica.

Hollister, C.D., Craddock, C. et al. 1976. Site 325. In: P. Worstell (ed.), *Initial reports of the DSDP* 35: 157–193, U.S. Gov. Printing Off., Washington.

Houtz, R. & Davey, F.J. 1973. Seismic profiler and sonobuoy measurements in Ross Sea, Antartica. *J. Geophys. Res.* 78: 3448–3468.

Hughes, T. 1977. West Antarctic ice streams. *Rev. Geophysics and Space Physics* 15: 1–46.

Johnsen, S.J., Dansgaard, W., Clausen, H.B. & Langway, C.C. 1972. Oxygen isotope profiles through the Antarctic and Greenland ice sheets. *Nature* 235: 429–434.

Keany, J. (in press). Paleoclimatic trends in Early and Middle Pliocene deep-sea sediments of the Antarctic.

Keany, J. & Kennett, J.P. 1972. Pliocene–early Pleistocene paleoclimatic history recorded in Antarctic–Subantarctic deep-sea cores. *Deep-Sea Res.* 19: 529–548.

Kemp, E.M. 1972. Reworked palynomorphs from the West Ice Shelf area, East Antarctica. and their possible geological and palaeoclimatological significance. *Marine Geology* 13: 145–154.

Kemp, E.M. 1975. Palynology of Leg 28 drill sites, Deep Sea Drilling Project. In: D.E. Hayes, L.A. Frakes et al. (eds.), *Initial reports of the DSDP* 28: 599–608.

Kemp, E.M. & Barrett, P.J. 1975. Antarctic glaciation and early Tertiary vegetation. *Nature* 258: 507–508.

Kemp, E.M., Frakes, L.A. & Hayes, D.E. 1975. Paleoclimatic significance of diachronous biogenic facies, Leg 28, Deep Sea Drilling Project. In: D.E. Hayes, L.A. Frakes et al. (eds.), *Initial reports of the DSDP* 28: 909–917.

Kennett, J.P. & Vella P. 1975. Late Cenozoic planktonic foraminifera and paleoceanography at DSDP Site 284 in the cool subtropical South Pacific. In: J.P. Kennett, R.E. Houtz et al (eds.), *Initial reports of the DSDP* 29: 769–782.

Kennett, J.P. & Watkins, N.D. 1974. Late Miocene–Early Pliocene paleomagnetic stratigraphy, paleoclimatology, and biostratigraphy in New Zealand. *Geol. Soc. Am. Bull.* 85: 1385–1398.

Kennett, J.P. and 10 others. 1975. Cenozoic paleoceanography in the southwest Pacific Ocean, Antarctic glaciation and the development of the circum-Antarctic current. In: J.P. Kennett, R.E. Houtz et al (eds.), *Initial reports of the DSDP* 29: 1155–1169.

LeMasurier, W.E. 1972. Volcanic record of Antarctic glacial history: implications with regard to Cenozoic sea levels. *Inst. British Geographers Spec. Publ.* 4: 59–74.

LeMasurier, W.E. & Rex, D.C. (in press). Migration of Cenozoic volcanic activity in Marie Byrd Land. *Symposium on Antarctic Geology and Geophysics, Madison, Wisconsin,* August 1977.

LeMasurier, W.E. & Wade, F.E. 1976. Volcanic history in Marie Byrd Land: implications with regard to Southern Hemisphere tectonic reconstructions. In: *Proc. International Symposium on Andean and Antarctic Volcanology Problems*, Santiago, Chile. 398–424.

Margolis, S.V. 1975. Paleoglacial history of Antarctica inferred from analysis of Leg 29 sediments by scanning-electron miscroscopy. In: J.P. Kennett, R.E. Houtz et al. (eds.), *Initial reports of the DSDP* 29: 1039–1043.

Margolis, S.V. & Kennett, J.P. 1971. Cenozoic paleoglacial history of Antarctica recorded in sub-antarctic deep-sea cores. *Am. J. Sci.* 271: 1–36.

Mayewski, P.A. 1972. Glacial geology near McMurdo Sound and comparison with the central Transantarctic Mountains. *Antarctic J. U.S.* 7: 103–106.

Mayewski, P.A. 1975. *Glacial geology and Late Cenozoic history of the Transantarctic Mountains, Antarctica.* Report No. 56, Inst. Polar Studies, Ohio State Univ., 168 pp.

McIntyre, D.J. & Wilson, G.J. 1966. Preliminary palynology of some Antarctic Tertiary erratics. *N.Z. J. Botany* 4: 315–321.

Menéndez, C.A. 1971. Floras terciárias de la Argentina. *Ameghiniana* 8: 357–371.

Menéndez, C.A. & Caccavari de Felice, M.A. 1975. Las especies de *Nothofagidites* (polen fosíl de *Nothofagus*) de sedimentos Terciários y Cretácicos de Estancia La Sara, Norte de Tierra del Fuego, Argentina. *Ameghiniana* 12: 165–183.

Mercer, J.H. 1968a. Antarctic ice and Sangamon sea level. *Intern. Assoc. Sci. Hydrology, Gen. Assembly of Berne, Publ.* 79: 217–225.

Mercer, J.H. 1968b. Glacial geology of the Reedy Glacier area, Antarctica. *Geol. Soc. Am. Bull.* 79: 471–486.

Mercer, J.H. 1970a. Some observations on the glacial geology of the Beardmore Glacier area. In: R.J. Adie (ed.), *Antarctic Geology and Geophysics,* Universitetsforlaget, Oslo: 427–433.

Mercer, J.H. 1970b. Variations of some Patagonian glaciers since the Late Glacial II. *Am. J. Science* 269: 1–29.

Mercer, J.H. 1973. Cainozoic temperature trends in the southern hemisphere: Antarctic and Andean glacial evidence. In: *Palaeoecology of Africa* 8: 85–114, Balkema, Cape Town.

Mercer, J.H. 1975. Southern Patagonia: glacial events between 4 m.y. and 1 m.y. ago. In: R.P. Suggate & M.M. Cresswell (eds.), *Quaternary Studies*, The Royal Soc. New Zealand, Wellington: 223–230.

Mercer, J.H. 1976. Glacial history of southernmost South America. *Quat. Res.* 6: 125–166.

Mercer, J.H. 1977. Radiocarbon dating of the last glaciation in Peru. *Geology*, 5: 600–604.

Mörner, N.-A. & Dreimanis, A. 1973. The Erie Interstade. In: R.F. Black, R.P. Goldthwait & H.B. Willman (eds.), The Wisconsinan Stage, *Geol. Soc. Am. Mem.* 136: 107–134.

Piper, D.J.W. & Brisco, C.D. 1975. Deep-water continental-margin sedimentation, DSDP Leg 28, Antarctica. In: D.E. Hayes, L.A. Frakes et al. (eds.), *Initial reports of the DSDP* 28: 727–755.

Robin, G. de Q. & Adie, R.J. 1964. The ice cover. In: R. Priestley, R.J. Adie & G. de Q. Robin (eds.), *Antarctic Research*, Butterworths, London: 100–117.

Rutford, R.H., Craddock, C., White, C.M. & Armstrong, R.L. 1972. Tertiary glaciation in the Jones Mountains. In: R.J. Adie (ed.), *Antarctic Geology and Geophysics,* Universitetsforlaget, Oslo: 239–243.

Ryan, W.B.F., Cita, M.B., Dreyfus-Rawson, M., Burckle, L.H. & Saito, T. 1974. A paleomagnetic assignment of Neogene stage boundaries and the development of isochronous datum planes between the Mediterranean, the Pacific and Indian oceans in order to investigate the response of the world oceans to the Mediterranean "Salinity Crisis." *Rivista Italiana de Paleontologia e Stratigrafia*, 80: 631–688.

Shackleton, N.J. & Kennett, J.P. 1975a. Paleotemperature history of the Cenozoic and the initiation of Antarctic glaciation: oxygen and carbon isotope analyses in DSDP sites 277, 279 and 281. In: J.P. Kennett, R.E. Houtz et al. (eds.), *Initial reports of the DSDP* 29: 743–755.

Shackleton, N.J. & Kennett, J.P. 1975b. Late Cenozoic oxygen and carbon isotopic changes at DSDP Site 284: implications for glacial history of the northern hemisphere and Antarctica. In: J.P. Kennett, R.E. Houtz et al. (eds.), *Initial reports of the DSDP* 29: 801–807.

Shackleton, N.J. & Opdyke, N.D. 1973. Oxygen isotope and palaeomagnetic stratigraphy of Equatorial Pacific core V28–238: oxygen isotope temperatures and ice volumes on a 10^5 year and 10^6 year scale. *Quat. Res.* 3: 39–55.

Weaver, F.M. 1973. Pliocene paleoclimatic and paleoglacial history of East Antarctica recorded in deep sea piston cores. M.Sc. thesis, Florida State University, Tallahassee, mimeo., 142 pp.

Weaver, F.M. (in preparation). Late Miocene and Pliocene paleoclimatology.

Weaver, F.M. & Dinkelman, M.G. 1976. Pliocene climatic and glacial history of Antarctica as revealed by southeast Indian Ocean deep-sea cores: discussion. *Geol. Soc. Am. Bull.* 87: 1529–1531.

Webb, P.N. 1972. Wright Fjord, Pliocene marine invasion of an Antarctic dry valley. *Antarctic J. U.S.* 7: 226–234.

Zinsmeister, W.J. (in preparation). Review of the Miocene molluscan faunas of Navidad, central Chile.

Fig. 2. Lateral moraines on the southern flank of Cook Glacier, St Andrews Bay. The T1 moraine is adjacent to the glacier edge. The T2 moraine is ca 100 m from the glacier edge.

Fig. 4. Terminal moraines crossing Hamberg Lakes Valley. The moraines belong to group T3. In the background T1 and T2 moraines lie between the glacier margin and a rock knoll.

Fig. 5. A fresh ice-scoured greywacke rock surface outside T3 limits on the peninsula east of Moraine Fjord.

Fig. 6. The upper and lower beach terraces lying at the base of degraded rock cliffs on the eastern shore of Cumberland East Bay. The beach surfaces are covered with tussock grass.

Fig. 7. An exposure of cemented beach deposit at an altitude of 30 m, Doubtful Bay. The beach rests on a bedrock platform and is overlain by till.

* 8 *

Glacier fluctuations in South Georgia and comparison with other island groups in the Scotia Sea

C.M. Clapperton, D.E. Sugden, R.V. Birnie, J.D. Hanson and G. Thom
Department of Geography, University of Aberdeen, Scotland

Manuscript received 1st September 1977

CONTENTS

ABSTRACT

Geomorphological evidence of the sequence and age of past glacier fluctuations in South Georgia is presented. At one stage (Early Wisconsin or earlier) an ice cap extended over the surrounding submarine platform to a depth of 200 m and was responsible for most glacial erosion. During a subsequent Interglacial/Interstadial raised beaches were deposited up to altitudes of 52 m and are now represented by isolated remnants of cemented beach deposits. Afterwards a glaciation which probably spanned the Late Wisconsin covered all lowlying parts of the island and may have extended offshore as far as the — 120 m submarine contour. The retreat of this ice cap was marked by a stillstand or readvance of valley glaciers at the mouths of troughs which occurred earlier than 9 000 years ago. Following retreat from this position, there have been two readvances, one a modest event 100–200 years ago and the other a very minor readvance in the early part of this century.

Comparison with other areas in the sub-Antarctic regions of the Scotia Sea reveals that the major glacial episodes seem to be duplicated elsewhere. However, some relatively minor readvances such as those of the Neoglacial/Little Ice Age are not always in phase. For example there is no South Georgia equivalent of a readvance which took place in the South Shetland Islands 500–700 years ago. The glacial history of the Falkland Islands, which are only 2° north of South Georgia but on the other side of the Antarctic Convergence is quite different.

INTRODUCTION

The aim of this article is to present evidence which helps to establish an approximate chronology of glacier fluctuations in South Georgia and to compare the sequence of events with that experienced elswhere in the Scotia Sea zone of the sub-Antarctic. Suggestions for relative chronology in South Georgia are based on the analysis of the distribution, form and textural characteristics of raised beach sediments and features of glacial deposition and erosion; the absolute dating of events since about 10 000 years BP is based on radiocarbon dating of organic materials.

South Georgia (lat. 54°–55° S, long. 36°–38° W) is an island ca 170 km long and 2–34 km wide which attains altitudes as high as 2 960 m in the majestic peaks of the Allardyce Range. Valley glaciers extend to the sea at the heads of many fjords and give the island its strongly indented coastline (Fig. 1). The island lies south of the Antarctic Convergence and experiences a sub-Antarctic climate. The waters around the island are affected by cold currents which may bring icebergs and very occasionally pack ice from the Weddell Sea to the south (Hardy, 1967).

Detailed fieldwork over several seasons has suggested a tentative chronology of glacial events and this is summarised in Table 1. The evidence for each phase of this chronology is discussed below.

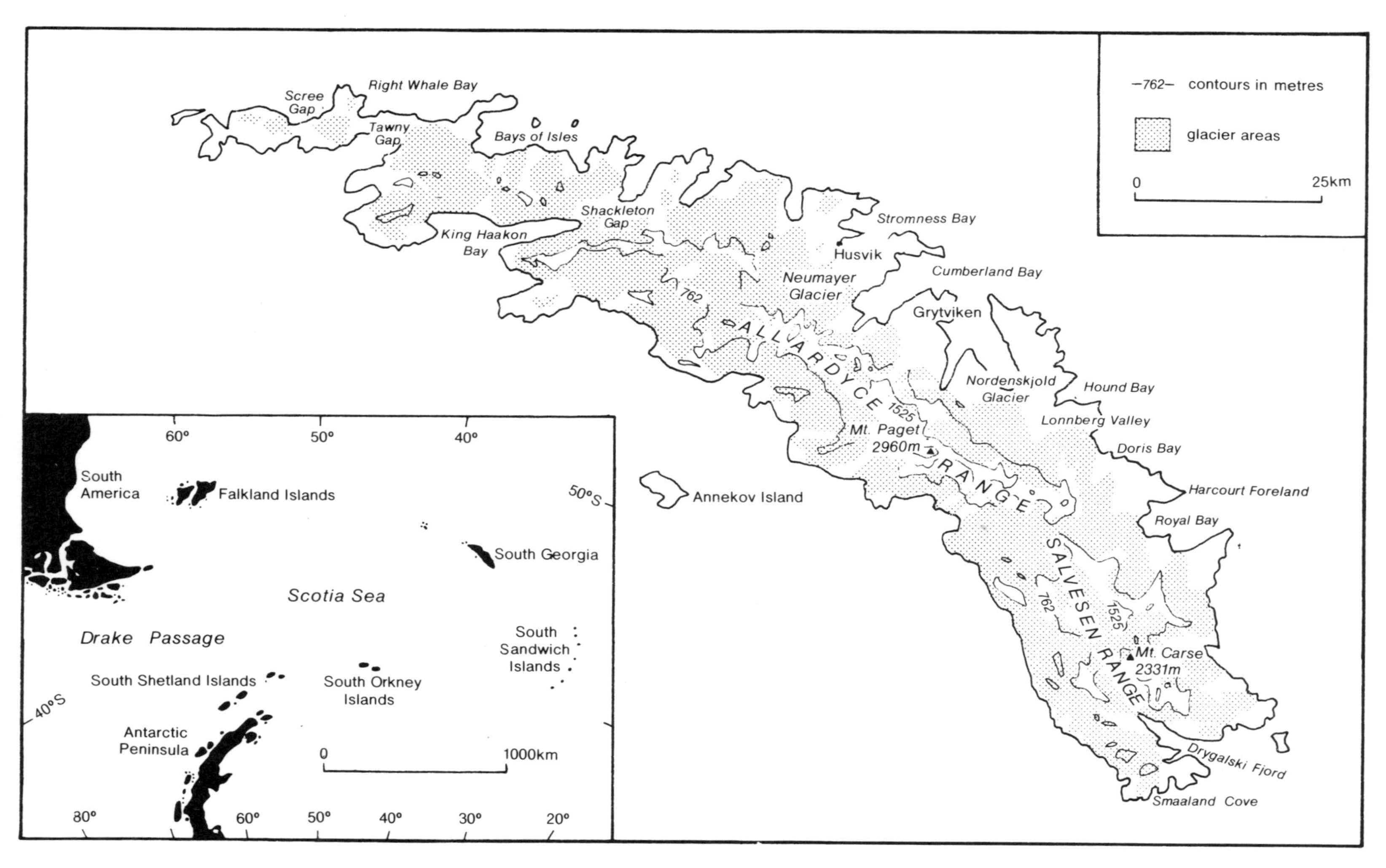

Fig. 1. Location map of South Georgia and other island groups around the Scotia Sea

THE EVIDENCE IN SOUTH GEORGIA

Historical records

It is now certain that most glaciers in South Georgia experienced a positive mass balance during the early part of the present century and advanced. This was most probably in response to the climatic deterioration, identified from climatic records kept at the whaling stations, which began in 1924 (Smith, 1960). Land-terminating glaciers, particularly the smaller cirque glaciers, advanced by as much as 500 m (e.g. southwest of Husvik), but the large sea-terminating glaciers advanced no more than 50 m beyond their present snouts. Smith (1960) concluded from photographic records that whereas sea-terminating glaciers reached their maximal positions between 1926 and 1929, a closely observed cirque glacier (probably Hodges Glacier) halted its advance some time between 1928 and 1936. The majority of glaciers seem to be still receding slowly from morainic limits (T1) associated with this advance, leaving an unstable mass of boulders, some of which are capped with till, and little vegetation cover (Fig. 2).

Moraines and till

It has also been confirmed that South Georgia glaciers responded to the global climatic deterioration of the last few centuries generally known as the Little Ice Age. Complexes of multiple terminal and lateral moraines (T2) normally occur more than 300 m beyond the snouts of present land-terminating glaciers (exceptionally 1 000 m southwest of Husvik)

Table 1

Tentative glacial chronology for South Georgia

Event	Extent (beyond present glacier margins)	Characteristics	Age	Name
Glacier advance	50–500 m	Multiple terminal/lateral moraines at margin of cirque glaciers. Single/double/ice cored moraines at margins of valley glaciers. Very little surface vegetation.	1926–36	T1
Glacier retreat	— ?			
Glacier advance	300–1 000 m	Multiple terminal/lateral moraines. Sparse cover of surface vegetation. Slight disturbance by cryogenic activity.	17th–19th centuries	T2
Glacier retreat	Within limits of T3	Raises beaches to altitude of 7.4 m a.s.l.	?	
Glacier stillstand or advance	2–6.5 km	Multiple terminal/lateral moraines. Good vegetation cover. Highly disturbed by cryogenic activity.	Post Late-Wisconsin maximum but before c. 10 000 BP	T3
Glaciation	Beyond present coast	Till sheet outside T3 moraines and offshore. Highly disturbed by cryogenic activity.	Late-Wisconsin maximum (c. 17 000 BP)	T4
Interstadial/interglacial	Recession to within T3 limits	Raised beach sediments (in situ) up to 52 m a.s.l.	Mid-Wisconsin or Sangamon	
Glaciation	?	Indurated till.	?	
Maximum glaciation(s)	To edge of submarine shelf (c. — 200 m depth)	Main erosional landforms troughs, areal scouring.	Early-Wisconsin? Illinoian or earlier?	

but less than 300 m beyond the snouts of sea-terminating glaciers. These T2 moraines are distinctly older than the contiguous T1 moraines; there is a sharp boundary between the two groups, clearly marked by factors such as density of plant cover, degree of discolouration by weathering, amount of disturbance by frost heaving and slope stability, suggesting that the two moraine groups represent distinct and separate events rather than a recessional age progression from one to the other. Conspicuous T2 moraines are also present in cirques now free of glacier ice and, in places, relict pro-talus ramparts and rock glaciers may also be equivalents (Birnie, 1977). The exact age and oscillatory nature of these Little Ice Age events in South Georgia have not yet been established in detail but there is one important date. The top layers of a peat buried by 60 cm of till and lying 60 m behind the T2 terminal moraine of the Heaney Glacier in St Andrews Bay gave a radiocarbon age of 155 ± 45 BP (SRR-738); this indicates that the Heaney Glacier reached its maximum Little Ice Age limit some time after 1795 A.D.

Beyond the limits of the Little Ice Age moraines there are four groups of phenomena related to more extensive glaciation; they are multiple moraines, older tills, raised beach deposits and glacial erosional landforms.

The terminal and lateral moraines (T3) comprise a group of distinctive landforms present in almost every valley, fjord and bay on the northeastern side of the island (Figs. 3 and 4). Terminal ridges are much more clearly developed in narrow valleys and fjords than in the open bays, where only lateral ridges may be present. Between and behind the moraines lies a discontinuous till sheet which is at least 10 m thick in places. Although the moraines vary in height from 1 to 35 m, they are generally larger features than the T1 and T2 moraines and are located much farther down-valley from existing glaciers (2 to 6.5 km). The forms are well preserved and sharply defined but the slopes and crests of the ridges are more degraded by gullying and slope movements than those of the T1 and T2 groups. Again in contrast to the T1 and T2 moraines, a dense cover of vegetation grows on many of the ridges and forms an almost continuous carpet in the low and sheltered coastal areas. Characteristics of the till can be seen where vegetation growth has been limited by exposure and altitude. In such localities the surface stones have normally been disturbed by frost action and are arranged into the form of conspicuous stripes, polygons and lobes. Many of the large surface boulders have disintegrated, apparently by frost shattering. Thus in terms of their relative position, morphology, vegetation cover and extent of cryogenic disturbance the T3 till and moraines are distinctly older than the T1 and T2 deposits.

The multiple terminal moraines are commonly interconnected by subsidiary cross ridges and thus seem to owe their origin to one phase of glaciation rather than to repeated advances separated by long time intervals. There is certainly no obvious field evidence which would indicate that the T3 moraines and associated till sheet relate to more than one glacial episode.

Because large numbers of glaciers in South Georgia advanced beyond the present shoreline when the T3 deposits were laid down, the end moraine marking the most distant position achieved is seldom visible. The only localities where clear maximum limits are continuously marked by lateral and end moraines are in the shallow reaches of Moraine Fjord, Hamberg Lakes valley (Fig. 4), Cumberland West Bay and Sea Leopard Fjord. Special topographic circumstances may have been responsible for the development and preservation of morainic landscapes at these localities. For example, in Moraine Fjord the magnificent and partly-submerged end-moraine complex rests on a rock threshold which is only 30 m deep. Similar relationships may also be present at the mouths of fjords occupied by Lyell and Geikie Glaciers in Cumberland West Bay.

In bays and fjords where T3 moraines occur mainly as lateral ridges and depositional trim lines sloping seawards out of the containing valleys, their possible continuations as terminal ridges on the sea floor of some shallow bays may be indicated by lines of kelp. Since thick belts of kelp grow all along the known submerged terminal moraines across Moraine Fjord and in front of the Lyell and Geikie Glaciers, it may be a reasonable assumption that similar patterns of kelp which are seaward continuations of land-based lateral moraines could also mark the positions of submerged terminal moraines. Such relationships exist in St Andrews Bay, Doris Bay and Kelp Bay. These relationships raise the possibility of topographic and sea-level control on the positioning and development of T3 moraines and this is considered more fully in the discussion section.

Glaciers in South Georgia appear to have responded to global climatic fluctuations recorded also by glaciers in most other parts of both hemispheres during the past 400 years. This encourages the prediction that most previous glacier oscillations which are well established global events, such as those earlier Neoglacial, Wisconsin and other Cainozoic episodes, may be expected to have occurred in South Georgia. It was in this belief that Clapperton (1971) tentatively correlated the T3 group of moraines with the Neoglacial advance identified in Chile and dated at between 4000 and 7000 years BP (Heusser, 1960; Mercer, 1968; 1970). However, more recent work in South Georgia on peat deposits lying within the T3 limits and the discovery of raised

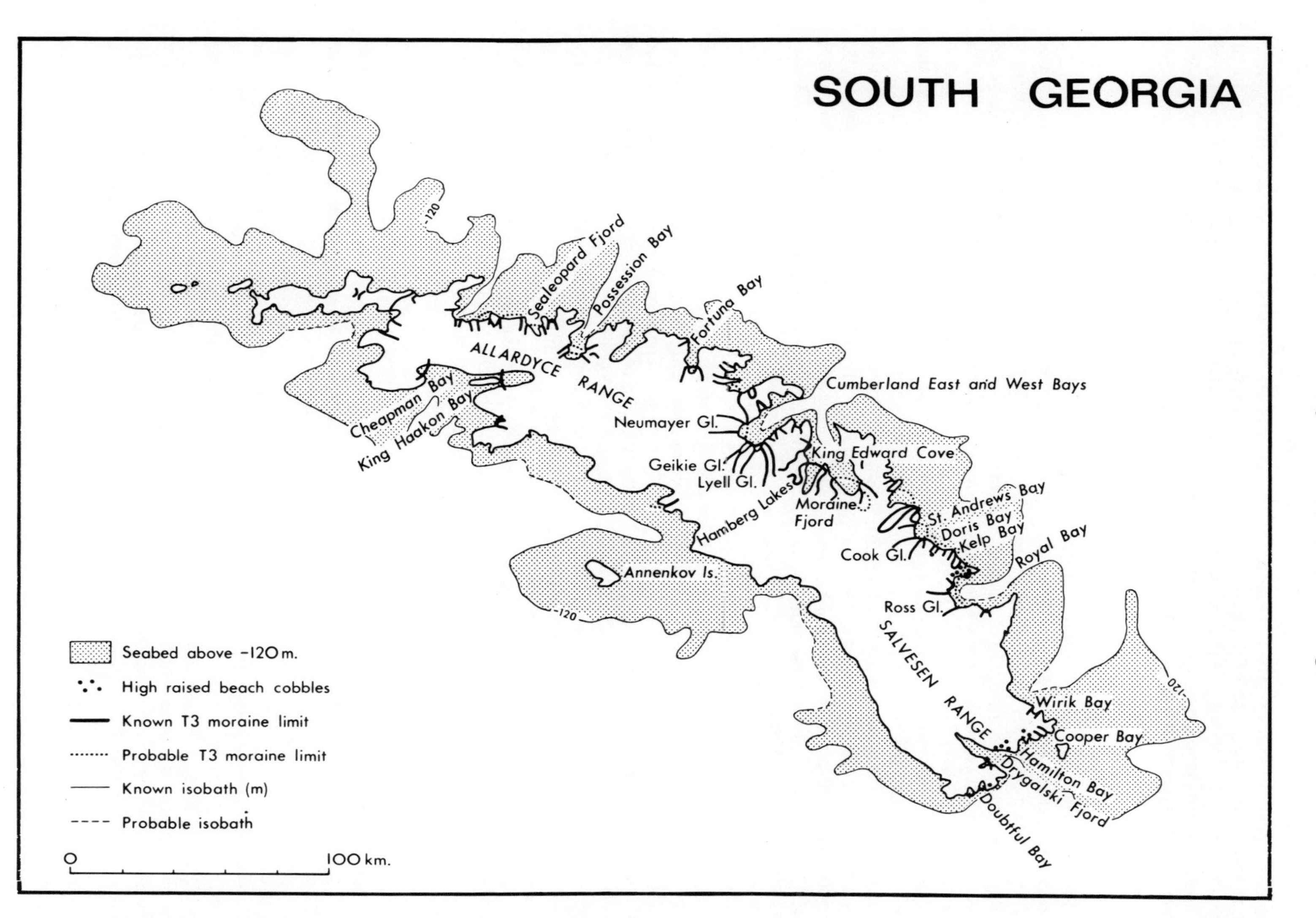

Fig. 3. Features of significance to the glacial chronology of South Georgia – T3 terminal and lateral moraine limits, sites of high raised beach cobbles (cemented beach deposits and cobbles in till) and submarine shelf areas shallower than 120 m. This latter shaded area is thought to represent areas covered by the Late Wisconsin ice cap.

beach deposits up to 52 m above sea level which have been buried and incorporated into T3 till, has made necessary a reappraisal of the age of this group of deposits and associated landforms.

The base of peat layers at two sites resting on T3 till within the multiple moraine complex on the south side of King Edward Cove gave radiocarbon ages of 9 493 ± 370 years BP (SRR–736) and 8 537 ± 65 years BP (SRR–582), clearly indicating that the area has been ice free for at least 9 500 years. It is also certain from radiocarbon dating (Harkness, in press) that peat was accumulating on T3 till in Cumberland West Bay 6 500 years ago; it has not since been covered by a glacier advance. This evidence suggests that T3 moraines cannot represent a Neoglacial advance of the glaciers, at least in the Cumberland West Bay area, and must be older. If Neoglacial advances did take place in South Georgia between 5 000 and 1 000 years BP, as in most other glacierized parts of the world (and notably in southern Chile–Patagonia 4 000–5 000 years BP and 2 700–2–200 years BP according to Heusser (1960) and Mercer (1968, 1970 and this volume respectively) then they must have been less extensive than those of the 17th to 19th centuries, which presumably overran any earlier moraines.

The till sheet (T4) which lies outside the T3 moraines occurs mainly on higher interfluves, such as that between Moraine Fjord and Hamberg Lakes valley, and on coastal forelands like those between the bays south of St Andrews Bay. Where the till surface is bare of vegetation it can be seen that the superficial layers have been disturbed by frost action and are similar in this respect to the till comprising the T3 moraines. Sedimentary analysis of the two tills suggests that the amount of frost disturbance of T4 till is only marginally greater than that of T3 till (Thom, Ph.D. thesis in preparation). Where bedrock is exposed in the same localities and has not yet disintegrated through frost action, it retains its freshly scoured surface with a fine glacial polish and frequent striations (Fig. 5).

These observations suggest that the T3 and T4 deposits may not be separated by a long period of time, any gap being of the order of a few hundred or thousand years rather than of tens of thousands of years. The location of the T4 deposit on high coastal peninsulas and on the very tip of prominent forelands, such as that on the north side of Royal Bay and on both sides of Wirik Bay, indicates a phase of ice cap glaciation extending beyond the present coastline.

Although for simplicity T4 till has been discussed as if it represents one till, it is likely than in places it comprises at least two components of different age. In the Cooper Bay area, for example, an indurated till deposit lies beneath unconsolidated T4 till on ice-scoured bedrock. The upper surface of the indurated till is itself ice moulded and indurated blocks were found incorporated in the overlying unconsolidated till. These relationships suggest that the indurated till is at least one glaciation older than the unconsolidated T4 till, but at this stage very little more can be said about its significance.

Raised beach deposits

Raised beach sediments are present in many localities in South Georgia and occur as distinct terrace forms, as *in situ* deposits beneat till and as scattered clasts within till. The terraces occur at two common levels, approximately 2–4 m and 6–7.5 m above present sea-level and are situated both inside and outside the T3 limits (Fig. 6). The upper terraces are nowhere found within the T3 limits, whilst in one location (Fortuna Bay) the lower terrace was found to be capped by a thin layer of T2 till at 3 m above present sea-level. The evidence for any local variations in altitude is limited for there appears to be no meaningful difference in terrace levels over the island, once allowance for exposure and wave energy is made. A minimum age for the low terrace comes from the St Andrews Bay radiocarbon date of 155 ± 45 BP. The base of a peat layer overlying the higher raised beach in Cumberland West Bay gave a radiocarbon age of 3 997 ± 85 years BP (SRR–597). In view of this latter date and the presence of the higher beach within T3 limits, it seems clear that the 6.0–7.2 m beach formed later than aproximately 9 500 years BP and earlier than 4 000 years BP.

Raised beach sediments, apparently *in situ*, have also been found beneath till at much higher altitudes (Fig. 7). They are present beneath T3 till at an altitude of 20 m in Kelp Bay and beneath T4 till at an altitude of 52 m on the north side of Royal Bay (Fig. 3). At both localities the deposit consists of from 2–4 m of sand and pebble layers containing water-rounded pebbles; stone roundness analysis showed the latter are similar in shape to those composing the beaches at present sea-level in the immediate vicinity. The upper 5–10 cm of the deposit at both localities is strongly cemented. The base of the deposit in Kelp Bay rests directly on glacially-polished and striated bedrock while that near Royal Bay nestles in a hollow in ice-scoured bedrock. Similar cemented beach sediments over 1.2 m thick have been found in six sites along the south coast of South Georgia where they overlie bedrock at altitudes up to 40 m and are buried by up to 25 m of till.

Cobbles and fragments of the cemented beach gravels, apparently derived from these high beach deposits, are contained in the T3 till of Kelp Bay up to an altitude of 90 m above sea-level and in T4 till on the northern and eastern sides of Harcourt Foreland; they have also been found in till at an altitude of 124 m in Hamilton Bay.

The significance of these cemented beach deposits in assessing a chronology of glacial events for South Georgia may be summarised as follows:

a) The sediments were deposited on glacially-scoured bedrock which is characteristic of lowlying parts of the island and the offshore submarine platform (Sugden & Clapperton, 1977). Thus they represent a marine phase following a period of glaciation sufficiently important to have eroded the main glacial features of South Georgia.

b) The position of the beach deposits on peninsulas and the fact that they have subsequently been covered by T3 and T4 till shows that they pre-date at least one glaciation which covered all lowlying parts of the island and extended onto the offshore submarine shelf. This last glaciation achieved little erosion in comparison to the earlier glaciation(s).

c) The altitude of 52 m for some of the deposits is far above the 7.5 m maximum of the beach terraces which followed the last phase of deglaciation. If the high altitude of the cemented beach deposits is explained by glacio-isostasy, then this implies that they followed either a more extensive glaciation than the lower younger beach terraces, one that disintegrated more rapidly, or both.

d) The occurrence of the cemented beach deposits within T3 limits suggests that they formed during a period when glaciers in South Georgia were more restricted than in T3 times and may have been as restricted as they are today. This indicates an Interglacial/Interstadial age for beach deposits.

DISCUSSION OF THE SOUTH GEORGIA CHRONOLOGY

From the distribution of glacial troughs and ice-scoured bedrock on the submarine platform surrounding South Georgia, it appears that the largest ice cap to have been centred on the island extended as far as the 200 m submarine contour, more or less at the edge of the shelf (Sugden & Clapperton, 1977). Assuming that world sea-level during that glacial period may not have fallen much below that of 17000 years ago, when it was around —120 m (Shackleton & Opdyke, 1973), the maximum ice cap must have built up sufficiently to have become grounded on the submarine platform and to have pushed outward into about 150 m of water (allowing for c 70 m isostatic depression at the ice cap margin). This would have been a potentially unstable situation (Weertman, 1974) and would have favoured very rapid collapse and shrinkage of the ice cap at the end of the glacial period as eustatic sea-level rose. The high rates of isostatic recovery that would have accompanied such a collapse could account for the high altitudes of the cemented beach deposits. The cementation of the deposits could have occurred during the subsequent interglacial period.

The expansion of glaciers which covered and partially removed the cemented beach deposits seems most likely to have occurred during the Late Wisconsin glaciation, judging from the position and limited weathering of the related till deposits and the age of peat deposits lying within the T3 moraines. The limits of this latter glaciation cannot be established firmly but if, as seems likely from the low altitude of recent raised beaches, isostatic recovery was more limited, then the ice cap may well have been smaller than at the full maximum. A possible reason for this could be that the ice cap was controlled in its extent by the position of contemporary sea-level. Shackleton & Opdyke (1973) suggested a figure of —120 m for world sea-level at 17000 years BP. This would have exposed the land area around South Georgia shown in Figure 3. It is immediately apparent that the deep troughs created by the earlier glaciation(s) would have played an important part in limiting the growth of the major outlet glaciers, since depths of over 200 m are common in major fjords such as the Bay of Isles, Possession Bay, Cumberland West Bay, Royal Bay and Drygalski Fjord. There is no clear indication of how restrictive these fjord basins were on the growth of any Wisconsin ice cap. However, the major outlet glaciers of the large fjords may not have advanced very much more than a few kilometres into the deep water of these inlets. Away from the inlets, glaciers and ice caps may have built up on parts of the island shelf that were exposed as dry land by the fall in sea-level during the Late Wisconsin. From Figure 3 it can be suggested that they may have extended 10–40 km from the present coastline.

This reconstruction of a possible Late Wisconsin ice cover is highly speculative. Nevertheless it does agree with several lines of evidence. On the one hand, the more restricted ice cover when compared to earlier glaciations helps to explain the contrast in the altitudes of the two sets of beach deposits. On the other hand it seems that the ice must have been more extensive than the T3 moraines, which implies that it extended beyond the present coast in many places. Evidence supporting this latter contention is the fact that the till sheet and associated rock surfaces immediately outside the T3 moraines do not appear to be much more weathered than those within the moraines, and in fact seem much less weathered than would be expected if they related to an early Wisconsin or earlier glaciation. Also, it seems most unlikely that South Georgian glaciers would have expanded no further than the site of the T3 moraines in response to such a major global event as the Late Wisconsin glaciation. In the absence of more conclusive evidence, it can be tentatively suggested that a Late Wisconsin ice cap, with the approximate

The radiocarbon ages of organic deposits within the T3 moraines of King Edward Cove indicate that South Georgian glaciers had probably receded from such limits by 9 500–10 000 years ago. If one accepts that the fresh glacial rock surfaces associated with T3 till cannot have survived exposure to maritime periglacial conditions for more than a few thousand years at most, then it seems most likely that the T3 moraines represent either a readvance or a recessional stage associated with the decay of the Late Wisconsin ice cap.

It is believed that by about 10 000 years BP world sea-level had risen to around —30 to —40 m (Thom, 1973). If one ignores isostatic adjustments for the moment, then in the Stromness Bay–Cumberland Bay areas such a sea-level would have exposed bedrock thresholds at the exits of rock basin fjords. It is interesting that prominent moraines are located in such positions; furthermore, in other areas declining lateral moraines which project out to sea do so in places where the sea floor of the bay is of the order of 30–40 m in depth, for example in Cheapman and King Haakon Bays on the southwest coast. Thus there is a distinct possibility that retreating sea-terminating glaciers first became grounded at or close to the mouths of fjords and bays, a situation which would have temporarily halted their general recession and permitted the accumulation of moraines. This means that the T3 moraines may reflect a recurring topographic sea-level relationship in South Georgia rather than a particular climatic event.

The multiplicity of T3 lateral and terminal moraines at such localities as Cumberland West Bay, Moraine Fjord and Kelp Bay, suggests that a period of stillstand of slow recession occurred before the final retreat which cleared the fjords of glacier ice and permitted the development of the raised beach terraces along the fjords.

Radiocarbon dates on several sites within the T3 moraines show that vegetation was well established by 9 500 years ago. It seems logical to anticipate that glaciers in South Georgia would have responded to global climatic oscillations which caused Neoglacial readvances around 5 000 and 2 500 years BP in other parts of the world. However, since no deposits or landforms clearly associated with a separate glacier advance lie between the moraines of the Little Ice Age expansion of the last few centuries and T3 moraines, it seems likely that any earlier Neoglacial readvances on the island were less extensive than those of the last few centuries, if they occurred at all.

COMPARISON BETWEEN SOUTH GEORGIA AND OTHER ISLAND GROUPS BORDERING THE SCOTIA SEA

The South Georgia chronology outlined above has many similarities with other sub-Antarctic islands lying to the south, but is very different to that of the Falkland Islands which is the nearest island group to the north (2° further north) but on the warm side of the Antarctic Convergence.

Evidence from the South Shetland Islands has been described in detail by John & Sugden (1971) and possible ages of the various glacial phases discussed at a former Antarctic Symposium (Sugden & John, 1973). Subsequent work in the islands has produced further radiocarbon dates on whalebones which fit in with these earlier results (Hansom, submitted). In the South Shetland Islands a maximum phase of glaciation extended out over the islands themselves and the surrounding submarine shelf and accomplished most glacial erosion. This glaciation(s) was followed by a marine phase where raised beach deposits, sometimes cemented, were deposited on glacially-scoured surfaces up to altitudes of 275 m. A subsequent glaciation eroded many of the marine deposits and incorporated many clasts in till. A lower suite of raised beaches extending as high as 54 m accompanied the withdrawal of the last glaciation. Radiocarbon dates on shells show that outlet glaciers had apparently withdrawn to near or within the present glacier limits by about 9 000 years BP. Since then one readvance, marked by moraines 1–3 km from outlet glacier snouts, has been tentatively dated to 500–750 radiocarbon years BP. In addition to this, relatively unstable moraines lie immediately adjacent to or outside glacier margins.

The main features of the glacial history of the South Shetland Islands and South Georgia are very similar. Both areas experienced two major phases of glaciation separated by a marine interval; in both cases the former ice cap(s) was the most extensive and accomplished most erosion. The main contrast between the two areas is that the raised beaches associated with each phase of deglaciation are higher in the case of the South Shetland Islands than South Georgia. This may reflect the presence of a larger ice mass in the South Shetland Islands during each phase because of the greater extent of shallow and undissected marine platform around the islands as well as the severer climate.

The dating of the main South Shetland Island glacial phases is far from clear. As with South Georgia, the radiocarbon dates that exist point to deglaciation resembling present day conditions by 9–10 000 years ago. Beyond this there seem to be the two main possibilities, as in South Georgia. If the last glaciation is assigned to the Late Wisconsin, then the earlier beach phase could represent an interstadial earlier in the Wisconsin or the Sangamon Interglacial. In the case of the former alternative the implication is that the maximum ice cap represents Early Wisconsin and/or earlier conditions. In the case of the second alternative the maximum ice cap represents Illinoian or earlier conditions, as suggested by John (1972).

Comparison of smaller-scale fluctuations between the South Shetland Islands and South Georgia raises problems. There is agreement that any Neoglacial maximum occurred in the last few centuries rather than earlier in the Holocene, unlike South America. But the recent readvance apparently took place 3–5 centuries earlier in the South Shetland Islands than in South Georgia and was more pronounced. At the moment it is difficult to pinpoint the regional climatic contrasts and/or the possible differences in glacier type and regime that could explain this apparent lack of correlation. Fluctuations over recent decades may be reflected in the South Shetland Islands by a moraine (often ice cored) which occurs adjacent to the margins of outlet glaciers and around the edge of ice caps. The moraine is actively slumping and yet stands above the adjacent ice surface. Although there are no dates, it could prove to be comparable to the T1 and T2 stages on South Georgia and relate to a minor expansion in the last few centuries.

A little information is availably from other sub-Antarctic Islands around the Scotia Sea. From Signy Island in the South Orkneys two moraines close to the existing ice cap have been dated at 1880 and 1837 (Lindsay, 1973), and appear to be Little Ice Age equivalents. It is also noteworthy that evidence of a former maximum glaciation extending over adjacent submarine shelves has been discovered in the South Orkney Islands and in the area around the tip of the Antarctic Peninsula (Sugden & Clapperton, 1977).

Comparison of the glacial history of the sub-Antarctic Islands with that of the Falkland Islands on the warmer, northern side of the Antarctic Convergence reveals dramatic differences. There is no direct evidence of the existence of an ice cap covering either the Falkland Islands or the surrounding submarine shelf. Instead the available evidence suggests that glaciers at their maximum extended only a few kilometres from a limited number of cirques (Clapperton & Sugden, 1976). One explanation of this contrast is that throughout the glacial age the Antarctic Convergence has remained in its approximate present position between South Georgia and the Falkland Islands. Perhaps the apparent stability of the Antarctic Convergence is due to the existence of strong submarine relief on the bed of the Scotia Sea. One culd go further and suggest that the oceanographic and climatic boundary of the Antarctic Convergence is perhaps one of the key controls on glacial history in this part of the Southern Hemisphere. Perhaps this is why there is no clear and obvious correlation between the Neoglacial history of southern South America and that of the adjacent sub-Antarctic.

ACKNOWLEDGEMENTS

This paper forms part of a programme of geomorphological work in Antarctica carried out by the Department of Geography, University of Aberdeen. We are very grateful to the British Antarctic Survey for support in the field. Chalmers Clapperton and David Sugden were given generous financial support by the Carnegie Trust for the Universities of Scotland and the Trans-Antarctic Association. Richard Birnie, James Hansom and Gordon Thom were financed by the Natural Environment Research Council research training studentships. We thank all these organisations for the opportunity to work in the Antarctic.

DISCUSSION

D.J. Drewry: What is the mechanism for producing the glaciated troughs on the shelf – could these be expanded outlet glaciers rather than ice streams within an enlarged ice cap?

D.E. Sugden: The evidence points to the creation of the troughs by ice streams within the ice cap, for areal scouring occurs on the submarine divides between the troughs and appears to extend as far out to sea as the troughs. If the troughs were cut by expanded outlet glaciers, then one would not expect areal scouring on the divides between the troughs.

R. Clark: What is the evidence in the reconstruction of South Shetland Island ice cap that Deception Island and Livingston Island were linked?

D.E. Sugden: The evidence here is the submarine trough morphology between Deception Island and Livingston Island which is most easily explained by postulating a continuous ice cap.

P.F. Barker: Is there any evidence on South Georgia for a climate significantly warmer than that of the present?

D.E. Sugden: We have found none. Most geomorphological evidence of such a warmer climate would lie beneath the existing glaciers and we have found no deposits beneath the sides of glaciers or in their moraines to indicate such an event. One possible approach to the problem is to examine the palaeobotanical evidence in the peat sections.

REFERENCES

Birnie, R.V. 1977. *Rock debris transport and deposition by glaciers in South Georgia.* Unpublished Ph.D. thesis, University of Aberdeen (submitted), 320 pp.

Clapperton, C.M. 1971. Geomorphology of the Stromness Bay – Cumberland Bay area, South Georgia. *British Antarctic Survey Sci. Rep.* 70, 25 pp.

Clapperton, C.M. & Sugden, D.E. 1976. The maximum extent of glaciers in part of West Falkland. *J. Glaciology* 17: 73–77.

Hanson, J.D. Radiocarbon dating of a raised beach at 10 m in the South Shetland Islands, Antarctica (submitted).

Hardy, A. 1967. *Great waters.* Collins.

Harkness (in press). Radiocarbon dates from Antarctica. *British Antarctic Survey Bull.*

Heusser, C.J. 1960. Late-Pleistocene environments of the Laguna de San Rafael area, Chile. *Geogr. Rev.* 50: 555–577.

John, B.S. 1972. Evidence from the South Shetland Islands towards a glacial history of West Antarctica. In: R.J. Price & D.E. Sugden (eds.), *Polar Geomorphology,* Inst. British Geographers Spec. Publ. 4: 75–92.

John, B.S. & Sugden, D.E. 1971. Raised marine features and phases of glaciation in the South Shetland Islands. *British Antarctic Survey Bull.* 24: 45–111.

Lindsay, D.C. 1973. Estimates of lichen growth rates in the maritime Antarctic. *Arctic and Alpine Research* 5(4): 341–346.

Mercer, J.H. 1968. Variations of some Patagonian glaciers since the Late-glacial. *Am. J. Sci.* 266: 91–109.

Mercer, J.H. 1970. Variations of some Patagonian glaciers since the Late-glacial, II. *Am. J. Sci.* 269: 1–25.

Shackleton, N.J. & Opdyke, N.O. 1973. Oxygen isotope temperatures and ice volumes on a 10^5 year and 10^6 year scale. *Quat. Res.* 3: 39–55.

Smith, J. 1960. Glacier problems in South Georgia. *J. Glaciology* 3(28): 705–714.

Sugden, D.E. & Clapperton, C.M. 1977. The maximum ice extent on island groups in the Scotia Sea, Antarctica. *Quat. Res.* 7: 268–282.

Sugden, D.E. & John, B.S. 1973. The ages of glacier fluctuations in the South Shetland Islands, Antarctica. *Palaeoecology of Africa* 8, Balkema, Cape Town: 141–159.

Thom, B.G. 1973. The dilemma of high interstadial sea levels during the last glaciation. *Progress in Geogr.* 5, Edward Arnold, London: 170–231.

Weertman, J. 1974. Stability of the junction of an ice sheet and an ice shelf. *J. Glaciology* 13: 3–11.

* 9 *

Aridification of the Namib Desert: Evidence from oceanic cores

William G. Siesser
Marine Geoscience, Department of Geology, University of Cape Town, Rondebosch, South Africa

Manuscript received 15th June 1977

CONTENTS

ABSTRACT

Extreme aridity in the Namib Desert is the result of several interacting atmospheric and oceanic phenomena. The presence immediately offshore of the Benguela Current is one of the major controlling factors. These cold, upwelled waters cool moisture-laden sea breezes, and combined with the atmospheric factors, prevent rain from falling in the Namib. If we can establish the time of the initiation of major cooling and upwelling in the Benguela Current System, we can approximate the time when aridification of the Namib was initiated, or at least greatly intensified.

Recent deep drilling on the Walvis Ridge Abutment has recovered a complete sequence of sediments ranging from Middle Eocene to Late Pleistocene. These open-ocean biogenic sediments provide a wealth of information on the history of the overlying waters in which they were formed: the Benguela Current.

Studies of sediment accumulation rates, diatom frustule abundance, planktonic Foraminifera and calcareous nannoplankton temperature preferences, primary productivity (expressed in C_{org}) and phosphorus incorporation in calcareous skeletons all suggest changes in the characteristics of the Benguela Current. These sedimentological, palaeontological and geochemical data suggest weak, spasmodic introduction of cool, upwelled waters along this coast from Middle or Late Oligocene until Middle Miocene times. In the early Late Miocene conditions changed markedly, strongly suggesting intensification of upwelling which brought cold, nutrient-rich waters to the surface along this coast.

Onshore faunal remains indicate that the Namib was mostly wooded-grasslands until Middle Miocene times, suggesting that the early spasmodic conditions of the Benguela did not cause significant aridification. It is suggested that the major cooling-upwelling of the Benguela in early Late Miocene times initiated aridification of the Namib Desert.

INTRODUCTION

Most of South West Africa has an arid or semi-arid climate. An extremely arid belt, the Namib Desert, extends along the entire coast from the Kunene River to the Orange River, eastwards at least to the Great Escarpment (Fig. 1). The reasons for the generally arid climate are two-fold: 1) the drying influence of the high-pressure (anticyclonic) cell located in the South Atlantic and 2) the effect of the cold Benguela Current and its intimately associated upwelling (van Zinderen Bakker, 1975a). The intensely arid Namib Desert adjacent to the coast is caused by several interacting atmospheric and oceanic phenomena, of which the Benguela Current is a major controlling factor. These cold waters cool moisture-laden sea breezes and, in combination with the atmospheric factors, prevent rain from falling in the Namib.

The 'Oldest Desert in the World'. This is a statement often heard when the Namib is described. But is it the oldest? What unequivocal evidence do we have which indicates the age of this desert? And in any case, how *do* we date a desert? The purpose of this paper is to present new evidence on the timing of aridification in South West Africa, with particular reference to the extreme aridification causing the Namib Desert.

TIMING OF THE INITIATION OF ARIDIFICATION

It is clear that if we can establish the time of the initiation of major cooling and upwelling in the Benguela Current System it will approximate the time of the initiation or, at least, intensification of aridification. Van Zinderen Bakker (1975b) made the first attempt to date the related Benguela Current–Namib Desert system. He presented a variety of evidence, part of which comes from Deep Sea Drilling Project (DSDP) cores raised from the sea floor off Antarctica. Oxygen and carbon isotope analyses of those cores showed that the temperature of high-southern-latitude bottom waters dropped markedly to the present low levels during Early Oligocene times (Shackleton & Kennett, 1975). On the basis of this and other evidence van Zinderen Bakker (1975b) concludes that '. . . in Early Oligocene times, when the cold Antarctic Intermediate Water could move northward, the stage was set for the origin of the Namib Desert'. But the initial availability of cold bottom waters in the high latitudes does not imply immediate development of cold, upwelled water off South West Africa. Certainly some sluggish, spasmodic upwelling may have been generated fairly soon (and there is evidence for this, as will be discussed in later sections), but major, intensive cooling/upwelling of waters began only in Late Miocene times, some 25 million years after the Early Oligocene origin of South Atlantic cold bottom waters.

DSDP SITE 362/362A

Evidence in support of this timing comes from DSDP cores collected off the northern coast of South West Africa during early 1975. Descriptions of these cores and relevant palaeo-environmental information are given by Bolli et al. (1975) and Siesser (in press, a).

The drill site most important to this study is Site 362/362A (362A is an offset hole drilled immediately adjacent to 362). The site is located on the western abutment of the Walvis Ridge (Fig. 1) in a water depth of 1 325 m. The total stratigraphic section penetrated was 1 081 m and a continuous sedimentary sequence from Upper Pleistocene to Middle Eocene was recovered. Sediment lithologies are remarkably consistent: they are overwhelmingly calcareous oozes, chalks and limestones dominantly composed of calcareous nannofossils with lesser amounts of planktic Foraminifera and other organisms. This has clearly been an open-ocean site from Eocene times onward. It is today under the direct influence of the Benguela Current (Moroshkin et al., 1970) and has been throughout all of the late Cainozoic. Thus the sediments at this site record the changing conditions of the overlying water mass throughout most of the Cainozoic times.

Sedimentological, palaeontological, and geochemical evidence have been collected from the sediments at this site in an attempt to elucidate the history of the Benguela Current.

SEDIMENTOLOGY

Sediment accumulation rates

Figure 2 shows the sediment accumulation rates at Site 362/362A for various Cainozoic ages and subages. These rates have been corrected for compaction and induration following the method of Schlanger et al. (1973). Eocene–Early Oligocene accumulation rates are low, although a definite increase occurs after Early Oligocene times. A rate between about 28 and 38 m/MY is maintained from Late Oligocene to Middle Miocene times. A dramatic increase in accumulation (72 m/MY) takes place in Late Miocene times, followed by an unexplained sharp decline in the Early Pliocene. The accumulation rate climbs again in Late Pliocene–Early Pleistocene times and by Late Pleistocene–Holocene, rates are almost back to Late Miocene levels.

The cold waters of the Benguela Current are rich in nutrients, which promote the growth of large numbers of planktic organisms in Benguela surface waters. Dead and discarded skeletons of these planktic organisms are the major components in the sediments at Site 362/362A. An onset or intensification of upwelling brings even more nutrients to the surface and should be reflected by vastly increased plankton abundance in surface waters and therefore an increased accumulation of pelagic sediment on the sea floor. It is plausible that the increase in sedimentation from Late Oligocene to Middle Miocene times may reflect the weak cooling/upwelling mentioned earlier. However, the accumulation rate in the Late Miocene almost doubles, strongly suggesting the production of more sediment-forming organisms and thus intense cooling/upwelling during that period. The brief Pliocene drop in accumulation is not supported by the other cooling/upwelling-

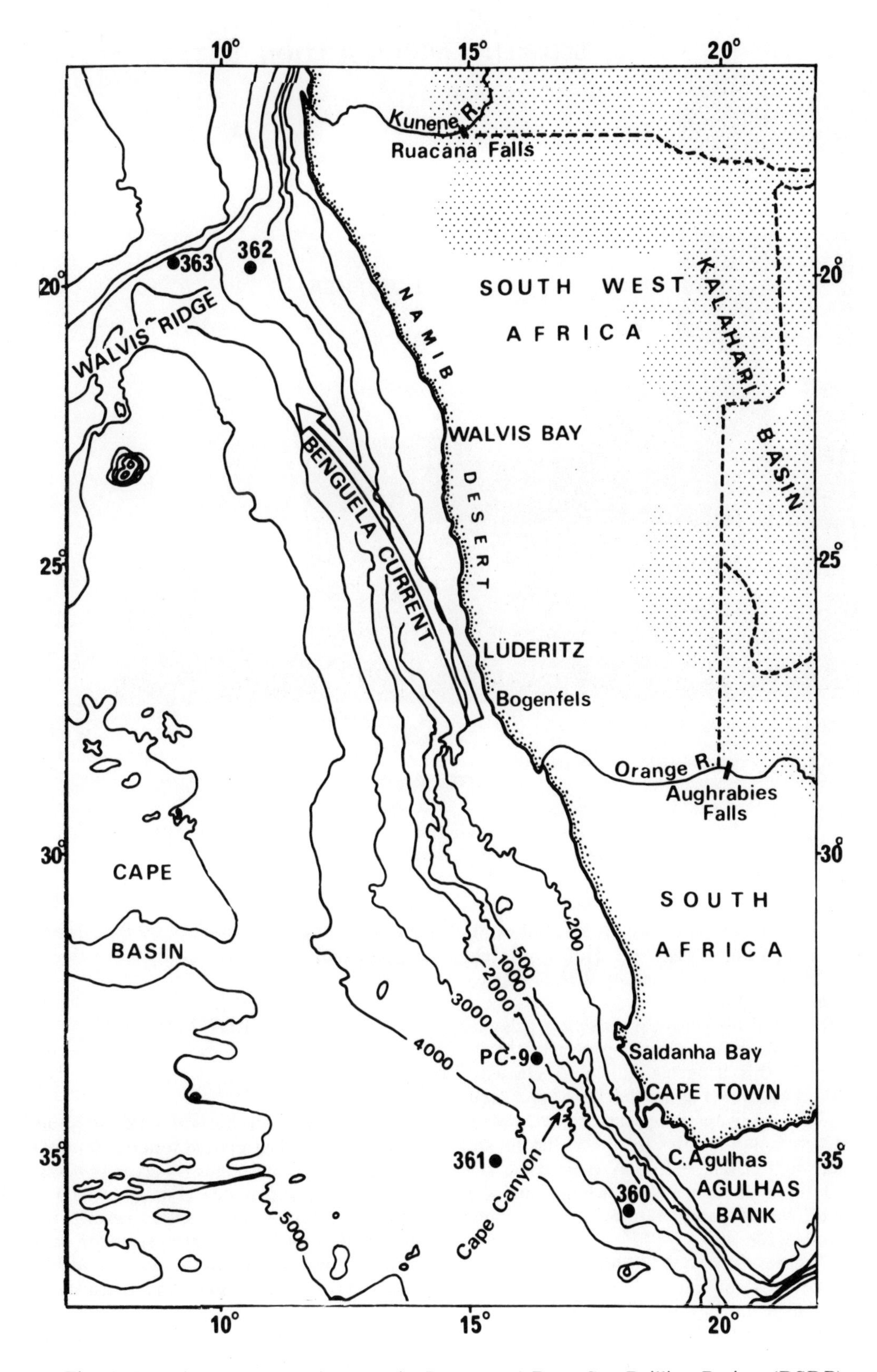

Fig. 1. Location map, showing Namib Desert and Deep Sea Drilling Project (DSDP) Sites. Isobaths are in metres

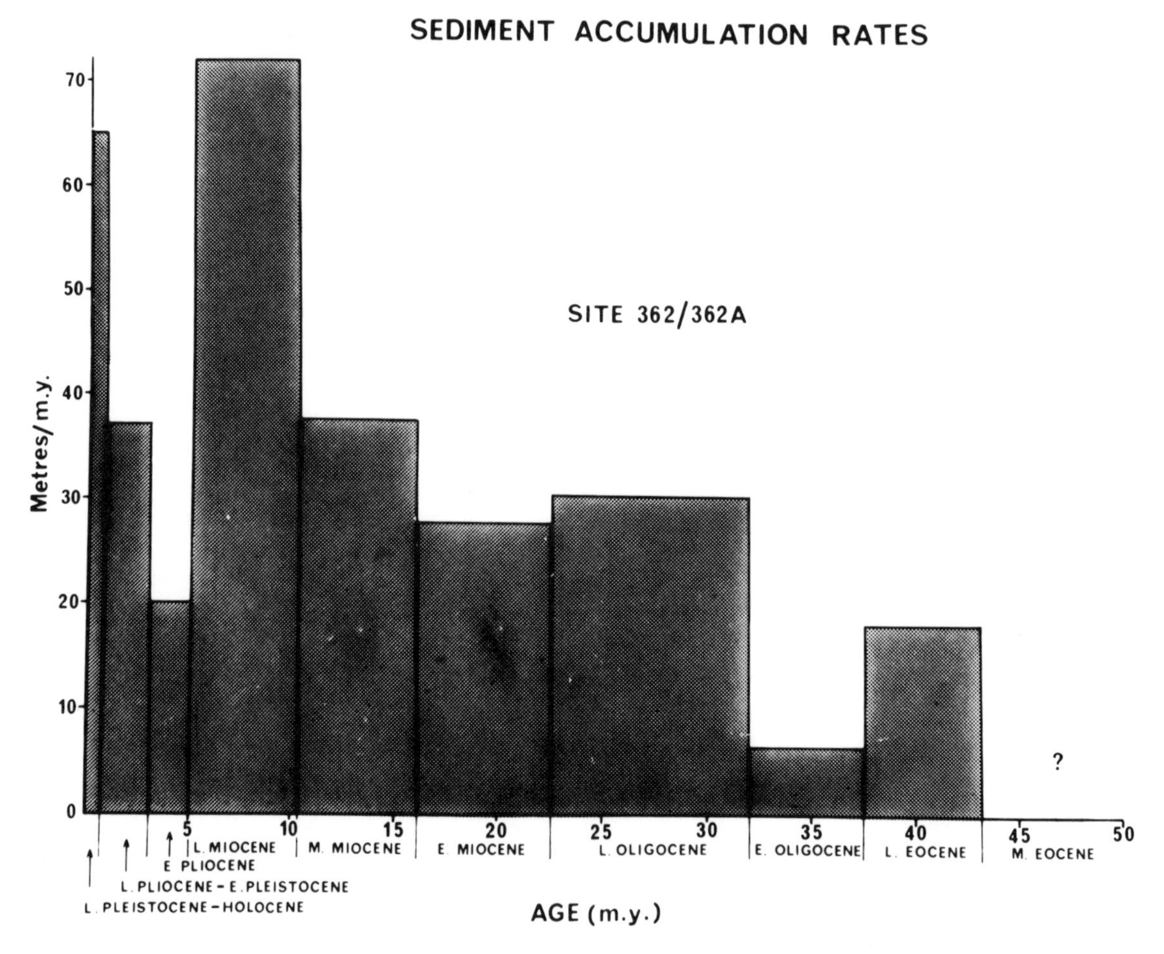

Fig. 2. Sediment accumulation rates in metres/million years at DSDP Site 362/362A. Rates are corrected for compaction and induration

indicative parameters, discussed in the following paragraphs. High Late Pleistocene–Holocene levels suggest near-modern accumulation rates similar to those in the Late Miocene.

The influence of the Benguela Current in pelagic-sediment production is clearly portrayed by a comparison of this Benguela-influenced site with sites (360/361 and 363 (Fig. 1)) that have never been directly influenced by the Benguela. Figure 3 shows sediment accumulation rates for these sites to the north and south of Site 362/362A. These data are calculated for each epoch as a whole, and are uncorrected, but still serve satisfactorily for comparison among the sites.

The differences are only a few m/MY at all sites during Palaeocene and Eocene times. Sites 360/361 and 363 still have similar rates during the Oligocene, but Site 362 shows the first slightly increased rate. This increase becomes dramatic in Miocene–Holocene times, almost doubling the rate at Site 360/361 (post-Miocene sediments were not recovered at Site 363), reflecting the influence of the Benguela Current over Site 362/362A.

PALAEONTOLOGY

Diatom frustules

Abundant diatom production is almost synonymous with cold, upwelled marine waters; diatom abundance is generally low in normal oceanic waters. These minute one-celled plants (Fig. 4), which build a skeleton out of opaline silica, extract nutrients (phosphates, nitrates, silicates) from sea water and form the lowest link in the food chain that makes the present-day waters off South and South West Africa one of the world's richest fishing grounds. Thus the relative abundance of these organisms, which are so dependent on upwelled nutrients for their growth, should record the history of the Benguela Current and its associated upwelling.

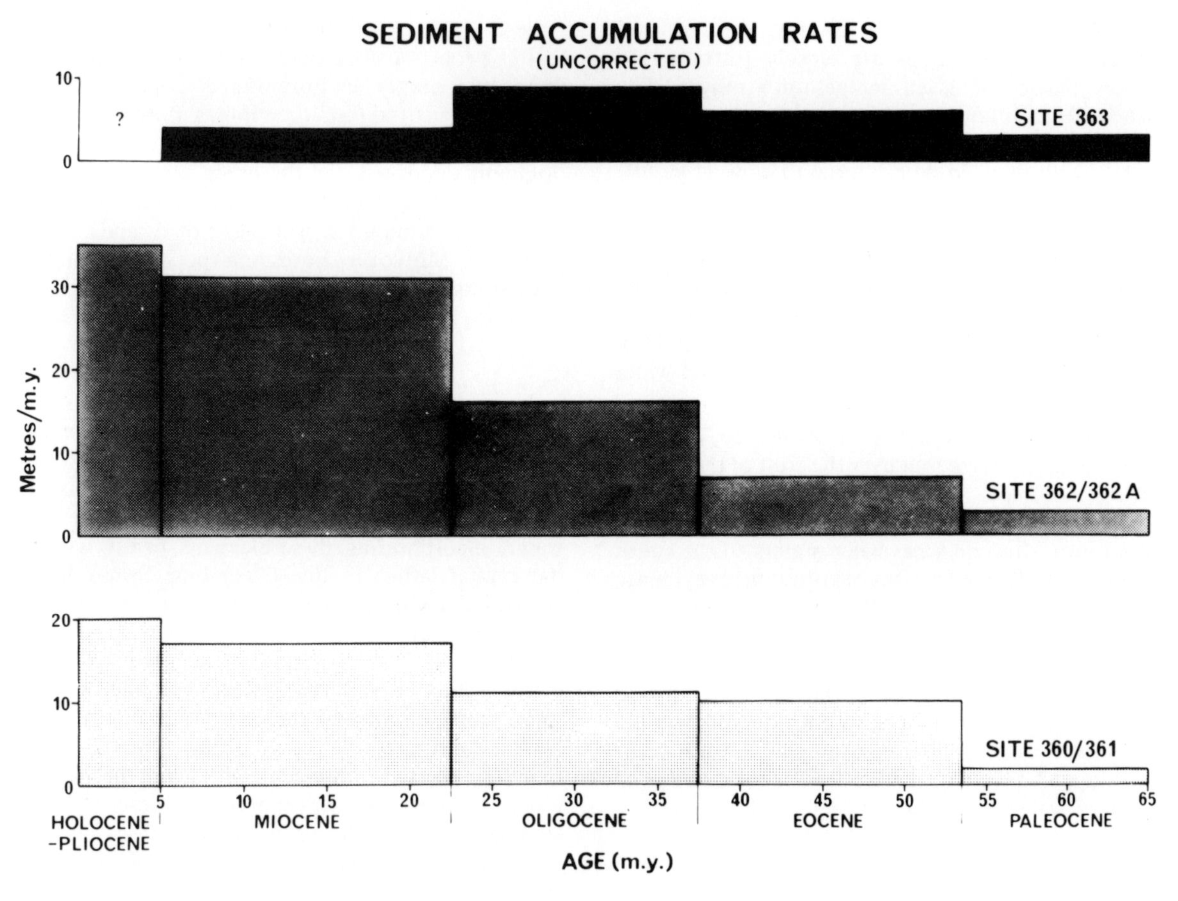

Fig. 3. Sediment accumulation rates in metres/million years at DSDP Site 360/361, 362/362A and 363. Rates are uncorrected

Figure 5 plots the average abundance of diatoms in each core against time. Negligible numbers of diatoms were found in sediments from Middle Eocene to earliest Late Miocene times. In the early Late Miocene the first recordable numbers of diatoms start to appear. Their abundances remain low until about the Late Miocene–Early Pliocene boundary, when a marked increase begins, which, although fluctuating, appears to continue increasing into the Pleistocene.

This is considered to be a real increase, and not, as might be suggested, resulting from progressive dissolution of buried diatom frustules. The only diatom frustules which normally reach the sea floor are those of robust species. Most of the weakly silicified, delicate diatoms are dissolved during their descent through the first few hundred metres of the water column. The robust species that accumulate at depths tend to be very stable, much more so than the calcareous skeletons of other planktic organisms. Thus, it is not uncommon to find pelagic red clays at depths greater than 4 000 m still containing diatom frustules, whereas all the skeletons of calcareous nannoplankton and Foraminifera have been dissolved.

The presence of solution-prone calcareous nannoplankton species with delicate skeletons in both cores where robust diatoms were and were not found indicates that solution has not removed robust diatoms from the older samples: they simply were not present in any abundance in the overlying waters.

Calcareous Nannoplankton and Foraminifera

Unlike diatoms, planktic Foraminifera and calcareous nannoplankton are found in great abundance in most oceanic water masses and not just in upwelling areas. Nevertheless, one would expect an increase in their abundance because of greater upwelling-induced nutrient production. However, this increase would be directly reflected by the sediment accumulation rates shown on Figure 2 and therefore has not been plotted separately.

But other valuable information can be obtained by an examination of the temperature preferences of species belonging to these groups, since certain species in both groups are closely restricted to water masses of given temperature ranges. The assemblages found at Site 362/362A indicate a very definite change in temperature of the overlying water with time. Middle Eocene to Lower/Middle Miocene sediments contain tropical-subtropical Foraminifera and 'warm-water' calcareous nannofossils. From Middle or Late Miocene times onward a marked cooling occurs, which is demonstrated by the dominance of cool-temperate-water planktic Foraminifera and calcareous nannofossils. By Pliocene times the assemblages are decidedly cold-water ones.

It is interesting to note that at Site 363, less than 150 km to the north, but over the crest of the Walvis Ridge, tropical faunas and floras were being deposited throughout the early and middle Tertiary, even into Miocene times when waters to the south were markedly cooler. This bears out other evidence (Bolli et al., 1975) that the Walvis Ridge, submerged though it may be, has acted as a substantial barrier to oceanic currents throughout the Tertiary. The Miocene cooling of the surface waters at Site 362/362A was caused by upwelling of Cold Atlantic Central Water. Large-scale upwelling did not occur at Site 363, owing to the Walvis Ridge barrier, and surface waters over that site remained warm.

GEOCHEMISTRY

Organic carbon

The high primary productivity of the cold Benguela Current and its associated upwelling has already been mentioned. Productivity is usually measured in terms of organic carbon (C_{org}). Foresman (in press) has measured the amounts of total C_{org} in these cores, and Figure 6 has been prepared from data presented by him.

C_{org} remains at very low levels from Middle Eocene to early Early Miocene times. A moderate increase occurs from late Early Miocene to middle Middle Miocene times, but again drops to negligible amounts near the Middle–Late Miocene boundary. However, in early Late Miocene times a marked increase in C_{org} begins, which persists into Late Pleistocene times. Extremely high values of C_{org} (3.6 and 4.2%) are recorded in the undifferentiated Late Pliocene/Early Pleistocene interval.

The curve shown on Fugure 6 suggests increased Benguela upwelling from Late Miocene times onward, corroborating the timing interpreted from data already presented. But, unlike the sediment accumulation rates, diatoms, and planktic assemblages – for all of which evidence can be presented showing that their fluctuations represent real trends – the C_{org} plot cannot be proven to be real. Organic matter progressively oxidizes with time, and what we see may simply be the amount of C_{org} that has not yet been destroyed (and therefore is more abundant in the younger cores) and not the amount that was originally deposited. On the other hand, oxidation of C_{org} is greatly retarded in environments of rapid sediment accumulation, and most of it tends to be preserved. Moreover, Foresman (pers. communic.) cites isotopic evidence which indicates that these C_{org} values do represent the original C_{org} content.

Phosphorus

A less equivocal chemical parameter that can be measured is phosphorus uptake in phytoplankton. It is well known that marine organisms extract minor and trace elements from the ambient sea water, incorporating these elements firmly within the crystal lattice in their skeletons. It has been stressed that Benguela Current waters are greatly enriched in nutrients such as phosphates, nitrates and silicates. Calcite skeletons of calcareous nannofossils were analysed using an electron microprobe to see what minor and trace elements might have been incorporated. The only element found, other than the expected components of calcium carbonate, was phosphorus. Siesser (in prep.) has described the techniques and results of this study.

Coccolithus pelagicus, a long-ranging Tertiary species was used throughout as a control species to avoid interspecific variation. The plot presented by Siesser (in prep.) shows that, after a Middle Eocene high, these organisms incorporated a low, fluctuating phosphorus content from Early Oligocene until Middle Miocene times. From Late Miocene times onward there is a slight, but steady increase in phosphorus uptake.

There are, of course, two possible explanations for this increase: 1) a physiological response on the part of the plants, reflecting progressively increasing need for, or ability to extract, phosphorus from the surrounding waters. Many marine organisms, for example, extract and incorporate Mg in direct proportion to ambient water temperatures. Is this, then, a reflection of warmer sea-surface waters? Information previously presented suggests this was in fact a time of cooling surface waters, and thus a warm-water environment factor seems unlikely.

The other explanation, 2) is simply increased availability of phosphorus in the water. This would obviously be the case if upwelling brought phosphate-rich waters into the zone where the plants lived, and this explanation seems best suited to explain the available facts.

The fluctuating values from Early Oligocene to Middle Miocene could represent the spasmodic introduction of phosphorus by irregular and discon-

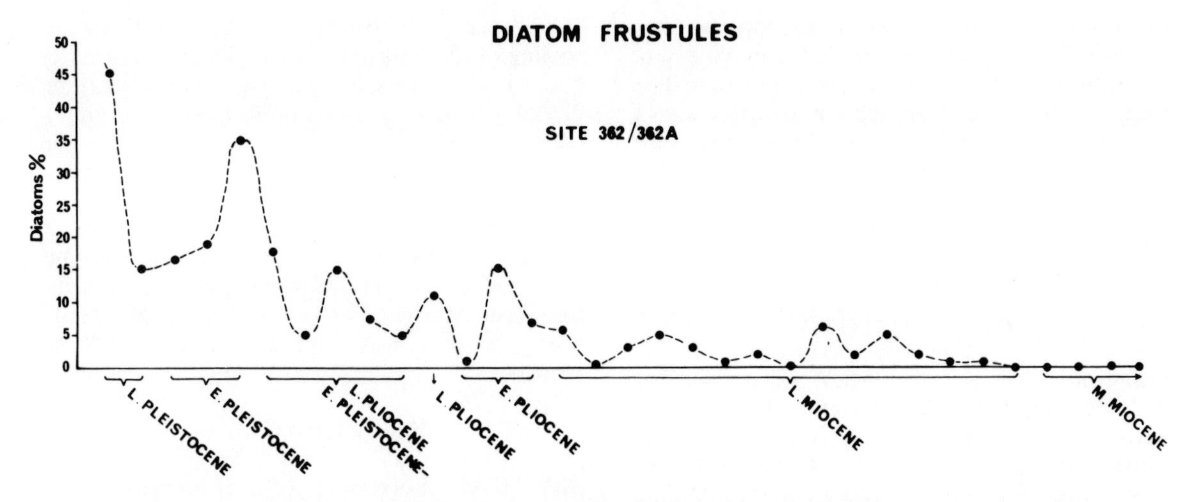

Fig. 5. Diatom frustule abundance at DSDP Site 362/362A. Diatom percentages are based on shipboard smear-slide estimates

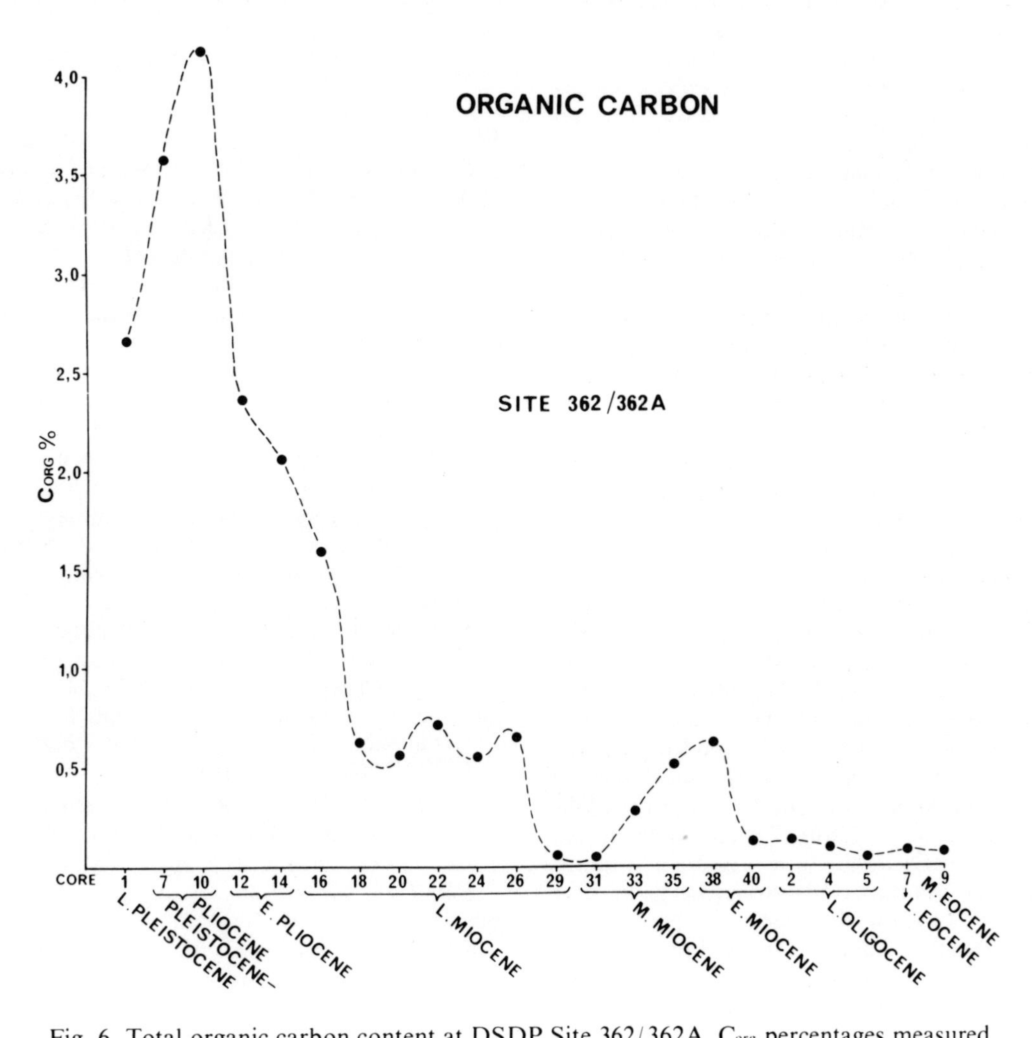

Fig. 6. Total organic carbon content at DSDP Site 362/362A. C_{org} percentages measured by combustion-gas chromatography

tinuous cells of upwelling. The steady increase after the Middle Miocene suggests that upwelling was established on a permanent basis after that time. The increased abundance of phosphorus in surface waters from Late Miocene times onward is strongly supported by the fact that the majority of firmly dated phosphorite rocks from the South African continental shelf are Pliocene in age (Siesser, in press, b).

CONCLUSIONS

Sedimentological, palaeontological and geochemical evidence suggests that the cold, nutrient-rich, northward-moving system we today call the Benguela Current first began to develop in Oligocene times.

Several lines of evidence suggest that from Middle or Late Oligocene until Middle Miocene times, cold, upwelled water was weakly and spasmodically introduced within this system. In the early Late Miocene times (~ 10 MY BP) a marked change may be seen in the Benguela Current: water temperature has dropped and nutrient content has increased sharply. These parameters strongly suggest intensification of upwelling, bringing more cold, nutrient-rich waters to the surface.

It is noteworthy that Savin et al. (1975) show a dramatic global cooling of bottom-water temperatures beginning in Middle Miocene, but reaching lowest values (lower, incidentally, than their calculated values for the Early Oligocene) only in Late Miocene times. This almost certainly corresponds to the development of the major Antarctic ice cap in Middle Miocene–early Late Miocene times (Kennett et al., 1975). Moreover, the circum-Antarctic current developed in Late Oligocene times, after the final separation of Australia and Antarctica, and undoubtedly had extensive influence on the spread of cold water northward.

It is difficult to assess the effect of the early, spasmodic nature of cold/upwelled water in the Benguela Current on aridification in South West Africa. Tankard & Rogers (in prep.) summarize onshore evidence for aridification along this coast. They quote Hopwood's (1929) study of Early Miocene deposits south of Lüderitz. He ascribed antelopes and jumping hares to a wooded-grassland (savanna) and tragulids to a riverine-woodland environment. Middle Miocene deposits north of the Orange River at Arrisdrift contain ruminant and rhinoceros fossils, which also suggest a wooded-grassland environment (Corvinus & Hendy, in press; quoted in Tankard & Rogers, in prep.). This suggests that the Namib was still well vegetated and watered up to at least Middle Miocene times, and that the current system offshore had not yet cooled sufficiently to promote significant aridification of the adjacent landmass.

Evidence presented here suggests that major cooling of the Benguela only became prominent in Late Miocene times, and rapid onshore desiccation probably followed. Tankard & Rogers (in prep.) reached the same general conclusions, suggesting that aridity on the subcontinent as a whole dates from the Pliocene. They further suggest that aridification was progressive, becoming fully developed during the Quaternary. Data presented here showing the overall increase in cold/upwelled offshore waters from Late Miocene to Pleistocene times also tends to support progressive aridification of South West Africa.

REFERENCES

Bolli, H.M., Foresman, J.B., Hotteman, W.E., Kagami, H., Longoria, J.R., McKnight, B.K., Melguen, M., Natland, J., Proto-Decima, F., Ryan, W.B.F. & Siesser, W.G. 1975. Basins and margins of the eastern South Atlantic. *Geotimes* 20: 22–24.

Corvinus, G. & Hendey, Q.B. (in press). A new Miocene vertebrate locality at Arrisdrift in Namibia (South West Africa), *S.A. J. Sci.*

Foresman, J.B. (in press). Organic geochemistry, DSDP Leg 40, continental rise of Southwest Africa. In: H.M. Bolli, W.B.F. Ryan et al. (eds.), *Initial reports of the DSDP* 40, U.S. Gov. Printing Office, Washington, D.C.

Hopwood, A.T. 1929. New and little-known mammals from the Miocene of Africa. *Am. Mus. Novitates* 344: 1–9.

Kennett, J.P., Houtz, R.E., Andrews, P.B., Edwards, A.R., Gostin, V.A., Hanos, M., Hampton, M., Jenins, D.G., Margolis, S.V., Ovenshine, A. T. & Perch-Nielsen, K. 1975. Cenozoic paleo-oceanography in the southwest Pacific Ocean, Antarctic glaciation, and the development of the Circum-Antarctic Current. In: J.P. Kennett, R.E. Houtz et al. (eds.), *Initial reports of the DSDP* 29: 1155–1169.

Moroshkin, K.V., Bubnov, V.A. & Bulatov, R.P. 1970. Water circulation in the eastern South Atlantic Ocean. *Oceanology* 10: 27–34.

Savin, S.M., Douglas R.G. & Stehli, F.G. 1975. Tertiary marine paleo-temperatures. *Bull. Geol. Soc. Am.* 86: 1499–1510.

Schlanger, S.O., Douglas, R.G., Lancelot, Y., Moore, T.C. & Roth, P.H. 1973. In: E.L. Winterer, J.L. Ewing et al. (eds.), *Initial reports of the DSDP* 29: 407–429.

Shackleton, N.J. & Kennett, J.P. 1975. Paleotemperature history of the Cenozoic and the initiation of Antarctic glaciation: oxygen and carbon isotope analyses in DSDP Sites 277, 279 and 281. In: J.P. Kennett, R.E. Houtz et al. (eds.), *Initial reports of the DSDP* 29: 743–755.

Siesser, W.G. (in press, a). Leg 40 results in relation to continental shelf and onshore geology. In: H. M. Bolli, W.B.F. Ryan et al. (eds.), *Initial reports of the DSDP* 40.

Siesser, W.G. (in press, b). Micropalaeontology and biostratigraphy of continental margin sediments. *Joint GS/UCT Marine Geoscience Group Tech. Rept.* 9, Dept. Geol. Univ. Cape Town.

Siesser W.G. (in prep.). Geochemical analysis of calcareous nannofossils.

Tankard, A.J. & Rogers, J. (in prep.). Progressive Late Cenozoic desiccation on the west coast of Southern Africa.

Zinderen Bakker Sr, E.M. van 1975a. Late Quaternary environmental changes in Southern Africa. *Ann. S. Afr. Mus.* 71: 141–152.

Zinderen Bakker Sr, E.M. van 1975b. The origin and palaeoenvironment of the Namib Desert biome. *J. Biogeography* 2: 65–73.

Fig. 4. Diatom frustule. Specimen is 0.13 mm in diameter

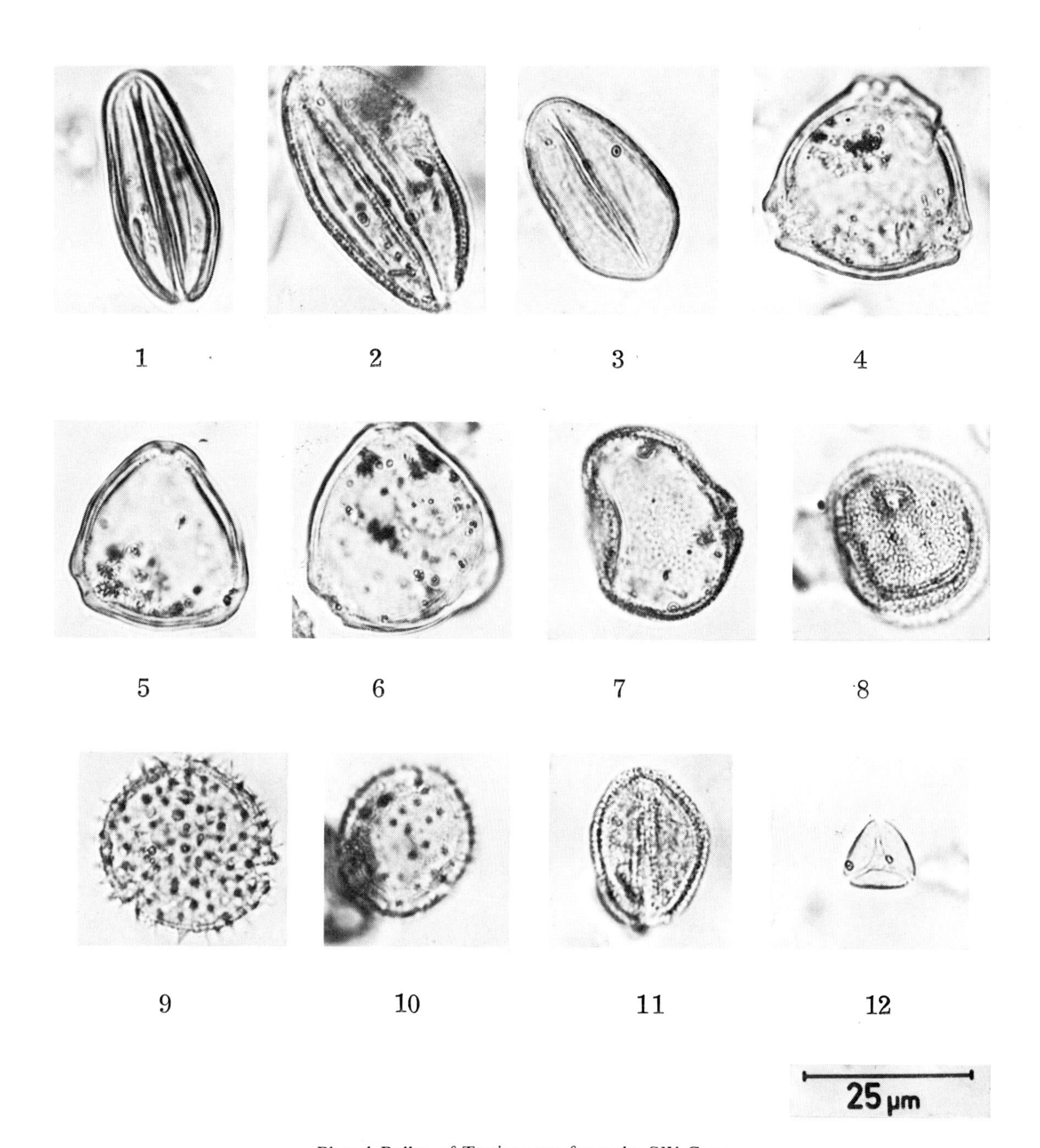

Plate 1. Pollen of Tertiary age from the SW Cape
1–3, Palmae; 4–6, cf. *Triorites harrisii* Couper; 7, 8, *Sparganiaceaepollenites* Thiergart; 9,10, *Echiperiporites* v.d. Hamm. & Wijmstra; 11, *Clavatipollenites* cf. *hughesii* Couper; 12, *Cupaniedites orthoteichus* Cookson & Pike

* 10 *

Late Cainozoic palaeoenvironments of southern Africa

J.A. Coetzee
Institute for Environmental Sciences, University of the O.F.S., Bloemfontein, South Africa

Manuscript received 17th August 1977

CONTENTS

ABSTRACT

Pollen bearing deposits of the southwestern Cape have been studied. In the oldest sections of the cores different types of Tertiary vegetation are represented since the Late Oligocene or Early Miocene until their ultimate extinction. The sporomorphae of the younger parts of the deposits are typical of the type of vegetation which exists in these regions at present and are considered to be of Quaternary age. Correlations with recent oceanographic data (Siesser, this volume) suggest that the Tertiary vegetation was probably eliminated in early Late Miocene times. The cold Benguela Current then originated together with the new climatic system for the sub-continent which is characterised by winter rainfall and dry summers. Data on the evolution of the Antarctic ice sheet (Mercer, this volume) could however point to a later extermination of this Tertiary vegetation during the Late Miocene to Early Pliocene.

Palm-dominated sub-tropical to tropical vegetation alternated with conifer forests during the Tertiary at the Cape which may be of significance with regard to the marked temperature oscillations of the Miocene (Mercer, this volume). Some of the Tertiary pollen types are of considerable botanical interest as they shed light on distribution patterns which could have existed before the separation of Africa from Antarctica between the Upper Jurassic and Lower Cretaceous.

Some evidence exists for the progressive aridification of the whole continent from Pliocene times onward and in particular the development of extreme aridity along the southwest coast. The results of research in different disciplines strongly indicate that the evolution of the Antarctic ice sheet and the subsequent Quaternary glaciations had a profound influence on these changes in climate and biogeography.

INTRODUCTION

The geological and biological background of southern Africa can only be fully appreciated if we go back in time to the Late Jurassic when the continent of Africa held a central position in Gondwanaland. In recent years much information has come forward from a wide array of disciplines which throw light on the old connections with other continents and on the processes of drift, separation and evolution which have taken place since the disruption of the old super-continent. The role southern Africa played in this complicated process is still obscure. The study of these pre-drift conditions is extremely important for our understanding of the location of ancestral groups of our flora and fauna and for the contacts their progeny could have in later times. The results of our micropalaeontological studies will throw some light on the palaeoenvironments which existed in southern Africa in Neogene times and will give indications on the old biological links which existed there before deterioration of the climate. There are strong indications that these final changes in southern Africa were closely correlated with the climatic evolution of Antarctica and especially with the history of the ice sheet.

The Antarcto-Tertiary Geoflora well known from southern South America, Australia, Tasmania and New Zealand had its counterpart in southern Africa, but the composition of this African assemblage was probably somewhat different indicating the early separation of Africa from the austral lands. The Mesozoic of southern Africa is practially unknown from this point of view and it will be shown in this study that a number of taxa will have existed here, but became extinct probably as a consequence of the considerable northward drift of the continent.

Africa separated from Antarctica between the Upper Jurassic and the Lower Cretaceous and may well have shared the first primitive Angiosperms with the other southern continents when Gondwanaland still formed an undivided unity. The biological connections with South America lasted much longer affording migratory routes until the Senonian (Jardine et al., 1974). These two continents share more of the Mesozoic Gondwana floral and faunal elements than any other two continents (Crackraft, 1975) as they were united for so long under very favourable climatic conditions. After the separation of Africa from the complex of Australia and Antarctica exchange of taxa between these lands could still take place via the link with South America. The ages of these migration routes are very important for the explanation of the contacts of Southern Africa.

After the final separation of the Gondwana sections the biota of the separate continents developed independently and culminated in southern Africa in later times in the evolution of the very rich Cape flora. During the Eocene and Oligocene these continents all shared a temperate to sub-tropical evergreen forest of Podocarpaceae, *Nothofagus*, Proteaceae and ferns. Forests of that age are not yet known from southern Africa, but have been described for central Australia from the Middle Eocene (Kemp, 1976a), from the Palaeocene to Oligocene from the islands of the Ninetyeast Ridge (Kemp & Harris, 1975), from the Late Eocene from the Black Island erratics (Antarctica), from the Late Oligocene in the Ross Sea region (Kemp & Barrett, 1975) and the Late Oligocene of McMurdo Sound (Cranwell, 1969). These forests were present in Antarctica in Late Cretaceous times before it rafted to its present position and indicate considerable warmth for that time (Cranwell, 1969; 1969a). They were probably completely eliminated in Miocene times before the major ice advance (Kemp & Barrett, 1975). Our own studies may give an indication of the influence these climatic events had on the vegetation of the southwestern part of South Africa.

EVALUATION OF THE PALYNOLOGICAL DATA OF THE SW CAPE DEPOSITS

Pollen zones

The stratigraphy of the deposits studied is based on biostratigraphic units or pollen zones representing definite units of vegetational history which can be correlated over wide areas. From the cores and other samples taken by the Geological Survey from deposits at Saldanha, Mamre, Noordhoek on the Cape Peninsula and from various sites on the Cape Flats (Fig. 1) two distinct pollen zones could be distinguished. These are demarcated on the basis of the presence or absence of pollen of extinct species. The oldest unit, characterised by the pollen of species which do not occur there at present, is designated Pollen Zone L and contains rich and diverse pollen types, many of which are typical of Tertiary periods in other continents. The other younger unit, Pollen Zone M, with less diverse pollen types is typified by pollen of vegetation which occurs in these regions at present. On these grounds it is considered to be of Quaternary age.

Tertiary vegetation at the Cape (Pollen Zone L) (Fig. 2)

The pollen spectra of this zone indicate the existence of sub-tropical–tropical and relatively cool–temperate vegetation types at different periods respectively in the southwestern Cape. These vegetation shifts are so marked that further subdivision of this unit is warranted.

The most significant information emerging from the pollen assemblages is that those pollen types indicating sub-tropical–tropical conditions at two different periods (Lii and Lv) are dominated by high percentages of Palmae pollen (33–34%). The associated pollen species suggest an open type of vegetation during these periods. No Palmae occur in these temperate regions at present and it is interesting that the fossil pollen types do not in the least resemble the pollen of any of the present Palmae of the sub-tropical–tropical parts of South Africa. Some of the fossil form species, however, show a resemblance to *Monosulcites waitakiensis* McInryre

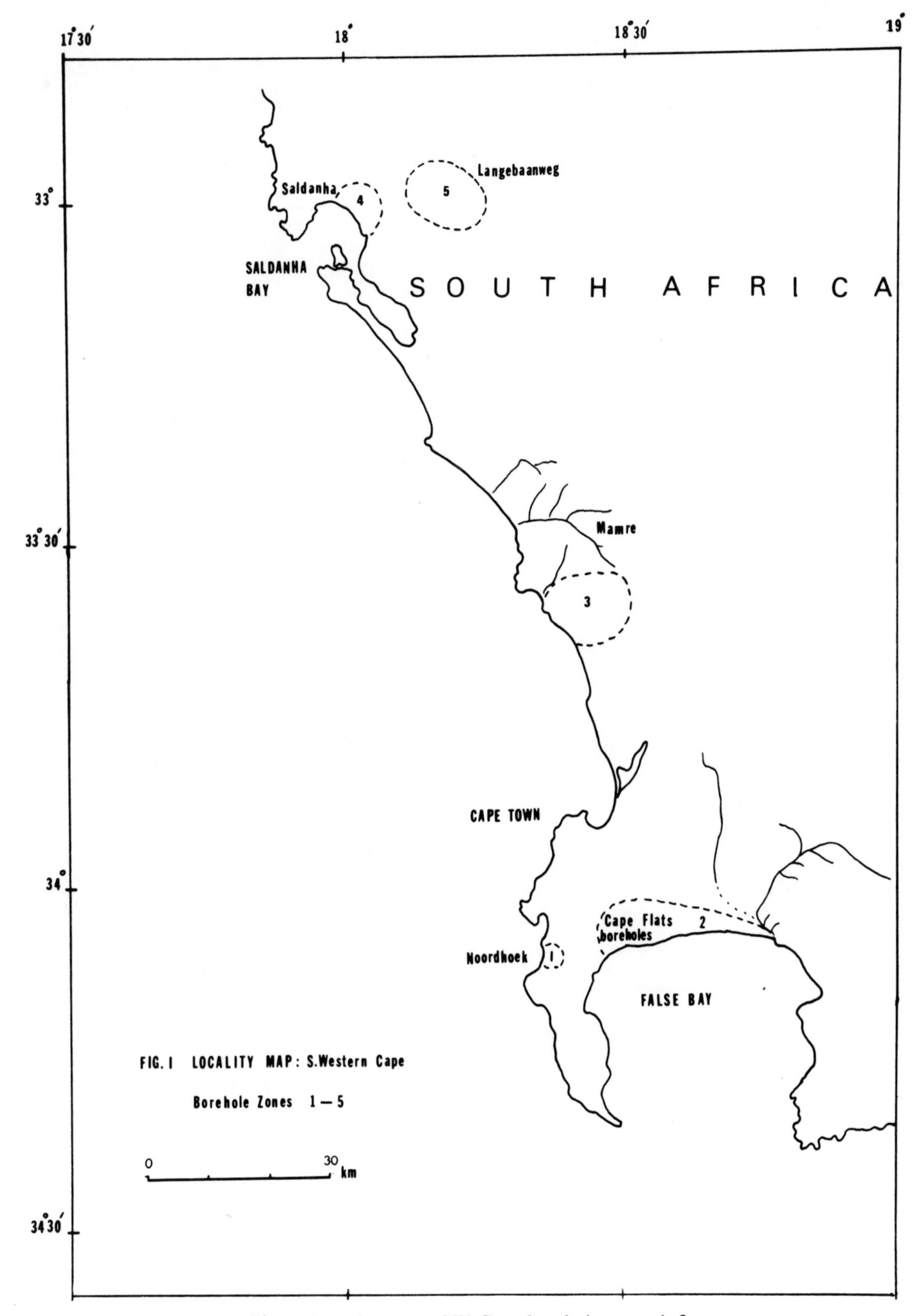

Fig. 1. Location map: SW Cape borehole zones 1–5

POLLEN ZONE	VEGETATION	CLIMATE	SUGGESTED STRATIGRAPHY	
M	Present Macchia	Present	Quaternary	
Lvii	First Strong Development of Macchia	Colder Drier	Tertiary	Pliocene
Lvi	Forest: Coniferae Casuarinaceae Cupaniedites	Cool Wet		Late Miocene
Lv	Palmae	Sub-Tropical Tropical		
Liv	Restionaceous Swamp	Temperate Locally Wet		
Liii	Forest: Coniferae First Compositae	Cool Wet		
Lii	Palmae	Sub-Tropical Tropical		Late Oligocene
Li	Forest: Coniferae	Cool Wet		

Fig. 2. Pollen zones of Late Cainozoic deposits in the SW Cape

(Palmae) from the Miocene of New Zealand and from the Palaeocene of Ninetyeast Ridge. Similar forms also occurred in Australia (Dr E.M. Kemp, pers. comm.).

The pollen spectra suggesting cooler and wetter conditions during periods alternating with the tropical vegetation are dominated by pollen of different Podocarpaceae while pollen of other forest species are also variously represented. The high pollen values of some of these associated species suggest that conifer forest of different composition (Li, Liii and Lvi) existed at different periods in close proximity to some of the sites. Very low pollen percentages of Podocarpaceae and other forest types were recorded during the time of the existence of the sub-tropical–tropical vegetation suggesting the survival of this forest in far distant niches, probably in the mountains of the ancient Cape Folded Belt. In the present contribution it is only possible to discuss the main Tertiary vegetation types recorded.

Assessment of the time-span of the Tertiary vegetation

In the light of the long time ranges of some of the extinct species found at the southwestern Cape and on other continents it is not possible to use them as stratigraphic markers. Moreover, these time ranges need not necessarily be applicable to South Africa. In this respect it may be of significance to mention that Africa was the first continent to leave the supercontinent of Gondwana in the Late Cretaceous, severing links with the other austral lands.

However, in considering the pollen values of the Compositae among the other typical Tertiary types, it may be possible to assign a Late Oligocene or Early Miocene age to these deposits. The first appearance of this family is an accepted global stratigraphic marker for this time span (Leopold, 1969). Compositae pollen is completely absent from the lowest levels while only an occasional grain occurs in the rest of the zone except for the uppermost level. It may be significant in this connection that pollen values of Ericaceae and Gramineae which range onward from the Palaeocene and Eocene respectively are also extremely low in the samples of the deeper levels and are likewise absent from the lowermost Sub Zone Li.

In the uppermost limit of the zone (Sub Zone Lvii) the first high pollen percentages of Compositae, Ericaceae, Gramineae and other typical species of the modern macchia vegetation of the Cape were recorded while the last traces of the pollen of the extinct Tertiary species were still evident. Among the latter are Palmae, *Triorites harrisii* and *Cupaniedites* pollen types. This pollen assemblage marks the occurrence of a very important palaeoclimatological event and the initiation of a complete change of vegetation to the type which exists at the Cape at present. The possible time of this episode will be discussed further on when data from changes in oceanic sediments from the southwest coast and other evidence will be considered. Tentative broad correlations of the vegetation changes indicated in the rest of the sub-zones will also be discussed in the light of this data. In the meantime the time ranges on some other continents of a few of the extinct species from the Cape listed below (Plate 1) can provide provisional clues to the periods of the major Tertiary vegetation changes and final extermination of this flora in these regions:

1. cf. *Triorites harrisii* Couper *(Haloragacidites trioratus* Couper) Casaurinaceae: Late Cretaceous to probably Pliocene in Australia (E.M. Kemp, pers. comm.); Late Cretaceous to Pliocene in New Zealand (Couper, 1960).
2. *Clavatipollenites* cf. *hughesii* Couper: Palaeocene and Oligocene at Ninetyeast Ridge (E.M. Kemp, Pers. comm.)
3. *Echiperiporites* v.d. Hamm & Wijmstra: Late Oligocene at Ninetyeast Ridge (E.M. Kemp, pers. comm.); Oligocene–Miocene, British Guyana (Potonié, 1970).
4. *Sparganiaceaepollenites* Thiergart: Miocene Germany (Jansonius & Hills, 1976), Oligocene at Ninetyeast Ridge (Kemp & Harris, 1975). Middle Eocene through Miocene, Australia (E.M. Kemp, pers. comm.).
5. *Cupaniedites orthoteichus* Cookson & Pike: Palaeocene–Upper Pliocene, New Zealand (Couper, 1960); Eocene–Pliocene, Australia (Cookson & Pike, 1954); Oligocene, Ninetyeast Ridge (Kemp & Harris, 1975).

It is evident that while a number of the above species range in other continents into the Pliocene others, according to present knowledge, apparently range to at least Miocene times or earlier. Among the latter group occurring in the Cape deposits, *Clavatipollenites* cf. *hughesii* and *Echiperiporites* are extinct together with almost all the Palmae after Sub-Zone Lv. On the other hand *Sparganiaceaepollenites* and *Microcachrydites* are not recorded in the deposits younger than Sub-Zone Lvi. It is, therefore, possible that at least Sub-Zones Lv and Lvi could be of Miocene age. The pollen types such as *Triorites harrisii* and *Cupaniedietes orthoteichus* which range into the Pliocene on other continents can on the other hand give possible indications of an approximate time of the final extinction of the Tertiary flora at the Cape.

Quaternary vegetation at the Cape (Zone M) (Fig. 2)

The pollen data, according to the site, show a progressive increase and diverse development of the different types of macchia vegetation which was already initiated at the time of the almost complete extermination of the Tertiary vegetation. Other evidence for Quaternary palaeoenvironments will be alluded to further on.

DISCUSSION OF THE PALAEOENVIRONMENTS OF SOUTHERN AFRICA IN RELATION TO THE ANTARCTIC GLACIATION

The Tertiary period

The above palynological evidence for a change from Tertiary to the more recent type of vegetation at the Cape must be related to palaeoclimatic events of great amplitude. A fundamental climatic change must have been responsible for the extermination of the Palmae together with other Tertiary forms in these regions.

Interesting evidence for the timing of this event can possibly be obtained from correlations with results from recent investigations of ocean sediments off the southwestern coast of South Africa (Siesser, this volume). Data on sediment accumulation rates, diatoms, foraminifera and calcareous nannoplankton with temperature preferences and other geochemical features give proof of warm oceanic conditions in these regions until early Late Miocene times. It is shown that at this time a marked drop in ocean temperature occurred intensifying the cold upwelling of nutrient rich waters which initiated the present cold Benguela Current. This upwelling was according to Siesser's evidence already weakly and spasmodically introduced from Middle to Late Oligocene times. It is suggested that this early proto-Benguela Current did, however, not affect the onshore fauna and vegetation during these times (Tankard & Rogers, in the press; Corvinus & Hendy in the press: quoted by Siesser, in this volume). Evidence for this conclusion comes from the occurrence of fossil jumping hares and antelopes of early Miocene age which points to the existence of savanna vegetation south of Luderitz and also from fossil tragulids which suggest riverine woodland in these regions during this period (Tankard & Rogers). Corvinus and Hendy provide evidence from fossil Miocene fauna for a similar environment at Arrisdrift north of the Orange River. All these manifestations imply that the upwelling was not yet strong enough to initiate the aridification of the present Namib Desert. This assertion is contrary to the original suggestion by van Zinderen Bakker (1975) that the aridification of the Namib Desert is of Oligocene age.

The Tertiary vegetation at the Cape, recorded by the recent palynological studies could substantiate the above oceanographic and offshore evidence for warmer and more humid conditions than at present along the south western coast from the Late Oligocene until the early Late Miocene. The final extermination of the Palmae together with the other forms sensitive to temperature and humidity changes must certainly have been related to the intense cooling of the coastal waters and the final establishment of the cold Benguela Current and its far reaching effects. It is interesting in this connection that Palmae existing in northwestern Europe together with rich flora of Asian affinity became extinct at the end of the Middle Pliocene. This event was clearly related to a cold period in these regions during that time (van der Hammen et al., 1971).

The powerful circum-Antarctic current which was initiated with the final separation of Antarctica and Australia in Middle to Late Eocene times (45–43 MY) (Kennett et al., 1974) was probably finally established by the beginning of the Miocene with the opening of the Drake Passage (Mercer, this volume). This important event brought about fundamental changes in the oceanic and climatic circulation patterns of the world which were intensified by the drop in ocean temperature resulting from the glaciation of Antarctica.

Palaeotemperature assessments of surface water at middle latitudes south of Tasmania and New Zealand have provided important evidence for the cooling which occurred in the Southern Ocean from the Eocene onward (Shackleton & Kennett, 1975). This information shows that this ocean changed from a warm sea into a polar water mass. In the Late Palaeocene and Early Eocene the surface water at these sites was 18–20° C. A dramatic drop in temperature occurred at the Eocene–Oligocene boundary to about 7° C. During the entire Oligocene the temperature regime was rather constant. Forest dominated by *Nothofagus fusca* with Proteaceae, Myrtaceae, Coniferae and ferns could exist in the Ross Sea area until 26 MY ago. This vegetation was eliminated at some time during the Miocene (Kemp & Barrett, 1975). Mercer has further concluded that during the Miocene two warm periods, with temperatures almost as high as in Late Eocene times, developed in Antarctica at 19 MY and 14 MY respectively.

After the East Antarctic ice sheet had accumulated between 14–10 MY a subsequent rise in temperature of 5° C occurred at 8 MY. This period was as warm as the Early Miocene. The rest of the Miocene and Early Pliocene was characterised by low temperatures while the West Antarctic ice sheet was accumulating.

The very marked alternations of subtropical-tropical and cool-wet forest vegetation types must be definite reflections of important climatic events from at least the Late Oligocene to more recent times. Although exact correlations cannot be made as yet it is very reasonable to assume that the above episodes of the climatic evolution of Antarctica affected the Tertiary vegetation in the southwestern Cape. The palm-dominated vegetation is clearly related to warm periods, possibly to certain of the Miocene warmer episodes. The conifer forests, on the other hand, belong to colder and probably wetter phases.

In the light of the more detailed record of climatic events provided by Mercer for Antarctica it is possible that the final elimination of the Tertiary flora at the Cape (Sub Zone Lvii) coincided with the final drop in temperature in Late Miocene and Early Pliocene times and not in the early Late Miocene as was considered earlier in the discussion. Nevertheless, the final establishment of the Benguela Current and the origin of the entirely new climatic system with dry summers and wet winters in these regions must have had a direct effect on the vegetation at the Cape.

These new conditions must have favoured the strong development of Macchia vegetation to the type which occurs there at present. The development of this flora at the expense of the Tertiary humid conifer forest (Sub Zone Lvi) could support the contention of Siesser (this volume) that the change in climate initiated progressive aridification of South West Africa, in particular of the Namib Desert. This phenomenon set in from the Late Miocene after the initiation of the cold Benguela Current. The palynological evidence could also further support the conclusions of Tankard & Rogers (in preparation: quoted by Siesser) that aridification of the subcontinent as a whole dates from the Pliocene.

In this connection the evolution of the Bovideae is of special interest as evidence for the progressive aridification of the continent (Vrba a, in press; b, in press). The more primitive representatives of the family belonging to the *Boodontia* which are characteristic of moist wooded country have been found in Miocene deposits in South West Africa (Gentry in: Vrba, a, in press). On the other hand the *Alcelaphini*, mostly wholly African, evolved about 5 MY BP and are highly adapted to arid country. Together with the *Antelopini* they are considered to be indicators for the Plio–Pleistocene opening up of the vegetation of the continent. Further evidence for aridification over the whole continent has been alluded to by Axelrod & Raven (in press) who mention fossil evidence in East and North Africa for the spread of savannas, thornbush and grasslands at the expense of forests since Late Miocene times.

It is interesting that similar independent correlations of the climatic evolution of the southeast Indian ocean and onshore Tertiary vegetation and climatic changes in Australia as well as Antarctica have also been made (Kemp, 1976b; Galloway & Kemp, in press). These authors likewise come to the conclusion that the final drop in temperature in the latest Miocene had a dramatic effect on the Australian climate which became more arid.

In connection with the above palynological and oceanographic studies it is significant to refer to the research of Zinsmeister (Mercer, this volume) done along the west coast of southern South America. He has indicated a molluscan faunal change of species typical of subtropical waters to cool-temperate types during the latest Miocene or earliest Pliocene. The change has been related to the development of the West Antarctic ice sheet and the intensification of the Humboldt Current.

Phytogeographical considerations of some of the Tertiary flora

The recent palynological studies of Tertiary vegetation at the Cape give an opportunity to include the South African subcontinent for the first time in discussions on some problems of Tertiary palaeophytogeography. The occurrence of certain microfossils in these regions may be of considerable botanical interest and warrants a brief digression in this field.

The oldest pollen assemblages show that the vegetation had much in common with that of other continents in the ancient past. Some of the floras shared with these continents are probably manifestations of pre-drift migratory routes (Axelrod & Raven, in press; Brundin, 1975; Cracraft, 1975; Moore, 1973) while others could provide evidence of land connections existing between some of the southern continents until Palaeocene times.

The occurrence in South Africa of the form-species *Triorites harrisii (Haloragacidites trioratus)* of the Casuarinaceae is of particular interest. According to E.M. Kemp (pers. comm.) the pollen strongly resembles that of *Casaurina* which fits the form-species *Triorites harrisii.* In Australia this pollen type ranges from the Late Cretaceous to the Pliocene and is considered to represent one of the most abundant and widely distributed of the Tertiary floras of that country (Cookson & Pike, 1954). It is also typical of the New Zealand Tertiary vegetation (Couper, 1960) and the early Tertiary of Ninetyeast Ridge (Kemp & Harris, 1975). These islands were emergent from possibly latest Cretaceous to Late Oligocene. *Casuarina* is also present in the early Palaeocene Deccan Trap Flora of India which was still situated at that time at 10° S against the now largely submerged Mascarene Plateau. Interconnections still existed during this period between this landmass, Madagascar and Africa. It is interesting in this connection that it was the Malagasy–Mascarene subcontinent which served as a migratory route for the Cretaceous dinosaurs common to Africa–

India–Madagascar and other regions (Axelrod & Raven, in the press).

The pollen of *Triorites harrisii*, however, is indistinguishable from that of *Myrica* and the problem is that while *Myrica* occurs at present along the eastern disjunct chain of mountains of Africa as far south as the Cape Peninsula, no Casuarinas are at present indigenous to South Africa. The question of whether Casuarinaceae are indigenous elsewhere in Africa is still equivocal. *C. equisetifolia* appears to be native to East Africa as a strand plant but there is no proof of it being indigeneous to these regions (J.P.M. Brenan (Kew) and J.B. Gillett (Nairobi), personal communication).

The possibility that this Tertiary pollen type belongs to *Myrica* may perhaps be ruled out on the grounds that *Myrica* is a boreal genus which could possibly only have reached South Africa from the north via the eastern chain of mountains. Most of the highlands and mountains of East Africa, however, only came into existence from the Miocene to Pliocene and even in mid-Pleistocene times as a result of faulting, downwarping and volcanism. It is possible, therefore, that the arrival of *Myrica* in South Africa was affected during Late Quaternary times and that it is a relatively recent component of the southern vegetation.

On the other hand *Casuarina* has a very interesting past distribution pattern and is at present a Southern Hemisphere genus particularly of Australia and New Caledonia but also occurs in southeast-tropical Asia, the Mascarene and Pacific Islands. Of particular interest in this connection is the pollen found in the Upper Cretaceous dysodil deposits at Banke in Namaqualand, which Kirchheimer (1932) assigned to *Myrica*. It is more likely, however, that this sporomorph belongs to *Triorites harrisii* and could give proof to the ancient occurrence of Casuarinaceae in South Africa as in Australia. *Triorites harrisii* is also reported from the Early Tertiary lignites of Knysna (Thiergart, Frantz & Raukopf, 1963).

The Late Cretaceous–Late Tertiary record of *Triorites harrisii* in South Africa testifies to an ancient distribution pattern either of the pre-drift period or of the time when migration routes between Africa and Australasia existed across the Indian Ocean via Ninetyeast Ridge–India–Madagascar until mid-Cretaceous times. It is therefore suggested that *Casuarina* rather than *Myrica* existed in South Africa during the Tertiary. *Triorites* pollen was a particularly abundant component in the cool wet temperate forest (Sub-Zone Lvi) which became eliminated subsequent to the extinction of the Palmae. According to Dr E.M. Kemp (personal communication) the Casuarinaceae must have belonged to an early Tertiary rainforest community judging from the associated fossil pollen. At present *Casuarina* does not occur in rainforest but has become more adapted to drier habitats in that continent.

The existence of Palmae at the southernmost tip of Africa during the Tertiary is also of particular significance. The different form species recovered from the deposits show great affinities to Tertiary Palmae in other continents but not to present Palmae of the sub-tropical–tropical regions of South Africa. This phenomenon and the fact that palms do not occur in the Cape at present, lend further support to the contention that this family had a much wider distribution in the past.

An interesting analysis by Moore (1973) of the present disjunct distribution patterns, relationships, diversity and centres of primitiveness as well as fossil evidence of Palmae has provided important clues as to their centre of origin and ancient distribution patterns. From this information he concluded that West Gondwanaland (South America and Africa) could have been the only likely centre of origin of this group during the Jurassic. However, as far as the time of the origin of the Palmae is concerned, a re-evaluation of so-called pre-Cretaceous angiosperm fossil records has shown that there is no unequivocal evidence pointing to a pre-Cretaceous origin for the Angiosperms (Wolfe et al., 1975). A possible Jurassic age for the origin of Palmae is, therefore, unacceptable.

Fossil evidence, according to Moore, points to two possible migratory routes. An austral dispersal is suggested from palm fossils of Cretaceous and Oligocene age from McMurdo Sound, which could explain the evidence for fossil Palmae in New Zealand and the occurrence of primitive groups in New Caledonia, New Guinea and the Indo-Pacific. Another distribution pattern through Africa to Laurasia and eastwards along the warm Tethys to Asia and Australasia is also indicated. An interesting disjunct pattern of ceroxyloid and chamaedoreoid palms in America, Madagascar and the Mascarenes, points to a massive extinction of these and other palms in Africa which is, according to Moore, feasible in the light of the Pleistocene history of this continent and the ecological demands of this group. They are mostly ill adapted to truly xeric conditions even in deserts and are often important components of the hydrosere.

Evidence for the probable origin of the Palmae in West Gondwanaland could point to a very ancient establishment of this group in South Africa.

Another interesting pollen type occurring in the Cape deposits is that of *Clavatipollenites* cf. *hughesii*. The form species *C. hughesii* is at present accepted to be the oldest definite angiosperm and was first described from rocks from England which are presumably older than the Barremian (Early Cretaceous) (Couper, 1958; Kemp, 1968 in: Wolfe et al., 1975). E.M. Kemp (pers. comm.) has reported *C.* cf.

hughesii from the Palaeocene and probable Late Oligocene of Ninetyeast Ridge, while in Australia it occurs in the Cretaceous, Palaeocene and Eocene.

The occurrence of the pollen of this most ancient angiosperm also in South Africa could, as in the case of the Palmae point to a distribution pattern which existed prior to the separation of Africa from Antarctica between the Upper Jurassic and Lower Cretaceous.

The Quaternary period

Since the origin of the circum-Antarctic current and the subsequent growth of the Antarctic ice sheet to its full extent by latest Miocene and early Pliocene times the climatic system of South Africa changed completely. As a consequence of the substantial cooling the Southern Ocean was divided into separate water masses, the Antarctic and Subtropical Convergences formed and the proto-Benguela Current was initiated. The effect of these changes was that the present day climatic system originated. This system is dominated by the high pressure belt in the middle latitudes. Summer rainfall of tropical origin can only penetrate the sub-continent when the anticyclone, which prevails above throughout the year at 2000 m altitude, is weakened. The south-western and southern tip of the continent receives winter rainfall from cyclones originating over the South Atlantic. This system is subjected to seasonal meridional displacement of about 4° latitude (Schulze, 1972). The consequence is that the Cape region does not receive rain in summer.

The stable anticyclone situated above the Atlantic Ocean at about 30° S and the Benguela Current which generates considerable cold upwelling are both responsible for the aridity of the Namib Desert and the dry climate of the interior of the sub-continent.

The Quaternary glaciations superimposed on the glaciation of Antarctica have had pronounced effects on the climatic system and consequently on the biogeography of the sub-continent. The last four Quaternary glaciations have been more severe than any of the previous cold periods, which have been recorded since the Pliocene (van der Hammen et al., 1971). Van Zinderen Bakker (1977) has given a review of the evidence for palaeoenvironments during these glacial and interglacial periods in southern Africa.

During glacial episodes the pressure gradient between the Antarctic Convergence and the equator was steepened with the result that the oceanic and atmospheric circulation was activiated. Cold polar air could penetrate the interior up to about 24° S latitude and the cold Benguela Current had more energy and moved further north. In the southern parts winters must have been very cold, wet and windy, while the northern parts were less cold and received limited summer rainfall. During warmer periods the climatic system moved south again.

MANIFESTATIONS OF LATE QUATERNARY CLIMATES

1. Changes in the oceanic environment

Evidence for changes in the oceans surrounding the subcontinent is very important for an evaluation of the Quaternary palaeoenvironments. The hypothesis postulating a northward displacement of the Antarctic Polar Front at the time of the last glaciation of Würmian age of Marion Island (46° 50′ S, 37° 40′ E) (van Zinderen Bakker, 1969) has been confirmed by palaeotemperature assessments of 34 ocean sediment cores from the sub-Antarctic region (Hays et al., 1976). It has been shown that in the western south Atlantic and northward shift was of the order of about 10° latitude while in the South Atlantic and Indian Oceans it was about 6° latitude. South of South Africa this Polar Front was almost stationary as a result of the east–west oceanic barrier ridge. The Subtropical Convergence remained almost stationary south of South Africa so that the sub-Antarctic region was narrowed especially during the severe winters when pack-ice reached almost to 50°. The summers were not so very different from present day conditions (Hays et al., 1976) but pack-ice occurred almost up to the latitude of 55°.

Confirmation of the northward shift and extension of the Benguela Current to as far as Gabon comes from the studies of Bornhold (1973) on the warm Angola basin. From foraminiferal studies by Vincent (1972) can be inferred that before 10000 BP the water temperature in the Moçambique channel at 25° S was 5° C lower than at present. McIntyre (1977) points out that at 18000 BP, as a consequence of the lowering in temperature in the Indian Ocean south of 10° S the Agulhas Current practically ceased to exist.

2. The South Cape coastal region

Reconstruction of the palaeoenvironments of this region is based on palaeontological, archaeological, palynological and geological evidence. The analyses of the fossil faunas found in many caves which occur in this region are supporting the conjecture that the coastal plain was covered with forest and bush during the last interglacial (Klein 1972; 1974; 1976). The fossil faunas of grazers from caves between Cape Town and Port Elizabeth suggest that the open grassland had replaced the forest during the glacial maxima (ibid.).

According to archaeological data the caves at

many sites on the coast and further inland were not occupied by prehistoric man for thousands of years (op. cit.). This hiatus between the habitation of MSA and LSA people is puzzling and could have been caused by the great distance of the caves from the principle food source, the sea, as a consequence of the substantial drop in sea-level. It is also feasible that prehistoric man had to leave the caves because of the adverse climate.

Some important palynological studies carried out in this region (Martin, 1968; Schalke, 1973) shed some light on the evolution of the vegetation in the Cape coastal region. Schalke (1973) concludes that *Podocarpus* forest existed in the southwest Cape during the last warm Kalambo interstadial and the cold last glacial maximum. The pollen percentages used as evidence are, however, too low (Coetzee, 1967) and macchia elements should be separated from the forest assemblages. Good proof for *Podocarpus* forest with pollen percentages of up to 40% is given for the time between 45 000 and 40 500 BP which coincides with the warmer Moershoofd interstadial of Europe and not with a cold period as suggested by Schalke.

The vegetation changes during a glacial- interglacial cycle are not yet well understood but fossil evidence especially of the studies by Martin (1968) could indicate the following succession. During the cold wet and windy last glacial maximum open grassland covered the coastal plain. When the climate ameliorated round 12 000 BP macchia vegetation and later forest with browsing animals replaced the open community. The forest expanded at the onset of the postglacial Hypsithermal period especially in areas where the optimum rainfall of over 750 mm per annum and a frost free climate occurred. When during the Climatic Optimum the climate was warmer than today the forest gave way to karroid vegetation. This stage coincides with a marine transgression which lasted from 6 870 ± 160 until 1 905 ± 60 BP (Martin, 1968). The forest returned when the climate became cooler and moister. The earlier hypothesis (van Zinderen Bakker, 1963) that forest spread during cold wet periods is now refuted on the basis of new evidence.

This explanation is not compatible with the views of Sarnthein & Hahn (1977) that during the Climatic Optimum summer rain spread as far south as 34° S in South Africa. No conclusive proof is yet available for this hypothesis.

3. *The Namib Desert*

On biogeographic, geomorphological and archaeological evidence van Zinderen Bakker concludes that during glacial maxima desert conditions must have shifted northwards along the Angolan coast. On the other hand, during interglacial periods, these arid conditions spread south of 17° S. It has now been concluded that the Namib Desert in its present geographical position is not as old as had previously been assumed (see discussion above). The mechanism by which the coastal desert originated had been well explained but new information on the age of the Antarctic ice sheet and the cooling of the ocean water have shown that the Namib Desert will probably be of terminal Miocene or Early Pliocene origin (Siesser; Mercer; and Kennett, this volume). This view is supported by our palynological data. The rich endemic fauna and flora testify to the antiquity of the core of the desert. It has also been suggested that during interglacials rains could influence the flow of present dry rivers from the catchment area through the desert as far as the ocean (Seely & Sandelowsky, 1974). During glacial periods winter rains could apparently penetrate the southern Namib as far as 24° S. (v. Zinderen Bakker, 1977).

4. *The interior of the southern part of Africa*

In broad terms it consists of a plateau separated from the coastal areas by an escarpment running round the sub-continent but which is most clearly defined in the east in the Natal Drakensberg where the highest point reaches 3 484 m. The Kalahari basin with lowlands of altitudes of 853–844 m forms a major part of the plateau while the South African Highveld, bounded on the east and south by the Great Escarpment, rises from 1 219 to 1 829 m.

Different types of vegetation and climates characterise various regions of the plateau which were greatly modified during glacial and interglacial phases. Van Zinderen Bakker postulates that during Late Pleistocene and Holocene periods the South African Highveld was treeless except for rocky outcrops and riverbanks. During glacial maxima alpine grassland from the eastern highlands could have invaded regions above 1 200 m which experienced very cold wet winters. At the same time the temperate and subtropical grasslands which occur at present on the plateau of the Orange Free State were driven to lower altitudes with less frost.

That cold wet climates existed in the interior during the maximum Würm glaciation is indicated by studies of Butzer et al. (1973a) at Alexandersfontein pan near Kimberley. A palaeolake existed there at 16 010 ± 185 BP which had a surface of 44 km² and was 17–19 m deep. A drop in temperature of 6° was calculated for this period. Geomorphological studies of the alluvial terraces of the Vaal River Basin (Butzer, 1972; Butzer et al., 1973b) indicate good grass cover during wet periods at a time of the coldest part of the Upper Pleniglacial. Further support for a substantial drop in temperature of 8–9° C during this period in the interior comes from the isotopic studies of speleothems of the Wolkberg cave in the Transvaal (Talma et al., 1974).

Palynological studies at Florisbad in the O.F.S. (van Zinderen Bakker, 1957) and at Aliwal North in the northern Cape (Coetzee, 1967) have shown that during cold wet periods of the Upper Pleniglacial and Late Glacial grassland probably of the Alpine type, spread into these regions replacing semi-arid karoo vegetation which existed there during very warm periods.

Cold climates in the high Drakensberg during the last glacial maximum can also be inferred from the age of the organic peaty deposits which occur above the altitude of 3 000 m in these mountains. These so-called sponges of the Orange River only started forming at the beginning of the Holocene (van Zinderen Bakker & Werger, 1974). Studies of the 'periglacial' features in these high mountains during glacial maxima have been reviewed by Butzer, 1973. A decrease in temperature of 5–9° C during the last glacial maximum must have depressed the tree line by 1 000 m (Harper, 1969; van Zinderen Bakker, 1977).

Interesting indications on how inhospitable the interior could have been between 9 500 and 4 600 BP has been provided by J. Deacon (1974). A study of 223 archaeological sites could give evidence of an occupation hiatus covering the Wilton and Smithfield cultures. The extreme aridity could have made it very difficult for hunter-gatherers to live on the semi-arid plateau where surface water was lacking through most of the year.

These surmises on semi-arid climate in the interior of South Africa during warm interglacials and the warm Climatic Optimum of Holocene age are supported by the palynological results obtained at Florisbad (van Zinderen Bakker, 1957) and at Aliwal North (Coetzee, 1967). Geomorphological studies by K.W. Butzer and palynological analyses by L. Scott in different parts may prove or disprove the former findings. The statement by Sarnthein & Hahn (1977) that during the Holocene Climatic Optimum at 6 000 BP 'monsoonal summer rains' reached as far south as 34° S in South Africa has not been confirmed by palaeobotanical evidence. If these conditions prevailed abundant tree growth should have spread into the present open grassveld.

The enormous semi-arid Kalahari Basin in the centre of southern Africa has also provided indications of Late Quaternary climatic changes. Evidence for redistribution and northward spread of the Kalahari sands during the Upper Pleniglacial show that a windy and arid climate existed here during glacial periods when the Benguela Current spread northwards. Laterite formation during the Hypsithermal period shows that the climate had rainy seasons during warm phases (Clark, 1963; 1968; van Zinderen Bakker, 1975). Vast palaeopans or fossil playas are also manifestations of former wetter climates in the Kalahari Basin (Grove, 1969).

CONCLUSIONS

Although much work still has to be done on deposits from more sites in the southwest Cape, the recent palynological data have thrown new light especially on the Tertiary palaeoenvironments of these regions. Marked subtropical–tropical vegetation dominated by Palmae alternated with cool–wet conifer forests. These results have also provided an insight, for the first time, into ancient distribution patterns which could have affected the southernmost tip of Africa before this continent separated from the other austral lands of Gondwana in the Late Jurassic.

Considerations of recent oceanographic data from the southwestern coast of South Africa (Siesser, this volume) could suggest that the extermination of certain elements of the Tertiary vegetation at the Cape possibly coincided with the origin of the cold Benguela Current in the early Late Miocene. Detailed evaluations of the data on the evolution of the Antarctic ice sheet by Mercer (this volume), however, could point to the later elimination of this vegetation during the Late Miocene to Early Pliocene. Temperature variations in the Antarctic regions (Shackleton & Kennett, 1975) during the Miocene could have had an effect on the profound changes of vegetation reflected by the palynological results.

The origin of the Benguela Current is directly related to the full glaciation of the Antarctic ice sheet which also resulted in an entirely new climatic system which had aridifying effects especially in the western half of the sub-continent.

Pollen evidence from the Cape can support other data which indicate a progressive aridification of the sub-continent from Pliocene times onward. Different lines of investigation on the palaeoenvironments of the Quaternary have also shown that the glacial events in the Northern and Southern Hemisphere profoundly affected the face of the sub-continent. The climatic changes forced the biota to extensive migrations while new habitats induced diversification of the flora and fauna.

ACKNOWLEDGEMENTS

Special thanks are due to the Geological Survey, Pretoria for initiating this project, for the generous financial support and for permission to publish the results. I am also very much indebted to Dr J.N.

Theron from the Geological Survey Cape Town for the care he took in supplying the valuable cores and samples. My sincere appreciation is furthermore extended to the following: Dr E.M. Kemp of the Bureau of Mineral Resources in Canberra, Australia who kindly confirmed the identification of the Palmae and other Tertiary pollen types and for her very useful comments; Dr M.A. Sowunmi, palynologist at the University of Ibadan, Nigeria who also confirmed the presence of Palmae pollen in the Tertiary deposits; Professor E.M. van Zinderen Bakker and Mr L. Scott who willingly assisted me with the microscopic analyses of some of the deposits.

REFERENCES

Axelrod, D.I. & Raven, P.H. 1978. Late Cretaceous and Tertiary vegetation history of Africa. In: M. J.A. Werger, *Biogeography and ecology of Southern Africa.* Junk, The Hague.

Bornhold, B.D. 1973. Late Quaternary sedimentation in the eastern Angola Basin, Woodshole Oceanogr. Inst. WHO 1: 73-80. Unpublished manuscript.

Brenan, J.P.M. & Greenway, F.L.S. 1949. *Check lists of the forest trees and shrubs of the British Empire* 5: Tanganyika Territory. Oxford Holywell Press. Part II, 194 pp.

Brundin, L. 1975. Circum-Antarctic distribution patterns and continental drift. *Mém. Mus. Nat. d'Histoire Naturelle. Sér. A. Zoologie* 68: 19-27.

Butzer, K.W. 1972. Fine alluvial fills in the Orange and Vaal Basins. *Palaeoecology of Africa* 6, Balkema, Cape Town: 149-151.

Butzer, K.W. 1973. Pleistocene "periglacial" phenomena in southern Africa. *Boreas* 2(1): 1-11.

Butzer, K.W., Fock, G.J., Stuckenrath, R. & Zilch, A. 1973a. Palaeohydrology of Late Pleistocene Lake Alexandersfontein, Kimberley, South Africa. *Nature* 243(5406): 328-330.

Butzer, K.W., Helgren, D.H., Fock, G.J. & Stuckenrath, R. 1973b. Alluvial terraces of the Lower Vaal River, South Africa: a re-appraisal and re-investigation. *J. Geology* 81: 341-362.

Clark, J. Desmond. 1963. Prehistoric cultures in north east Angola and their significance in tropical Africa. *Museo do Dundo, Diameng, Pub. Cult.* 62(1): 225.

Clark, J. Desmond. 1968. Further palaeo-anthropological studies in Northern Lunda, *ibid* 78: 196.

Coetzee, J.A. 1967. Pollen analytical studies in East and Southern Africa. *Palaeoecology of Africa* 3, Balkema, Cape Town: 1-146.

Cookson, I.C. & Pike, K.M. 1954. Some dicotyledonous pollen types from Cainozoic deposits in the Australian Region. *Austr. J. Bot.* 2(1): 197-219.

Couper, R.A. 1958. British Mesozoic microspores and pollen grains. *Palaeontographica Abt. B. Paläophytol.* 103: 75-179.

Couper, R.A. 1960. New Zealand Mesozoic and Cainozoic plant microfossils. *N.Z. Geol. Survey Palaeont. Bull.* 32: 87 pp.

Cracraft, J. 1975. Mesozoic dispersal of terrestrial faunas around the southern end of the world. *Mém. Mus. Nat. d'Histoire Naturelle Sér. A. Zoologie* 68: 29-52.

Cranwell, L.M. 1969. Antarctic and Circum-Antarctic palynological contributions. *Antarctic J. U.S.* 4: 197-198.

Cranwell, L.M. 1969a. Palynological intimations of some pre-Oligocene Antarctic climates. *Palaeoecology of Africa* 5, Balkema, Cape Town: 1-7.

Dale, I.R. & Greenway, P.J. 1961. *Kenya trees and shrubs.* Glasgow, Univ. Press, 654 pp.

Deason, J. 1974. Patterning in the radiocarbon dates for the Wilton/Smithfield complex in Southern Africa. *S.A. Archaeol. Bull.* 29(113 + 114): 3-18.

Drewrey, D.J. 1975. Initiation and growth of the East Antarctic ice sheet. *J. Geol. Soc. Lond.* 131: 255-273.

Galloway, R.W. & Kemp, E.M. (in press). Late Cainozoic environments in Australia. In: A. Keast (ed.), *Ecological biogeography in Australia*, Junk, The Hague.

Grove, A.T. 1969. Landforms and climatic change in the Kalahari and Ngamiland. *Geographical J.* 135(2).

Hammen, T. van der, Wijmstra, T.A. & Zagwijn, W.H. 1971. The floral record of the Late Cenozoic of Europe. In: K.K. Turekian (ed.), *The Late Cenozoic glacial ages*, New Haven, Yale University press; 391-424.

Harper, G. 1969. Periglacial evidence in Southern Africa during the Pleistocene epoch. *Palaeoecology of Africa* 4, Balkema, Cape Town: 71-101.

Hays, J.D., Lozano, J.A. & Shackleton, N. 1976. Reconstruction of the Atlantic and Western Indian Ocean sectors of the 18 000 BP Antarctic Ocean. *Geol. Soc. Am. Mem.* 145: 337-372.

Jansonius, J. & Hills, L.V. 1976. *Genera file of fossil spores.* Spec Publ., Dept. Geol, Univ. Calgary.

Jardiné, S., Biens, P. & Doerenkamp, A. 1974. *Dicheiropollis etruscus*, un pollen caractéristique du Crétacé inférieur Afro-Sudaméricain. *Sci. Géol. Bull.* 27(1-2): 87-100.

Kemp, E.M. 1968. Probable angiosperm pollen from British Barremian to Albian strata. *Palaeontology* 11: 421-434.

Kemp, E.M. 1976a. Early Tertiary pollen from Napperby, Central Australia. *BMRJ Aust. Geol. Geophys.* 1: 109-114.

Kemp, E.M. 1976b. Tertiary climatic change in the southern Indian Ocean Region and its influence on continental vegetation. *Abstr. 25th Intern. Geol. Congress.*

Kemp, E.M. & Barrett, P.J. 1975. Antarctic glaciation and early Tertiary vegetation. *Nature* 258 (5535): 507–508.

Kemp, E.M. & Harris, W.K. 1975. The vegetation of Tertiary islands on the Ninetyeast Ridge. *Nature* 258 (5533): 303–307.

Kennett, J.P. et al. 1974. Development of the Circum-Antarctic Current. *Science* 186 (4159): 144–147.

Kershaw, A.P. 1970. Pollen morphological variation within the Casuarinaceae *Pollen et Spores* 12(2): 145–161.

Klein, R.G. 1972. The Late-Quaternary mammalian fauna of Nelson Bay Cave (Cape Province, South Africa): Its implications for megafaunal extinctions and environmental and cultural change. *Quat. Res.* 2(2): 135–142.

Klein, R.G. 1974. Environment and subsistence of prehistoric man in the southern Cape Province, South Africa. *World Archaeol.* 5(3): 249–284.

Klein, R.G. 1976. The mammalian fauna of the Klasies River Mouth sites, Southern Cape Province, South Africa. *S.A. Archaeol. Bull.* 31: 75–98.

Leopold, E. 1969. Late Cenozoic Palynology. In: R.H. Tschudy & R.A. Scott (eds.), *Aspects of Palynology*, Wiley, N.Y.: 377–438.

Martin, A.R.H. 1968. Pollen analysis of Groenvlei Lake sediments, Knysna (South Africa). *Rev. Palaeobot. Palynol.* 7: 107–144.

McIntyre, A. 1977. Winter and summer reconstructions of the ocean's surface 18 000 years ago by CLIMAP. *Abstracts X INQUA Congress, Birmingham*: 280.

Moore, Jr, H.E. 1973. Palms in the tropical forest ecosystems of Africa and South America. In: B.J. Heggers, E.S. Ayensu & W.D. Duckworth (eds.), *Tropical forest ecosystems in Africa and South America: A comparative review*, Smithsonian Inst. Press., Washington: 63–88.

Potonie, R. 1970. Synopsis der Gattungen der sporae dispersae V. *Beih. geol. Jb.* 87: 222 pp.

Sarnthein, M. & Hahn, D.G. 1977. Active sand deserts: Present, 18 000 BP and 6 000 BP (Reconstructed and simulated aridity). *Abstracts X INQUA Congress, Birmingham*: 398.

Schalke, H.J.W.G. 1973. The Upper Quaternary of the Cape Flats area (Cape Province, South Africa). *Scripta Geol.* 15: 1–57.

Schulze, B.R. 1972. South Africa. In: J.F. Griffiths (ed.), *World survey of climatology* 10, Elsevier, Amsterdam: 501–586.

Seely, M.K. & Sandelowsky, B.H. 1974. Dating the regression of a river's end point. *S.A. Archaeol. Bull. Goodwin Ser.* 2: 61–64.

Shackleton, N.J. & Kennett, J.P. 1975. Palaeotemperature history of the Cenozoic and the initiation of Antarctic glaciation. *Initial Reports of the DSDP* 29: 743–755.

Talma, A.S., Vogel, J.C. & Partridge, T.C. 1974. Isotopic contents of some Transvaal speleothems and their palaeoclimatic significance. *S.A. J. Sci.* 70: 135–140.

Thiergart, F., Frantz, U. & Raukopf, K. 1963. Palynologische Untersuchungen von Tertiärkohlen und einer Oberflächenprobe nahe Knysna, Südafrika. *Advancing Frontiers Plant Sciences* 4: 151–178.

Vincent, E. 1972. Climatic change at the Pleistocene–Holocene boundary in the southwestern Indian Ocean. *Palaeoecology of Africa* 6, Balkema, Cape Town: 45–54.

Vrba, E.S. (in press, a). The significance of Bovid remains as indicators of environment and predation patterns. Burgwartenstein Symposium 69, 1976.

Vrba, E.S. (in press, b). Phylogenetic analysis and classification of fossil and recent *Alcelaphini* (family Bovidae, Mammalia).

Wolfe, J.A., Doyle, J.A. & Page, V.M. 1975. The bases of angiosperm phylogeny: Palaeobotany. *Anns. Miss. Bot. Gard.* 62(3): 801–824.

Zinderen Bakker Sr, E.M. van. 1957. A pollen analytical investigation of the Florisbad deposits (South Africa). *Proc. 3rd Pan-Afr. Congr. Prehist.*, Livingstone 1955: 56–67.

Zinderen Bakker Sr, E.M. van. 1963. Palaeobotanical studies. In: Symposium on early man and his environments in southern Africa. *S.A. J. Sci.* 59: 332–340.

Zinderen Bakker Sr, E.M. van. 1969. Quaternary pollen analytical studies in the southern hemisphere with special reference to the sub-Antarctic. *Palaeoecology of Africa* 5, Balkema, Cape Town: 175–212.

Zinderen Bakker Sr, E.M. van. 1975. The origin and palaeoenvironment of the Namib Desert biome. *J. Biogeography* 2: 65–73.

Zinderen Bakker Sr, E.M. van. 1977. The evolution of the Late-Quaternary palaeoclimates of Southern Africa. *Palaeoecology of Africa* 9, Balkema, Cape Town: 160–202.

Zinderen Bakker Sr, E.M. van & Werger, M.J.A. 1974. Environment, vegetation and phytogeography of the high altitude bogs of Lesotho. *Vegetatio* 29: 37–49.

* 11 *

Late–Mesozoic and Tertiary palaeoenvironments of the Sahara region

E.M. van Zinderen Bakker Sr
Institute for Environmental Sciences, University of the O.F.S., Bloemfontein, South Africa

Manuscript received 15th October 1977

CONTENTS

ABSTRACT

The present Sahara region has undergone dramatic changes in its environment which can be correlated with the palaeogeographic position of the African continent and the continuous variations in world climate. Mainly micropalaeontological studies have shown that the humid tropical climate of the Late Cretaceous became progressively drier. Savanna and woodlands replaced the tropical forest in Early Tertiary times. This process is the consequence of the northward shift of the continent, the diminishing of the oceanic influence and the cooling of the Earth since the Early Tertiary.

So far no well dated close correlations are available between the lowering in temperature in the Antarctic and the gradual desiccation of North Africa but the parallel trend of these events cannot be denied. The abrupt cooling which has been described from the Southern Ocean at the Eocene–Oligocene bundary must have affected the climate of the southern continents and even aggravated the aridification of the North African region. When in the late-Miocene the *Nothofagus* flora died out in Antarctica as a consequence of the severe polar climate, desert conditions developed in the Sahara. The complete desiccation of the Mediterranean Sea between 6.2 and 5.3 MY ago was probably closely correlated with the glacial evolution of Antarctica. This geological catastrophe intensified the arid conditions in the whole of Northern Africa.

Little is known about the Pliocene environments of the Sahara region, but an impressive accumulation of data has in recent times become available on the climatic changes of the desert region during the Quaternary. It is not yet possible to correlate these changes with events in the Antarctic region. The alternation of more humid and hyperarid periods inferred from geomorphological and palaeontological studies is strongly influenced by the glacial history of the Northern Nemisphere.

INTRODUCTION

The African continent has undergone many drastic changes since it emerged as a separate entity from Gondwanaland. The evolution of the old continent is marked by drifting, tectonic movements, internal rifting and large scale volcanism which changed the palaeogeography and topography considerably. The mainly northward drift of the continent affected the climatic setting and the biological contacts. The surrounding oceans enlarged gradually in Late Mesozoic and Cainozoic times and brought Africa under the influence of tropical and polar climatic systems. The warm tropical Tethyan Ocean in the north was replaced at a later stage, as a consequence of plate tectonics by the much smaller Mediterranean which dried up completely at the end of the Miocene.

The separation of Africa from the austral lands of Gondwana from the Late Jurassic onward isolated the continent from the biota of these southern continents after the development of the first primitive Angiosperms. The African continent could, however, for a considerable time even until the Palaeocene, exchange floral and faunal elements to the east and the west. After its final biological separation from all the other continents the African biota evolved in complete isolation during Neogene times until the northward drifting African shield made contact in Miocene times with the Eurasian landmass. This contact enriched its biota with plant and animal species but this migration route was cut off in Pliocene times when the Saharan–Arabian desert belt developed. This separation was, however, not complete and this route was partly opened during more humid periods of the Pleistocene.

The palaeoenvironmental evolution of Africa during the late Cainozoic is marked by two important climatological processes, viz. a gradual lowering in world temperature and a decrease in humidity. These factors are closely correlated with the global energy budget. The incoming solar energy shows fluctuations of the Milankovitch type (Hays, this volume) and the storage and reflection of this energy by the Earth depends very much on the distribution of land and sea and on the Earth's albedo (Adam, 1973; 1975). The changes in humidity compared with those of temperature, are of a secondary nature and depend also on many other factors, such as the climatic system of a certain region, the orography, etc.

The important changes in world temperature which affected Africa were closely correlated with the building up of the vast ice sheets on Antarctica and Greenland in late Cainozoic times. The first indication of Cainozoic cooling of the Earth was found in the Bottom Water in the SW Pacific in the early Oligocene (Shackleton & Kennett, 1975). As the decrease in temperature affected the oceans and the atmosphere, glaciers moved down the mountains on the Antarctic continent and reached the ocean at the beginning of the Oligocene. The cooling process occurred haltingly between very long periods of a more or less even temperature. The full development of the East Antarctic ice sheet was completed in latest Miocene times, while the West Antarctic ice sheet is of latest Miocene to Early Pliocene age (Drewry, 1975 and this volume: Drewry; Mercer and Wilson). These events had a considerable impact on the whole of Africa.

The world energy budget deteriorated further in Pliocene–Pleistocene times when a long sequence of temperature oscillations with increasing amplitude affected the Earth. Evidence for this impressive range of cycles with a duration of 100 000 to 40 000 years has been found with oxygen isotopic analyses of deep-sea cores from the Pacific and the Caribbean Ocean (Shackleton & Opdyke, 1973; 1976; Emiliani, 1966) and in the loess stratigraphy of the lowlands of Czechoslovakia (Kukla, 1975). The last of these oscillations led to the glaciations of the Northern Hemisphere. These colder periods completely changed the face of the African continent, as is well known from the last or Kenya Glaciation of Africa (Coetzee, 1967). The drastic cooling accompanying these glaciations was superimposed on the global lowering in temperature caused by the Antarctic ice cap.

In this chapter a general evaluation will be given of the Cainozoic palaeoenvironments of the Sahara region with special attention to possible correlations with glacial episodes of a global nature. In this brief account attention will mainly be paid to climatic and geological changes and no detailed consideration will be given to biological or archaeological arguments.

The history of the Nile and its sediments which is very ably described by Butzer (in press) will not be included in these discussions as this river is "exotic" to the desert belt which it traverses.

The Sahara, the largest desert of the world, with an area of 9 million km^2 is roughly situated between 16 and 34° N lat. In this review the entire northern part of the continent will be included stretching from the Mediterranean Sea to the Sahel at about 14° N, and from the Atlantic Ocean to the Red Sea (Figure 1). The environmental history of this enormous area is very varied and includes episodes of marine transgression, mountain building, and volcanism, while the bioclimatic setting changed from the extreme of tropical rainforest to hyperarid desert. The processes involved in this evolution reach far back into history and for a full appreciation of these events we have to digress shortly into the Mesozoic.

THE MESOZOIC

During the Mesozoic the climatic and palaeogeographic situation of the north African region was entirely different from that of today. The high Atlas mountains and the Alps did not yet exist and a large shallow sea which had developed in the Tethyan intercratonic geosyncline separated Laurasia from Gondwanaland. This Tethyan Ocean covered central and southern Europe, the Near East and southern Asia as far as Burma. In Triassic times this tropical sea in which coral reefs occurred in northern Italy, was situated just north of the equator. The present desert belt was therefore idealy situated for receiving heavy high-sun rainfall.

As a consequence of the northward displacement of the continent the equator moved geographically southward and during the Middle Cretaceous it probably crossed the Sahara at the present latitude of about 14° N (Jardiné et al., 1974). From the replacement during the Barremian of the Early Cretaceous Gymnosperm Flora by the drier type of vegetation with *Ephedripites* it can be concluded that the climate must have been warm and dry (ibid.). During this important period in the geology of the continent, when the first Angiosperms appeared in the tropical parts, Africa started to

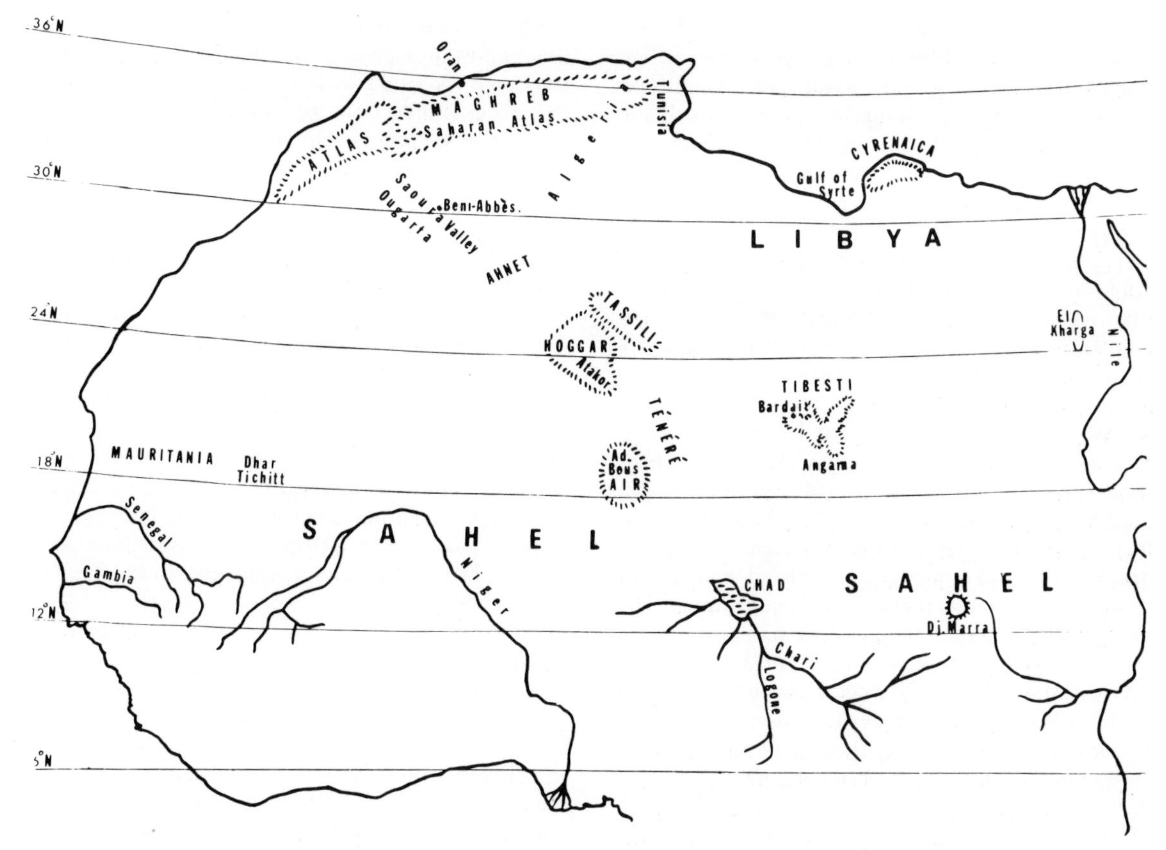

Fig. 1. Locality map of the Sahara region

separate from South America. The two continents, however, remained in fairly close contact for a considerable period of time and the Southern Atlantic Ocean only opened up in the Senonian during the Late Cretaceous. The consequence of this process was biological isolation and concomitant enhancement of the maritime influence.

At the same time several marine transgressions invaded the northern and central Sahara and attained their largest extent during the Late Cretaceous when they covered the northern Sahara as far south as the present Hoggar. The final regression set in during the Middle Miocene. Saad & Ghazaly (1976) have described the coastlands of one of these transgressions from the Kharga Oasis in Egypt and have shown that in Albian–Cenomanian times northern Africa formed part of the warm and dry belt which stretched from South America to Arabia.

During the Senonian the *Ephedripites* flora was replaced by *Proteacidites* which was succeeded by palm trees *(Spinozonocolpites)* and other species indicating a tropical climate in the last part of the Cretaceous (Jardiné et al., 1974). Humid tropical conditions must for instance have existed along the north coast in the Syrte basin where the growth of tropical rainforest has been recorded (Jordi, 1974).

Axelrod & Raven (1978) have given a very convincing description of this humid warm episode and mention some fossil floras which indicate the occurrence of rainforest in late Cretaceous up to Eocene times in Libya, Egypt and along the Red Sea coast. They also refer to fossil woods of tropical affinity which have been described by Kräusel and others from the present Sahara proper.

THE PALAEOGENE

In Palaeocene times these tropical humid conditions in the Sahara were favoured by the Tethyan and the enlarging Atlantic Oceans. The tropical and subtropical flora could in Eocene times even have spread far into Europe and reached southern England and Germany. Roche (1974) implies from the distribution of the *Nipa* palm that the equator could have crossed the Tethys at 35° N lat. during this period. This inference was made from the very high percentage of 39% of pollen of *Nipa (Spinozonocolpites)* attained in Spain. Although this hypothesis seems extraordinary, the facts prove that the climate round the Tethyan Ocean was of a tropical nature.

It can be assumed that in early Neogene times the

north and west coastal regions of Africa were covered with a humid tropical vegetation which graded further inland into subtropical woodland and savanna. The old mountain complexes in the central Sahara had only very low elevations and could perhaps have supported montane forest islands surrounded by plains covered with drier types of vegetation.

These favourable conditions deteriorated from Oligocene times onward mainly as a consequence of two factors, viz. the continued drop in world temperature and the changes in the palaeogeography of the region. Oxygen and carbon isotopic studies of ocean cores of Eocene to Miocene age from south of Australia and New Zealand by Shackleton & Kennett (1975) reveal declines in temperature which were rather rapid and show that the intervening periods were marked by a relatively stable temperature. At the beginning of the Oligocene (38 MY) a dramatic drop in temperature occurred, similar to a sudden decrease much earlier in Eocene times. The Oligocene lowering in temperature was of the order of 5° (Shackleton & Kennett, 1975) to 10° C (Mercer, this volume) and since then the temperature of the Antarctic Deep Water has been as low as at present. The Antarctic ice sheet may already have existed but would still have been of limited size. This decrease in temperature had a considerable impact on the Sahara region. As a consequence of continental drift this zone was at the same time moving out of the intertropical climatic belt and came under the influence of the high pressure system of middle latitudes which was strengthened as a result of the decrease in temperature.

THE NEOGENE

Very important changes were also taking place in the palaeogeography of the surroundings of the present desert belt. The marine transgressions had come to an end in Eocene times and the Tethyan Ocean, which had been an important source of humidity, disappeared in the late Early Miocene (Hsü et al., 1977) as the consequence of the orogeny of the Alps and the Atlas in the Oligocene–Miocene. These revolutions in the geographic and climatic setting of the north African region were caused by the compression of the plate margins of the European and Afro-Arabian cratons with the result that the Mediterranean basis was left over as a last relic of former oceanic crust.

The disappearance of the Tethys, which had formed a definite migration barrier for land faunas, had considerable consequences in that floras and faunas from Europe and Asia could henceforth be exchanged. This phenomenon is supposed to have been one of the 'major palaeobiogeographical events' of the Tertiary (Hsü et al., 1977). The new migration routes now possibly enabled the eastern Mediterranean mammalian fauna to invade the Iberian Peninsula and vast numbers of tragoceres, giraffids, *Hipparion mediterraneus*, gazelles, etc. populated the grasslands of Spain (Hsü, 1974).

Another very important phenomenon of world-wide importance was the lowering in temperature which has been recorded in the Southern Ocean during the Miocene (Mercer, Kennett, this volume). This cooling process was interrupted during three different periods by temporary warming, viz. at 19 MY, 14 MY and 8 MY ago. The general decrease in temperature between 14 MY and 10 MY ago resulted in the build-up of the East Antarctic ice sheet which reached its largest mass in the terminal Miocene. These temperature changes had a world-wide effect and diminished the evaporation from the tropical oceans, strengthened the anticyclonic systems and caused greater aridity, also in the present Sahara region.

The Miocene also witnessed important tectonic changes in the Sahara which resulted in the development of the large basins and ridges which still mark the topography of the region. The vast Chad basin has subsided since the Pliocene at a rate of 50 m 10^6 y^{-1} (Faure, 1971; 1977). The relief was further accentuated by the uplift of the old centres of the Saharan mountains at the same time as the Ethiopian highlands evolved. The massives of the Hoggar, Aïr and Tibesti were uplifted at a mean rate of 100 m 10^6 y^{-1} since the beginning of the Miocene (25 MY) (personal communication Prof. H. Faure). These mountains probably gained ca 2 500 m in altitude, so that new biotopes with a higher humidity and lower temperature originated. This process must have favoured the spread of montane forest and high altitude vegetation which will have formed a strong contrast with the vast surrounding plains where the climate became progressively drier. Savanna and dry scrub vegetation will have invaded the former woodland areas, while the coastal humid forest may have been replaced by evergreen laurel forest related to the present Canarian forest (Axelrod & Raven, 1978).

In the Late Miocene the evolution of the Mediterranean basin led to extremely radical changes in the wide surroundings. As a consequence of the Alpine orogeny the Mediterranean Sea became more and more isolated and was first cut off from its contact with the Indo-Pacific Ocean via the eastern Paratethys in the late Middle Miocene (14–15 MY ago) so that tropical water could no longer reach the inland sea from the east (Hsü et al., 1977). The Mediterranean was still connected with the Atlantic Ocean through two gateways, the deep Betic Strait in southern Spain and the Riff Strait in Morocco. The first connection was closed at the end of the Middle

Miocene. The ocean link through the Riff Strait was finally severed at the end of Late Miocene during the Messinian. The consequences of this event were of a very dramatic nature and led to the complete desiccation and to the so-called 'salinity crisis' of the Mediterranean which lasted one million years, from 6.2 to 5.3 MY ago (Smith, 1977).

The evidence for the Messinian salinity crisis is indisputable (Nesteroff et al., 1971; Hsü, 1974; Hsü et al., 1977). It was discovered during cruise Leg 13 of the Deep Sea Drilling Project in 1970 and re-examined during leg 42 A cruise of the Glomar Challenger in 1975. The deposition of 2–3 km thick salt layers started in deep water and continued until the basin had changed into a plain covered with sebkhras in which nodular anhydrite was precipitated near the ground-water table at temperatures above 35° C (Hsü, 1974). It appears that repeated marine influx was responsible for recycling of the older evaporites and that finally a nearly freshwater lake, the 'Lago Mare', developed.

The countries round the Mediterranean basin were left high and dry during the Messinian event and rivers flowing into the basin formed very deep canyons while the desiccation according to Hsü (1974) caused karst formation. The same author describes the climatic consequences this event had as far away as in the Vienna basin where forest was replaced by warm dry savanna in which antelopes and gazelles grazed. The implications for the Sahara region have not yet been studied, but this dramatic change which occurred in the old sea basin with an area of over 2½ million km² must have caused catastrophic consequences and resulted in much drier and also colder conditions and increased continentality in the Sahara region. Axelrod & Raven (1978) stress that selection for drought resistance was intensified considerably in Late Miocene times. The elevated areas, such as the Atlas and the emerging Aïr, Hoggar and Tibesti volcanoes could perhaps still have received orographic rainfall so that an impoverished montane forest could survive here.

It is of interest in the context of this volume to speculate on the possible correlations between the Messinian salinity crisis and important glacial events. These possibilities have been discussed in literature (Hsü, 1974; Hsü et al., 1977; Mercer, this volume) and also during the Second Messinian Seminar held in Italy in September 1976 (Smith, 1977). Lowering of the sea-level as a consequence of large scale glaciation could have been one of the causes of the isolation of the Mediterranean basin. The build-up of the Antarctic ice sheet (Kennett, this volume) might together with tectonic uplifts have closed the Rift Strait and severed the last connection with the Atlantic Ocean. At present it is not possible to assess whether the Antarctic glaciation was at least partly responsible for the Messinian salinity crisis or whether the completely opposite argument is valid. It has in fact been argued that the Messinian event suddenly lowered the salinity of the world oceans and could as a consequence have triggered the formation of the West Antarctic ice sheet (Mercer, this volume). Another possible correlation between the Antarctic and the Mediterranean could have been extensive melting of the southern ice sheet near the Miocene–Pliocene boundary which caused a higher sea-level and consequently the flooding of the dry Mediterranean basin (Anonymus, 1973). The sudden deluge of Atlantic water ended the dry Messinian in the early Pliocene.

The filling up of the Mediterranean basin will have ameliorated climatic conditions in the northern Sahara and it is well possible that, favoured by this climatic change, sclerophyll vegetation and open woodland again invaded the northern part of the desert. Very little evidence of former vegetation of Pliocene age exists for this region. One or two cold periods of uncertain age have been postulated for the Pliocene and it is therefore possible that colder-drier and warmer-wetter episodes alternated in the present desert region. Jordi (1974) mentions for the Upper Pliocene the occurrence of silicified wood of palms and also some animals which will have lived in isolated waterbodies. Evaporites in the eastern, northern and southern parts of the Sahara point to semi-arid conditions during later Pliocene times. Along the Red Sea coast similar conditions will have prevailed 3.5–4 MY ago when the Indian Ocean penetrated the Strait of Bab el Mandeb and inundated that basin (Hsü et al., 1977).

SUMMARY

The changes in world climate were in late-Cainozoic times closely correlated with glacial events especially of the Antarctic region. These changes had a considerable influence on the geological and biological processes in the vast North African region. The present Sahara region was also strongly affected by the northward drift of the African continent which caused the opening up and closing of surrounding ocean basins.

The changes in the oceanic environment of northern Africa were tremendous as the Atlantic Ocean to the west opened up from the Senonian onward and large parts of the present northern and central Sahara suffered extensive marine transgressions which ended in the Eocene. As a consequence of the northward drift the Atlas mountain ranges were pushed up during the alpine orogeny in the Miocene when the African plate collided with the Eurasian craton. In this process the wide tropical Tethyan Ocean, which had separated Gondwanaland from

Laurasia, disappeared and was replaced by the much smaller Mediterranean. All these processes had a great impact on the climatic setting of the northern sub-continent.

Since the beginning of the Miocene the topography of the Sahara region was changed by uplift and faulting which caused the development of large basins and intervening ridges. The large massifs of the central Saharan mountains gradually rose to their present altitude creating new climatic conditions and new habitats for flora and fauna. The end of the Miocene also witnessed the complete closure of the Mediterranean basin and the dessication of this sea of 2½ million km² between 6.2 and 5.3 MY ago. This geological catastrophe caused widespread aridification in Europe and northern Africa and can probably partly be correlated with the building up of large ice masses in Antarctica. The flooding of this large sea basin at the end of the Messinian ameliorated the aridity just as the drowning of the newly formed Red Sea valley during the later Pliocene.

All these geological processes were accompanied by a lowering in temperature of the Earth which set in in the Eocene. This decrease in temperature manifested itself in the evolution of polar conditions in the Antarctic region. There exists a parallel trend between the cooling of the Antarctic and the progressive aridification of northern Africa. The evolution of the palaeoenvironments of the Sahara region as a consequence of these glacial events is described in terms of climate and vegetation.

ACKNOWLEDGEMENT

I wish to thank my colleague Dr J.A. Coetzee for fruitful discussions and for reading the manuscript critically.

LITERATURE

Adam, D.P. 1973. Ice ages and the thermal equilibrium of the Earth. *J. Res. U.S. Geol. Survey* 1(5): 587–596.

Adam, D.P. 1975. Ice ages and the thermal equilibrium of the Earth, II. *Quat. Res.* 5: 161–171.

Anonymus. 1973. Leg 28 deep-sea drilling in the Southern Ocean. *Geotimes* 18(6): 19–24.

Axelrod, D.I. & Raven, P.H. 1978. Late Cretaceous and Tertiary vegetation history of Africa. In: M.J.A. Werger (ed.), *Biogeography and ecology of Southern Africa*, Junk, The Hague.

Butzer, K.W. Pleistocene history of the Nile valley in Egypt and Lower Nubia. In: M.A.J. Williams & H. Faure (eds.), *The Sahara and the Nile*, Balkema, Rotterdam (in press).

Climap Project Members. 1976. The surface of the Ice-Age Earth. *Science* 191 (4232): 1131–1137.

Coetzee, J.A. 1967. Pollen analytical studies in East and Southern Africa. *Palaeoecology of Africa* 3: 1–146.

Dansgaard, W., Johnson, S.J., Clausen, H.B. & Langway, C.C. 1971. Climatic record revealed by the Camp Century ice core. In: K.K. Turekian (ed.), *Late Cenozoic glacial ages*, Yale Univ. Press, New Haven: 37–56.

Dott, R.H. & Batten, R.L. 1971. *Evolution of the Earth*, McGraw-Hill, New York. 649 pp.

Drewry, D.J. 1975. Initiation and growth of the East Antarctic ice sheet. *J. Geol. Soc.* 131: 255–273.

Emiliani, C. 1966. Palaeotemperature analysis of Caribbean cores P6304–8 and P6304–9 and a generalised temperature curve from the last 425,000 years. *J. Geology* 74: 109–126.

Fairbridge, R.W. 1971. Quaternary sedimentation in the Mediterranean region controlled by tectonics, palaeoclimates and sea level. In: D.J. Stanley (ed.), *The Mediterranean Sea: A Natural Sedimentation Laboratory,* Dowden, Hutchinson & Ross, Stroudsburg, Pa.: 99–113.

Faure, H. 1971. Rélations dynamiques entre la croûte et le manteau d'après l'étude de l'évolution paléogéographique des bassins sédimentaires. *C.R. Acad. Sc. Paris* 272: 3239–3242.

Faure, H. 1977. Late Cenozoic vertical movements in Africa. *Symposium: Earth rheology and late Cenozoic isostatic movements Stockholm* (manuscript).

Hsü, K.J. 1974. The Miocene desiccation of the Mediterranean and its climatical and zoogeographical implications. *Naturwissenschaften* 61 (4): 137–142.

Hsü, K.J. et al. 1977. History of the Mediterranean salinity crisis. *Nature* 267: 399–403.

Jardiné, S., Kieser, G. & Reyre, Y. 1974. L'individualisation progressive du continent Africain vue à travers les données palynologiques de l'ère secondaire. *Sci. Géol. Bull.* 27(1–2): 69–85.

Jordi, U. 1974. Geologie. In: B. Messerli (ed.), *Sahara Exkursion 1973*, Geogr. Inst. Univ. Bern: 93–107.

Kennett, J.P. et al. 1974. Development of the Circum-Antarctic Current. *Science* 186: 144–147.

Kukla, W.G. 1975. Loess stratigraphy of Central Europe. In: K.W. Butzer & Gl. Isaac (eds.), *After the Australopithecines*: 99–188.

Nesteroff, W.D. et al. 1971. Evolution de la sédimentation pendant le Néogène en Méditerranée d'après les forages JOIDES – DSDP. In: D.J. Stanley (ed.), *The Mediterranean Sea: A Natural Sedimentation Laboratory*, Dowden, Hutchinson & Ross, Stroudsburg, Pa.: 47–62.

Roche, E. 1974. Paléobotanie, paléoclimatologie et dérive des continents. *Sci. Géol. Bull.* 27(1–2): 9–24.

Saad, S.I. & Ghazaly, G. 1975. Palynological studies in Nubia sandstone from Kharga Oasis. *Pollen et Spores* 18(3): 407–470.

Shackleton, N.J. & Kennett, J.P. 1975. Palaeotemperature history of the Cenozoic and the initiation of Antarctic glaciation: Oxygen and carbon isotope analyses in DSDP sites 277, 279 and 281. In: J.P. Kennett et al. (eds.), *Initial reports of the DSDP* 19: 743–755.

Shackleton, N.J. & Opdyke, N.D. 1973. Oxygen isotope and palaeomagnetic stratigraphy of equatorial pacific Core V28–238: Oxygen isotope temperatures and ice volumes on a 10^5 year and 10^6 year scale. *Quat. Res.* 3: 39–55.

Shackleton, N.J. & Opdyke, N.D. 1976. Oxygen-isotope and paleomagnetic stratigraphy of Pacific Core V28–239. Late Pliocene to Latest Pleistocene. *Geol. Soc. Am. Mem.* 145: 449–464.

Smith, L.A. 1977. Messinian event. *Geotimes*: 20–23.

* 12 *

Evidence for Quaternary glaciation of Marion Island (sub-Antarctic) and some implications

K.J. Hall

Institute for Environmental Sciences, University of the O.F.S., Bloemfontein, South Africa

Manuscript received 15th May 1977

CONTENTS

ABSTRACT

Marion Island is at present located only 2° latitude north of the Antarctic Convergence. Besides former palynological and geological proof for a glaciation of Würmian–Wisconsin age, further evidence has been found to indicate that Marion has been subjected to glaciations of both Würm and Riss age. Large areas of drift have been recognised for the first time together with a number of glacial landforms. Work so far undertaken suggests that there was a central ice cap from which a number of glaciers radiated. Till fabric analysis, striation observations and landforms have enabled a reconstruction of the glacier distribution on the northeastern side of the island to be undertaken.

The evidence from Marion is compared with that from other sub-Antarctic islands and from ocean floor sediments. The relationship of the findings to the Quaternary climate of the island is suggested. Evidence for sea level changes is also given and compared to that found on other sub-Antarctic islands.

INTRODUCTION

Marion Island (lat. 46° 54′ S, long. 37° 45′ E) is a roughly oval volcano of some 290 square kilometres, rising to 1 230 m, located in the southern Indian Ocean only some 2° latitude north of the Polar Front (Fig. 1). The radially faulted island is composed of older grey basaltic lavas dating from 276 000 BP (± 30 000) and younger black lavas, and associated scoria cones, dating from 15 000 BP (± 8 000) (McDougall, 1971). Considering the age of the oldest grey lavas Verwoerd (1971) suggested that the island could have been subjected to the southern equivalents of the Riss and Würm glaciations. Pollen analysis by Schalke and van Zinderen Bakker Sr (1967) showed that the island had indeed been subjected to a cold phase which was approximately coeval with the Würm of the Northern Hemisphere. Physical evidence for the existence of glaciers during this cold phase was found by van Zinderen Bakker Jr and Huntley in the form of striated pavements and glacially moulded grey lava outcrops (Verwoerd, 1971). The black lavas show no signs of glacial action and so are considered to post-date the glacial episode. Verwoerd (1971) suggested that any moraines that had been formed were at present below sea level and that the diamicts found at several places on the coast were more likely of volcanic, rather than glacial, origin. Thus although it was known that Marion had been glaciated at least once, the nature and extent of that glaciation was unkown, as too was evidence for the earlier glacial period.

The special research programme to investigate the glacial history of Marion Island was initiated in December, 1975 and the first results give evidence to:

1. substantiate two major periods of ice growth,
2. show the distribution of glaciers for part of the island, and
3. describe the climatic variations experienced in this area during the last glaciations as shown by glacial landforms and deposits.

The post-glacial faulting and lava flows have obliterated all evidence of the glacial episodes from the northern and western sections of the island. The

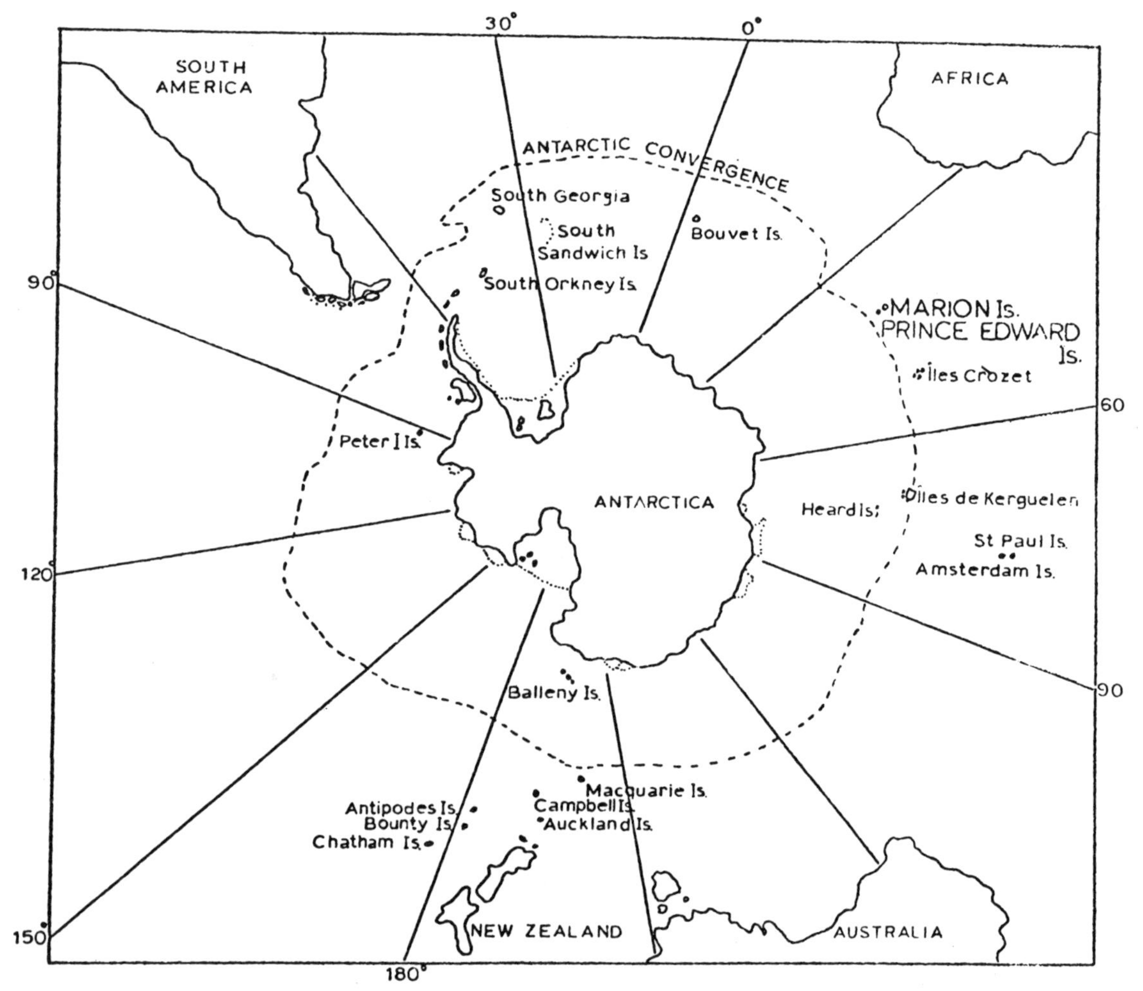

Fig. 1. Locality map, showing the position of Marion Island with respect to the Antarctic Convergence.

main area in which glacial deposits and landforms are exposed stretches from the northeast, at Long Ridge, down to the southern coast at Greyheaded Albatross Ridge. By reference to several specific locations within this area comments on the glacial history of Marion Island can now be given.

TECHNIQUES USED

A number of simple but complementary field techniques were employed in the study. Till fabric analysis was undertaken on many coastal cliff sections and along the banks of streams that were incised in till deposits. A high density of sample points was used with measurement of 50 stones at each point. The orientation and dip of the a-axis of stones longer than 0.02 m and shorter than 0.25 m with a minimum a : b ratio of 2 : 1 were measured. At each area one sample location was used for a detailed study of the stones measured and a number of parameters were monitored. Each fabric was subjected to a Chi-Squared test and accepted at the 95% level. Vertical and horizontal comparisons of fabrics were undertaken by means of the Kolmogorov-Smirnov test and all fabrics were subjected to vector analysis to obtain the resultant fabric vector and strength. At each fabric point the composition of the till with regard to the number of pyroclasts, striated stones and grey lava stones were noted and expressed as a percentage (Table 1). This was used as an aid to mapping of the beds and to noting the occurrence of volcanic phases of the island.

Wherever solid grey lava outcrops occurred measurements of striations and moulded forms were undertaken and the resulting mean directions obtained. Surveyed profiles of the moraines and other depositional landforms were studied as an aid to understanding their genesis and formative ice-flow directions. Surface stone fabrics were also obtained from the moraines for additional information on ice

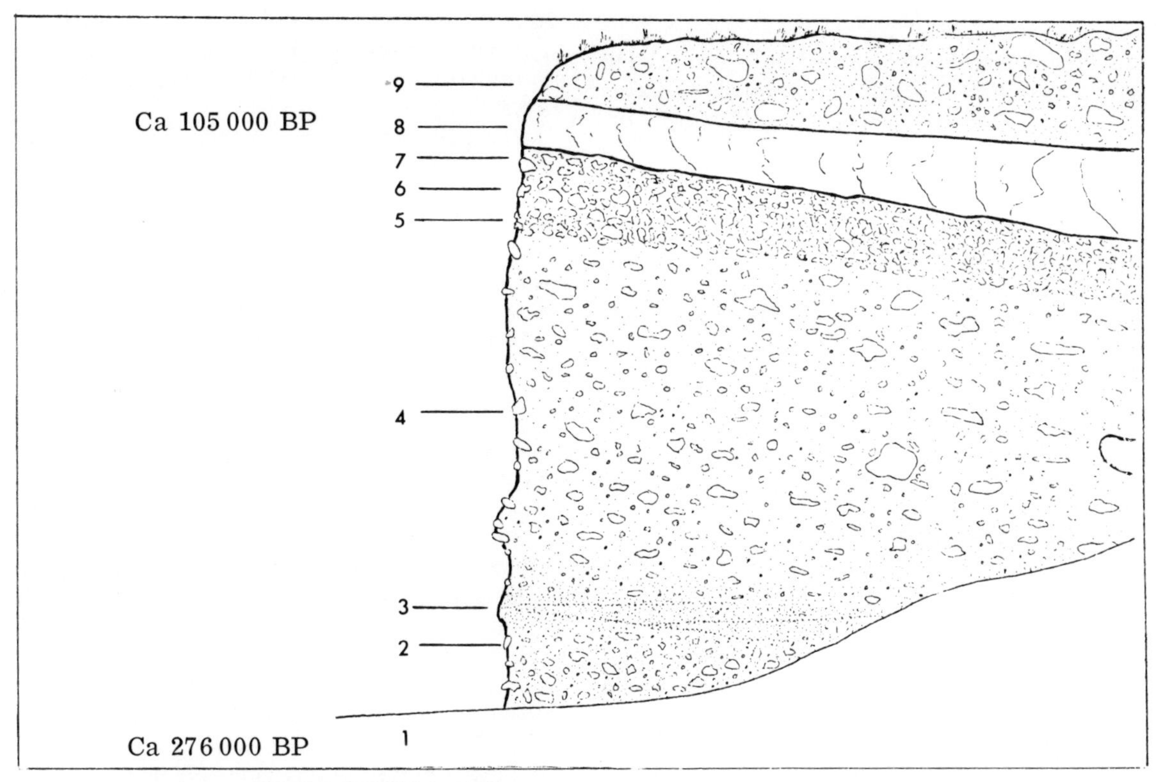

1 = Lava c. 220 000–276 000 yrs BP
2 = Till, largely composed of pyroclastic material, overlying fluvial sediments (total thickness = 3 m)
3 = Interstadial melt-out till with ablation till capping (in wedge structures due to subsequent shearing)
4 = Main till (11–15 m in thickness)
5 = Capping ablation till disrupted by volcanic bombs (6)
7 = Red pyroclasts
8 = Grey lava c. 105 000 yrs BP of Riss-Würm interglacial age
9 = Till of last glacial (Würm)

Fig. 2. Simplified section along the sea-eroded fault at Ships Cove

flow directions. Samples of till from both inland and coastal locations were collected and subjected to grain size analysis.

Alone each of these simple techniques was of limited value but together their complementary evidence assumes some degree of power. The combination of striated surfaces, surface stone fabrics, and asymmetry of moraines plus the coastal till fabrics allow the identification of the former glaciers with a reasonable degree of certainty.

EVIDENCE FOR AND NATURE OF THE OLDEST GLACIATION

Evidence for the earlier glaciation which Verwoerd (1971) had speculated about has been found at four coastal exposures, namely Kildalkey Bay, Macaroni Bay, Ships Cove and Goodhope Bay. At these locations the occurrence of the earlier glaciation was indicated by a vertical sequence of – till, grey lava, till – of which the upper till was known to be from the last glacial. The lava dividing the two tills had been dated at two outcrops (McDougall, 1971) at about 105 000 BP (± 25 000), thus the underlying till had to predate this lava and postdate the oldest lava (276 000 ± 30 000) thereby locating the lower till at approximately the same age as the Riss of the Northern Hemisphere. The intervening lavas, where they occurred, made good stratigraphic junctions but additional lines of evidence to denote two till sequences were also found. A location near Kildalkey Bay which lacked the intervening lava flow showed instead a distinct palaeosol developed in the

Table 1

To show the percentage of pyroclasts, striated and grey lava stones in successive tills from the Kildalkey area

Location	a	b	c
Top of till sequence	18	0	82
Top of till sequence	12	1	88
Top of till on top of palaeosol	23	2	77
Intermediate locations in till on top of palaeosol	58	4	42
	82	1	18
Base of till on top of palaeosol	89	3	11
Palaeosol	10	1(?)	90
Base of till below palaeosol	86	0	14

a = % pyroclastic stones
b = % striated stones
c = % of non-pyroclastic stones

lower till. Till fabric analysis for the lower and upper till at this point, and at a similar situation at Macaroni Bay, showed distinctly different preferred stone orientations between the tills. This can be explained by the interglacial outpourings of lava that affected the ensueing ice-flow directions. In addition, at Macaroni Bay, where the oldest grey lavas occur, the striations found on the lower lava agree with the preferred fabric orientation of the till resting directly on top but differ from the fabrics of the upper tills, whose preferred orientation agrees with the striations found on the more recent grey lava outcrops. Finally the lower tills were found to be well consolidated compared to the upper sequences.

At Ships Cove, along a sea-eroded fault, occurs the largest exposure of deposits from the earlier glacial (Fig. 2). The lowest till in the sequence is composed largely of pyroclastic material and only near its top do sub-rounded and angular blocks of grey lava occur and begin to predominate. This reflects the stripping of the unconsolidated surficial pyroclastic cover of the island, by the ice, and then the later incorporation of the frost-shattered underlying grey lava. This till is covered by what appears to be a melt-out till topped by a thin, angular, ablation till. The tills occur in a series of wedges which thin towards the sea. The wedging is thought to result from either slight movement during the melt-out process or due to subsequent overriding by ice of the succeeding stade. The melt-out – ablation till sequence is considered to represent an interstadial for it is covered by a further till. No other major oscillations of this nature have been observed in the 15 m of till but several distinct changes in lithology have been recognised and are still under study. Volcanic bombs can be seen at the top of the till. They disrupt the capping ablation till and gradually phase into a purely pyroclastic layer (very different in composition to the lowest pyroclastic till), upon which the lava rests.

The palaeosol, which was found on the coast where the interglacial lava capping was missing, is some 2 m thick and reddy-brown at the top, grading with depth, back to the grey of the till beneath. There is some evidence to suggest frost-sorting of the surface layers of the palaeosol prior to the return of the ice but no ice-wedge structures, suggesting permafrost, have been found. The depth and degree of weathering suggest that the interglacial may have been warm and humid. Table 2 gives data of the thicknesses of weathering rinds found on stones in the palaeosol and the tills beneath and above thereby indicating the distinct weathered layer in the upper part of the palaeosol. It is considered that the weathered stones found in the lower part of the capping till result from disturbance of the underlying soil and the inclusion of some of the already weathered stones. Samples of organic material have been collected from this weathered layer and it is hoped that they will be useful for pollen analysis which may give indications of the climatic conditions. The evidence for the earlier glacial episode is limited and much of the information is still being analysed but it is possible to state that Marion Island was ice covered sometime during the period after approximately 276 000 BP and before 105 000 BP. The dates of the bracketing lavas locates the glaciation as broadly coeval with the Riss of the Northern Hemisphere. It would appear that there was an interstade early on the 'Riss' sequence and that fans of outwash material were produced which were later

Table 2

Thicknesses of weathering rind observations from palaeosol at Kildalkey Bay

Location	a	b	c	d	
Base of till on top of palaeosol	15	2	0.6	0.57	
Top 0.3 m of palaeosol	15	15	3.14	1.93	n = 24
0.5 m–0.3 m depth in palaeosol	15	9	1.83	0.71	$\bar{x}$ = 2.65 s = 1.69
Top of till below palaeosol	20	0	—	—	

a = number of stones sampled
b = number of stones with weathering rinds
c = mean ($\bar{x}$) thickness of measured rinds
d = sample standard deviation (s) of measured rinds

covered by the till of the re-advance. The ensuing interglacial was a period of volcanic activity with lavas and pyroclasts capping much of the till, but uncovered till areas developed a reddy-brown soil under a possibly warm, humid climate. Prior to their being covered by ice during the youngest glacial the soils were subject to frost sorting but not to permafrost.

YOUNGEST GLACIATION

a) Mapping of the ice cover

Estimation of the area and distribution of the glaciers during the last glacial has been undertaken by means of till fabric analysis from coastal and inland exposures, striation measurements, and the mapping of moraines and other landforms. At the time of writing some 200 fabrics have been completed and striations measured on all known grey lava outcrops. Contrary to the expectations of Verwoerd (1971) large numbers of moraines have been found on the eastern and southern sides of the island. Surveyed profiles have been completed on four moraines and distal and proximal slopes measured on fourteen others; surface stone fabrics have been undertaken on six of the larger moraines. Both 'push' and 'dump' type moraines have been recognised.

At Skua Ridge and Albatross Lakes the ridges of debris are distinctly asymmetrical with shallow (5°–9°) proximal slopes and steep (17°–28°) distals. At both locations the moraine profile is broken by a bench 5–10 m in width on the distal side. This is thought to represent the top of a former moraine which has been partially overriden by readvancing ice thus producing a multiple moraine (Fig. 3). The preferred orientation of the larger surface stones found on the proximal slope of these moraines is at right angles to the moraine crest suggesting additional evidence for overriding ice. The much larger (±50 m high, 200–300 m broad) moraines found in the southeast at Stony Ridge to Kildalkey Bay are of the dump variety, often showing steeper ice-contact proximal slopes.

From the moraines, striations, and till fabrics the

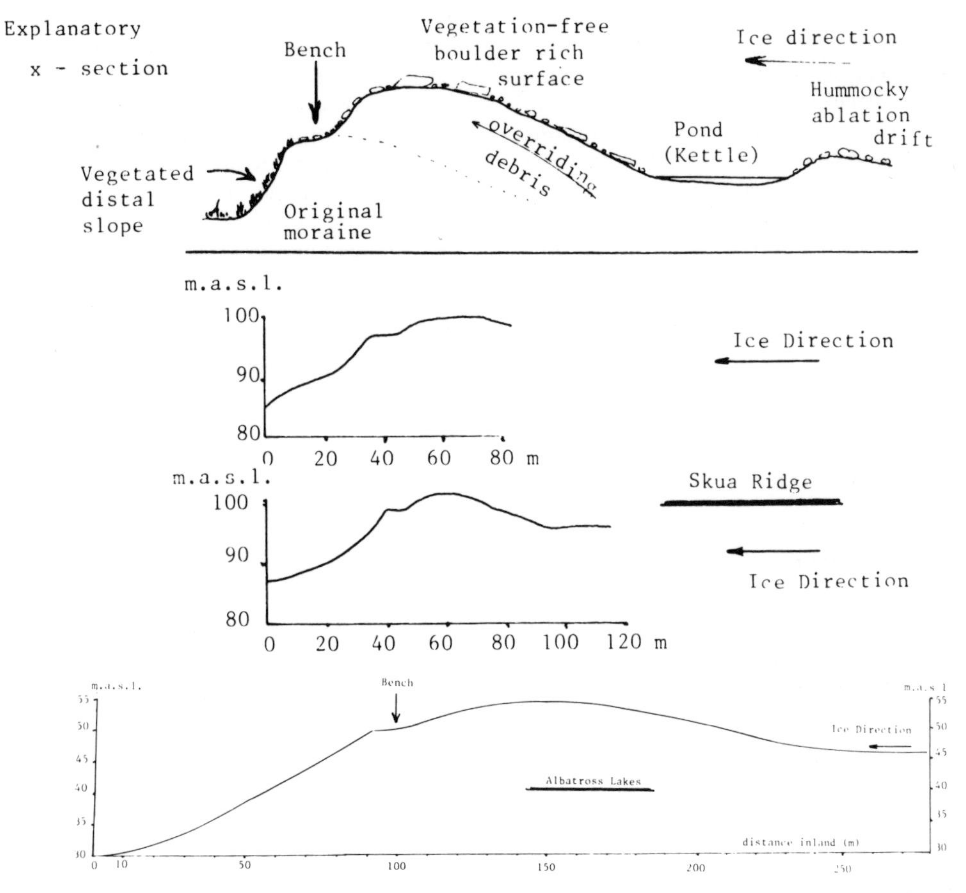

Fig. 3. Levelled x-sections from two locations of moraines exhibiting evidence for minor readvances (with simplistic explanatory diagramme)

locations of five former glaciers have been determined (Fig. 4):

1. to the north of Long Ridge (northward extension uncertain; fed from the Katedraalkrans–Middelman area)
2. Table Mountain to Skua Ridge (being fed from the Piew Crags to upper Table Mountain area)
3. Macaroni Bay–Albatross Lakes Area (being fed from the region of Freds Hill)
4. Stony Ridge to Soft Plume River (being fed from Tates to just north of Hoë Rooikop)
5. Soft Plume River to Kildalkey Bay (being fed from Hoë Rooikop to Beret area)

It was found that many of the post-glacial lava flows follow the intermoraine areas and that these are now the locations of the major stream courses.

The large laterals in this south-eastern area start at approximately 200 m a.s.l. and thus it is considered that this equates to the approximate altitude of the equilibrium line of the glaciers (see Andrews, 1975, p. 54). Using 200 m as the equilibrium line and the width of the glacier as equating to the distance between the main laterals the ablation area of four of the glaciers has been calculated:

1) Kildalkey Glacier = 6.2 km^2
2) Stony Ridge Glacier = 5.9 km^2
3) Albatross Lakes Glacier = 5.9 km^2
4) Skua Ridge Glacier (minimum size) = 4.9 km^2

When surveying the profiles across the equilibrium line have been completed it is anticipated that the volume of the ice in the ablation area and the approximate rates of flow can be calculated.

In addition to the moraines areas of disintegration ridges (and kettle holes) and two marginal stream channels have been recognised. Disintegration ridges and kettle holes have been found at three locations (Long Ridge, Skua Ridge and Albatross Lakes) within the moraine sequences. At Long Ridge, far back from the maximum ice extent, two parallel former marginal stream channels, which were formed during the main retreat phase, occur. The significance of these features and the moraines will be put in context after the earlier glacial history, as shown by the till deposits, has been described.

b) The sequence of the last glaciation as shown by the till deposits

The sequence of the youngest glacial will be described in detail from only one site, that of Long Ridge, but additional data from other locations will be commented on.

At the seaward end of Long Ridge a clear exposure of glacial debris upwards from a grey lava exists (Fig. 5). Immediately on top of the lava occurs a 2 m thick deposit of bedded fines and gravels which is capped by a distinctive 3 m thick till. The till has a reddish-brown colour and is composed entirely of pyroclastic debris of mainly clay to gravel size with few boulders or cobbles. It is a homogeneous deposit with only one variation occurring at its top where there is a distinct layer of platy, grey lava blocks. This is interpreted as an ice advance and retreat sequence. The advancing ice cleared the surficial pyroclastic debris, which has a few large blocks and weathers rapidly, thus producing the reddish-brown till. The occurrence of angular, platy grey debris at the top of the till suggests a supraglacial or englacial deposit that has been let down, by the melting ice, as an ablation till. If the material had been subglacially derived it would have been more rounded and less angular than was found.

This lower till is covered by a thick (up to 2 m) sequence of sands and gravels (with organic material) which grades upwards into a rhythmite sequence and then back to fluvial beds. This represents an interstade during which the ice retreated extensively

Fig. 4. Topographic map of Marion Island with moraines and glacier positions (etc.) marked

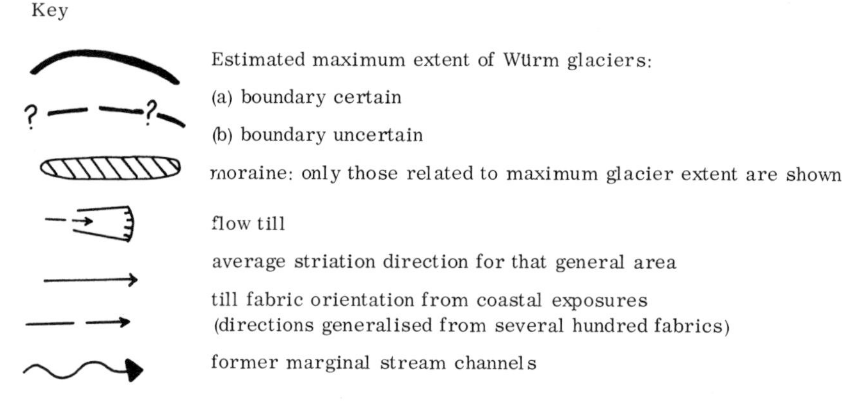

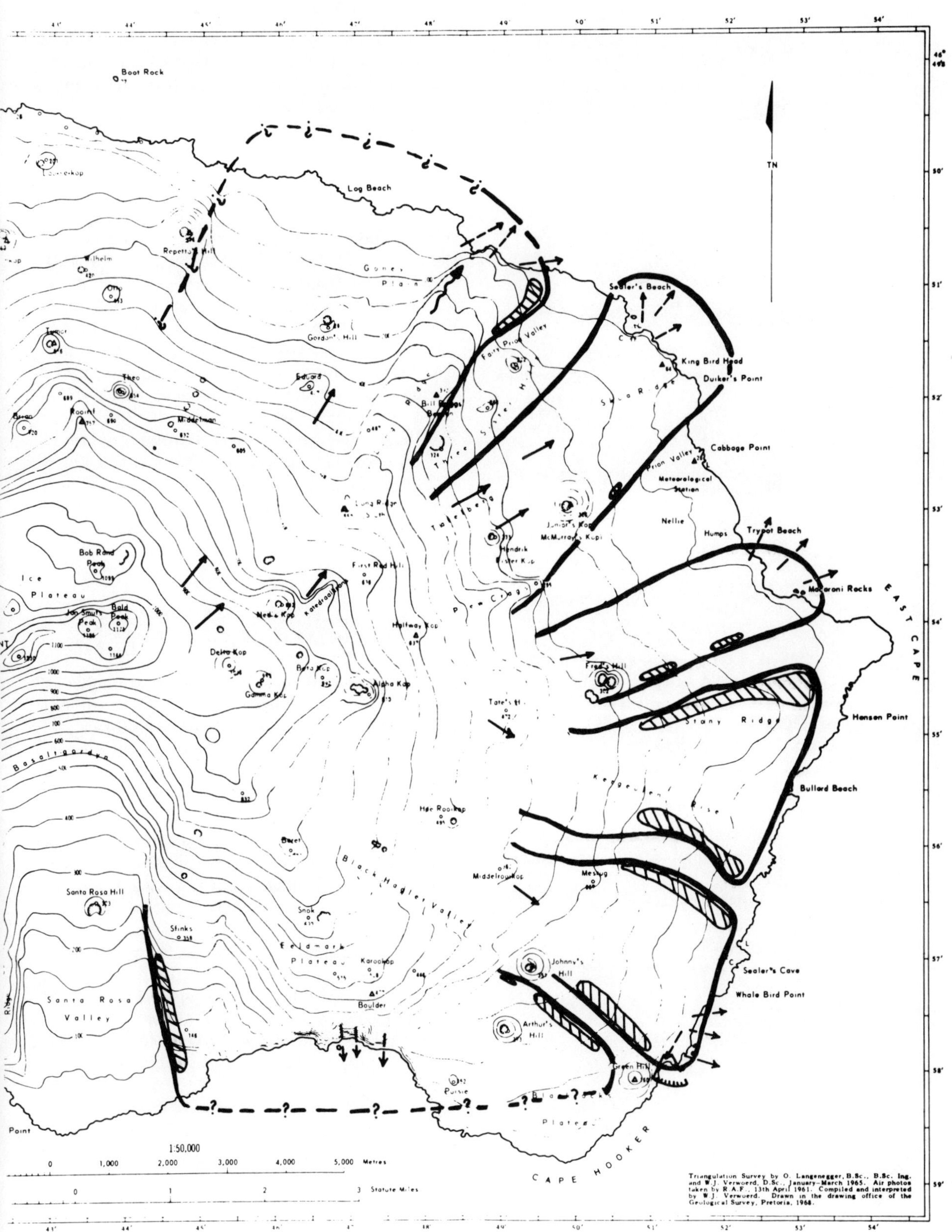

Form lines every 50 m.
The western side of the island is not shown as postglacial volcanics have destroyed nearly all evidence of former glacial activity.
The information is given on the map of Marion Island produced by Langenegger and Verwoerd, 1968.

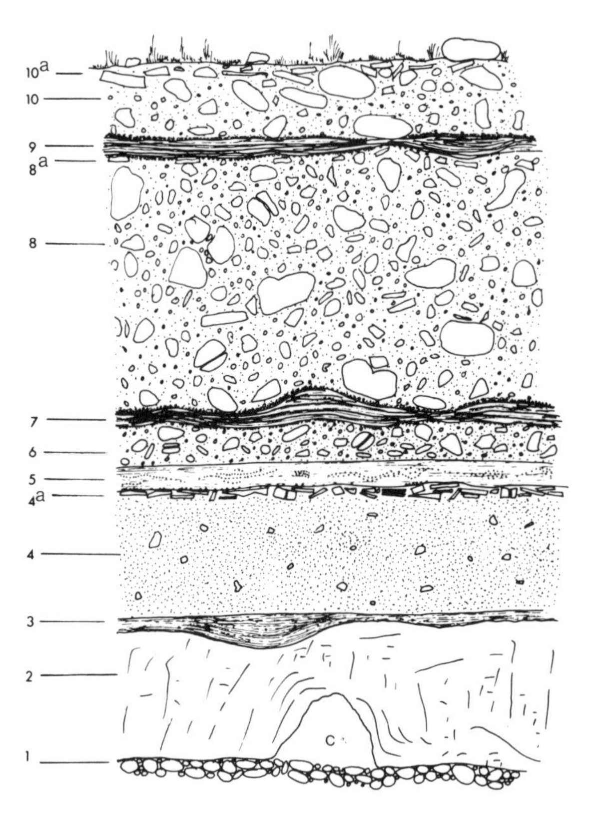

Fig. 5. Simplified section of upper (Würm) till sequence from Long Ridge.
1 = Beach
2 = Grey lava (C = cave)
3 = Bedded fines and gravels
4 = Till, reddish-Brown, composed of pyroclastic debris (3 m)
4a = Platy grey lava blocks
5 = Interstadial fluvial deposit of sand and gravel, containing a rhythmnite sequence (up to 2 m)
6 = Till, grey, c. 0.5 m thick
7 = Rhythmite-with-dropstones sequence, faulted and folded
8 = Till, 10–12 m thick
9 = Rhythmites (c. 1 m) covered by gravels
10 = Homogenous grey till
10a = Platy ablation till
Total height of profile c. 21 m

prior to a readvance. The covering till, which is only (approx.) 0.5 m thick, is grey in colour and has few pyroclast inclusions. Towards the top this till becomes very gravel rich and grades into a rhythmite-with-dropstones sequence. The rhythmites become gravel rich at their upper boundary and show faulting and folding, probably produced by the readvance of the ice which deposited a thick (10–12 m) till on top. This sequence is seen as indicating a glacial advance of relatively short duration followed by a minor retreat during which a proglacial lake was formed. The subsequent stade which disturbed the rhythmites was of long duration and a thick homogeneous till was produced.

The only variation found in the rest of the sequence is a further bed of rhythmites close to the cliff top. Possibly there was a minor retreat and readvance prior to the major retreat sequence and during this oscillation a proglacial lake occurred. A comparable sequence of events can be found at the southern end of the island at Kildalkey (Table 3). It can be seen from Table 2 that the initial stages of the two separate areas are very similar. The thin till is missing from the Kildalkey area which is probably due to the ice not having readvanced to its former maximum position prior to its subsequent retreat. At the southern end a distinct flow occurs which could be attributed to either the earlier or the later interstade. The flow till extends beyond the limits of the Kildalkey ice and is a most distinctive feature. Considering the small size of the island and the enormous oceanic influence it is considered acceptable that the glaciers should be fairly synchronous in their responses to climatic variations.

BRIEF OBSERVATIONS ON FORMER SEA LEVELS

True raised beaches were observed at a number of locations around the island – all of which showed wave-smoothed rocks, wave rounded pebbles and boulders, and macro-cliffs at the back of the beach. At Transvaal Cove and Trypot Beach raised beaches were surveyed and heights of +3.4 m and +2.9 m were found respectively. Just to the north of Cabbage Point a sequence of two beaches was found with the lower one at a height of +3.3 m and the upper one at +6.1 m. Stone roundness (Pi) (using Cailleux's equation) and flatness (Fi) indices were obtained for both levels:

+3.3 m beach	+6.1 m beach
Pi = 422.9 (s = 125.6)	Pi = 381.2 (s = 121.3)
Fi = 448.1 (s = 61.0)	Fi = 492.9 (s = 117.6)
n = 100	n = 100

The difference in indices between the beaches is further emphasised by the values obtained for the present day beach and the +2.9 m beach at Trypot where the following values were obtained:

Present Beach	+2.9 m Beach
Pi = 557.5 (s = 120.3)	Pi = 417.1 (s = 119.5)

Thus the three beach levels (present day, ca +3 m and ca +6 m) can be differentiated in terms of the stone roundness indices and it is interesting to note that very similar values for Pi were found for the two ca +3 m levels at Cabbage Point and Trypot Beach (422.9 and 417.1). These levels of ca +3.0 m and ca +6.0 m were observed at several locations

Table 3

Comparison of events during the last glacial at Long Ridge and Kildalkey Bay

Long Ridge	Kildalkey area
Ablation till	Ablation till
Thin till	Till
Rhythmites	Sands & gravels
Ablation till	Ablation till
Thick grey till	Thick bouldary till
Rhythmites (+ fluvials)	Flow till sheets
Thin grey till	Thin grey till
Fluvials (+ rhythmites)	Fluvials
Ablation till (grey plates)	Ablation till
Pyroclastic till	Pyroclastic till
Fluvials	Fluvials
Lava	Palaeosol

around the island, notably Macaroni Bay, Water Tunnel Stream, Goodhope Bay, Fur Seal Bay and Cape Davis.

In addition to direct observation of raised beach levels the long profile of a river, the Van den Boogaard, was surveyed and levels extrapolated from the nick points in this. From the long profile obtained a number of distinct nick points were discernable and whilst an exponential curve with a coefficient of determination (r^2) of 0.83 best fitted the whole profile, linear extrapolation was found to best fit the level above each nick point. A line generated for the seaward segment indicated a former level of +5.9 m with an r^2 of 0.98. The inland segment showed an extrapolated level of +10.86 m with an r^2 of 0.95. Thus the extrapolated level of +5.9 m is in close agreement with the level of the true raised beaches found around the coast. Whilst several localities show what may possibly be raised beaches at ca +10.9 m level none have actually been surveyed as such. However, it is very likely that a beach level occurred at a ca +10.9 m level.

CONCLUSIONS

Combining the information from the glacial landforms and till sequences a first approximation of the glacial history of Marion Island can be made (Table 4). It is hoped that the major events can be dated before long – notably the first interstade of the last glacial. Unfortunately the counting of varve couplets has been precluded by their disturbed nature although a set for the minor retreat at Long Ridge indicate a minimum duration of 78 years. Thus the information

Table 4

A first aproximation of the sequence of events during the last glacial as shown by the deposits and landforms

Interglacial	Lava flows – scoria palaeosols developed
Interglacial – onset of glaciation	Frost action in soils fluvial sequences
Stade	Pyroclastic tills produced from clearing of surficial deposits Deformation of earlier fluvials
Ice retreat	Distinct ablation till – angular grey blocks capping brown till beneath
Interstade	Outwash sands and gravels capped by rhythmites (thick)
Stade	Deforms underlying rhythmites Thin grey till produced
Minor interstade	Proglacial lake rhythmites Flow tills
Stade	Extensive deformation of rhythmites Thick grey till
Retreat	Distinct level of ablation till
Interstade	Outwash gravels followed by preglacial rhythmite deposit
Stade	Grey bouldary till
Retreat	Capping of ablation till
End of glacial Ice retreating	Ice begins major retreat at end of glacial
Minor readvance	Push moraine formed
Minor readvance	Overriding of earlier moraine
Period of relative stillstand	Dump moraine formed inland Marginal stream channels develop
Ice continues retreat	Ice retreats rapidly Thin till covering of striated pavements inland

obtained, given here in very brief outline, shows that the island was subject to two major glaciations and that each was composed of a series of stades and interstades.

The implication of the glaciations of Marion Island is that the Polar Front moved northwards to encompass Marion and the onset of the glacials is seen as a result of this. The northward shift of the Front would lower the mean annual temperature of the sea water by approximately 2° C and in addition put Marion within a zone of more southerly winds which are 3–4° C colder than those experienced at present (Schulze, 1971). Also the precipitation may have increased as the dry anticyclones would have passed further to the north. The increased precipitation together with the lower temperatures would result in a greater annual snowfall with less summer ablation which would account for the growth of the ice-cap and glaciers. Hays et al. (1976) showed that the Polar Front did not move very far north in the region to the south of Africa, so Marion would have only just come within its influence. Thus the island would have been in a sensitive marginal position with respect to the movement of the Front and it can be assumed that there could have been rapid build-ups and losses of ice as the island moved in or out of the influence of the Polar Front.

Hays et al. (1976) suggest that there was a lowering of temperature up to 3.5° C. Using the calculated level of the equilibrium line of the glaciers and from this calculating the approximate height of the snowline, it is considered that the minimum decrease in temperature on Marion was 3.5° C. This gives a mean annual temperature of 1.5° C at sea level. Thus the temperatures calculated from the information available on the island agree very closely with those found by Hays et al. from the ocean floor sediments and van Zinderen Bakker (1973) from palynological studies.

The results do not indicate that the island has been completely ice covered which is in accordance with the suggestions of Schalke & van Zinderen Bakker (1967). They suggest that as certain of the island biota had 'overwintered' the glacial, there must have been ice-free areas, as has also been suggested for Kerguelen by Young & Schofield (1973). The distribution of moraines and the occurrence of localities (not covered by postglacial lava flows) which show no signs of glacial deposits indicate, that a number of sections of the island were indeed ice-free. Clapperton & Sugden (1976) show that the Falkland Islands lacked extensive glaciers during the last glacial. However, the glaciation of Marion is seen as far more extensive than that of the Falklands. The glaciers of Marion are indicated as being 6–7 km long and 3 km wide flowing out from a central ice-cap, whilst Clapperton & Sugden (1976) suggest a maximum cirque glacier length of 2.7 km for the Falklands. The more extensive glaciation of Marion is seen as being due, in part, to the far greater precipitation. Hays et al. (1976) also showed that the Falkland Islands remained north of the Polar Front and were thus not subject to this colder influence.

The raised beach levels found on Marion appear to reflect those described at other Antarctic–sub-Antarctic locations. Nougier (1971) has noted a +3.0 m level on Kerguelen whilst in the South Shetlands Sugden & John (1973) have found levels of approximately +3.0 m and 6.0 m. On Livingston Island Everett (1971) has found levels of +10.6 m and +6.1 m. Thus there would seem to be some evidence from Marion to show that the +3.0 m, +6.0 m and +10.6 m sea levels occurred in the sub-Antarctic region.

The implications of the glaciations of Marion Island are far reaching, for not only do they assume significance for the sub-Antarctic and Antarctic region but also for the African continent. The possible northward shift of the climatic belts, and their oscillations, are pertinent to the understanding of the Quaternary history of Africa. In the light of the work on oceanic sediments to reconstruct the Quaternary climatic conditions the availability of terrestrial results from Marion helps clarify the situation. Much data still awaits analysis and so it is hoped that more detailed information will shortly be forthcoming. It is also hoped that future work undertaken in the Borga nunatak area of Antarctica will provide complementary information to link with that of Marion. Thus this information is seen as a step in the completion of an overall pattern of data collection from Africa through the sub-Antarctic to Antarctica proper from which a better understanding of the glacial episodes in the Southern Hemisphere may be obtained.

SUMMARY

Detailed investigation of Marion Island deposits has shown that the island was subject to two glacial episodes roughly coeval with the Riss and Würm on the Northern Hemisphere. Till deposits have been found at a number of coastal cliff exposures. The results of till fabric analysis together with striation observations has indicated the main ice-flow directions. Mapping and surveying the moraines has enabled the boundaries of the glaciers to be delimited for the southern and eastern sides of the island. Dates from the lava flows interbedded with the tills have shown that there were two major glacial phases which broadly coincide with the Riss and Würm. Where the intervening lava flows are absent palaeosols have been found.

Within both glacial episodes sequences of stades and interstades have been recognised. Evidence has been found for the occurrence of proglacial lakes, flow tills and multiple moraines. Two former marginal stream channels have been recognised together with kettle holes and hummocky ablation drift. Using the altitude of the start of lateral moraines the approximate height of the glacier equilibrium lines have been calculated together with the ablation area of the glacier. Palaeotemperatures for ± 18 000 BP have been calculated from the estimation of the palaeosnowline, which agrees with evidence from ocean floor sediments and fossil pollen.

A number of raised beaches have been found and surveyed at several locations on the island. Complementary evidence for the raised beach levels has been obtained from extrapolation of a long profile of a river.

ACKNOWLEDGEMENTS

Financial and logistic support was provided by the Department of Transport (Pretoria). The research programme was initiated by Professor van Zinderen Bakker to whom go sincere thanks. The help and companionship in the field of Doug Langley is most gratefully acknowledged for without his help much less would have been accomplished.

REFERENCES

Andrews, J.T. 1975. *Glacial systems*, Duxbury Press. 191 pp.

Clapperton, C.M. & Sugden, D.E. 1976. Maximum extent of glaciers in part of West Falkland. *J. Glaciology* 18 (75): 73–77.

Everett, K.R. 1971. Observations on the glacial history of Livingston Island. *Arctic* 24 (1): 41–50.

Hays, J.D., Lozano, J.A., Shackleton, N. & Irving, G. 1976. Reconstruction of the Atlantic and Western Indian Ocean sectors of the 18 000 BP Antarctic Ocean. *Geol. Soc. Am. Mem.* 145: 337–372.

McDougall, I, 1971. Geochronology. In: E.M. van Zinderen Bakker, J.M. Winterbottom & R.A. Dyer (eds.), *Marion and Prince Edward Islands*, Balkema, Cape Town: 22–77.

Nougier, J. 1971. *Contribution a l'étude géologique et géomorphologique des Isles Kerguelen.* French National Committee Antarctic Res. Publ. 27. 440 pp.

Schalke, H.J.W.G. & van Zinderen Bakker, E.M. 1967. A preliminary report on palynological research on Marion Island (sub-Antarctic). *S.A. J. Sci.* 63 (5): 254–260.

Schulze, B.R. 1971. The climate of Marion Island. In: E.M. van Zinderen Bakker, J.M. Winterbottom & R.A. Dyer (eds.), *Marion and Prince Edward Islands*, Balkema, Cape Town: 16–31.

Sugden, D.T. & John, B.S. 1973. The ages of glacier fluctuations in the South Shetlands Islands, Antarctica. *Palaeoecology of Africa* 8: 139–159.

Van Zinderen Bakker Sr, E.M. 1973. The glaciation(s) of Marion Island (sub-Antarctic). *Palaeoecology of Africa* 8: 161–178.

Verwoerd, W.J. 1971. Geology. In: E.M. van Zinderen Bakker, J.M. Winterbottom & R.A. Dyer (eds.), *Marion and Prince Edward Islands*, Balkema, Cape Town: 40–62.

Young, S.B. & Schofield, E.K. 1973. Palynological evidence for the late glacial occurrence of *Pringlea* and *Lyallia* on Kerguelen Islands. *Rhodora* 78 (802): 239–247.

* 13 *

Glacial age aeolian events at high and low latitudes: A Southern Hemisphere perspective

J.M. Bowler

Department of Biogeography and Geomorphology, Australian National University, Canberra, A.C.T., Australia

Manuscript received 17th August 1977

CONTENTS

ABSTRACT

Reconstruction of glacial age environments in semi-arid southern Australia has demonstrated the reality there of a wet phase (45 000 to 25 000 BP) followed by a long dry interval (25 000 to 14 000 BP). The transition from wet to dry is identified by a fall in lake levels, but the construction of lakeshore clay-rich dunes (lunettes) and by renewed activation of longitudinal desert quartz dunes. Measured by the maximum aeolian activity, the peak aridity occurred between 18 000 and 16 000 BP coincident with the maximum advance of global ice sheets.

The two-phase nature of southern Australian glacial age environments is evident also in the chronologies of other low latitude dry regions, especially in Lake Chad and Lake Abhé in northern Africa.

Loess sequences from cold Pleistocene semi-arid environments of Europe, USSR and North America also reflect a two-stage cycle; the transition there from pedogenic to aeolian environments occurred about the same time as comparable low latitude changes. After 25 000 BP the quantity of long-distance dust reaching Antarctica increased markedly.

The association of intensified aeolian activity with cold glacial conditions can be validated for the past 700 000 years, both in Atlantic deep sea cores, and in the continental loess sequences of Czechoslovakia and China. The consistency of the transformation from wet to dry environments preceding each glacial maximum, points to a major change in atmospheric circulation about that time. However, the last such change about 25 000 BP is not matched by changes of equivalent magnitude either in oxygen isotope or sea surface temperature reconstructions.

Whilst the cause of the changes that occurred about 25 000 BP is not well understood, the Australian evidence suggests intensification in the frequency and velocity of winds emanating from the continental interior. In glacial summers, sites downwind of such winds would have experienced hot dry conditions in which enhanced seasonal evaporation helped produce low lake levels; these seasonal changes amplified the general trend towards aridity already evident due to decreased precipitation.

Widespread dusty environments at high and low latitudes with consequent effects of both increased advective heat transfer and dust-induced thermal modifications had characterised each major glacial episode. The repetition of such conditions preceding each glacial-interglacial transition suggests a possible two-way interaction. In helping to produce dry, dusty environments on a global scale, the expanded ice sheets and sea ice of high latitudes may well have been agents instrumental in their own destruction.

INTRODUCTION

The climate of southern Australia was dominated in the past, as it is today, by the circulation of air masses that originate in the Southern Ocean and sub-Antarctic region (Gentilli, 1971). In winter the easterly travelling cyclonic depressions with their associated cold fronts that spiral northward from the polar front penetrate deeply into the southern continental margin bringing cold winter rains. Conversely the summer season, dominated by the subtropical anticyclone, is hot and dry. At such times the centres of the cyclonic depression lie well south of the continent. Any change in the position of the polar front (Hays et al., 1976a), the generating area of the southern cyclones, would affect their influence and depth of penetration over the Australian continent (Fig. 1). Furthermore any increase in the meridional temperature gradient consequent on such a shift may be reflected in intensified pressure systems with increased velocity of the geostrophic winds (Lamb, 1972; Barry & Chorley, 1971). The evidence from Pleistocene aeolian deposits may indicate if such changes did indeed take place in the past.

The influence of Cainozoic Antarctic glaciation on the continents of the Southern Hemisphere has been explored elsewhere (Flohn, 1973; Mercer, 1973; Kemp, 1977). However, the nature and influence of detailed oscillations of Quaternary age between different latitudinal zones and between sea and southern land masses remain to be specified.

If further justification is needed for including details from southern Australia in a volume devoted more specifically to Antarctica it is found in the nature of the evidence available. Whilst much of the Antarctic palaeoclimatic record is reconstructed from the surrounding oceans (Hays et al., 1976a and 1976b), and from sub-Antarctic islands (Bellair, 1965; van Zinderen Bakker, 1973; Sugden & Clapperton, 1977), data from the continent itself are inevitably sparse and less diverse than from other continents that escaped the great erosive influence of ice sheets. In these non-glacial regions, such as southern Australia, the imprint of climatic oscillations may be both more diverse in the legacy of landforms and sediments it expresses and relatively more continuous in the length of records preserved. Moreover, our understanding of modern climates is based on parameters observed and measured mainly from such continental regions. The detailed palaeoclimatic reconstructions now becoming available from ocean sediments require confirmation from equally reliable documentation on land. The regional and interzonal reconstructions made possible through such integrated projects as CLIMAP have yet to be repeated onshore.

This paper is concerned largely with aeolian deposits. Because they occur synchronously over wide regions extending from continents to ocean basins and because they provide important evidence of provenance, such deposits are of considerable importance in dating and reconstructing Quaternary environmental changes. Moreover, their lateral extent provides a means of extending correlations beyond continental margins and across different climatic zones, a correlation that has yet to be fully explored and whose palaeoclimatic implications have still to be spelled out.

In this paper, details of dated events in southern Australia are presented first, then compared with related events in other parts of the world. Whilst the proposition of 'curve-matching' may invite facile and irrelevant comparisons, as dangerous in their simplicity as in the complexity they may ignore, nevertheless we reach a stage when changes, identified on a continental scale, must be evaluated against data collected from regions of equivalent dimensions in other parts of the globe. Thus, in proceeding from local and regional data to inter-zonal and intercontinental comparisons, we may be led to identify primary causes which affect circulation changes on a global scale.

EVENTS OF GLACIAL AND DEGLACIAL AGE IN SOUTHERN AUSTRALIA

The general sequence of glacial age environments in southern Australia (Fig. 2) has been summarized elsewhere (Bowler et al., 1976) whilst episodes of aridity have received specific attention (Bowler, 1976). The data record two contrasting hydrologic episodes. the first relates to that phase in which the availability of water in the landscape was greater than it is today. Identified first from high lake levels dating from before 40 000 to about 25 000 BP in the Willandra Lakes of western New South Wales (Bowler, 1971), it finds supporting evidence from southeastern Australia in Lake George (Coventry, 1976; Singh, in Bowler et al., 1976) and Lake Leake in the southeast of south Australia (Dodson, 1975). The implication drawn from lakes is that large quantities of surface water were present throughout large areas of southeastern Australia, with profound effects on stream discharge and regional water-

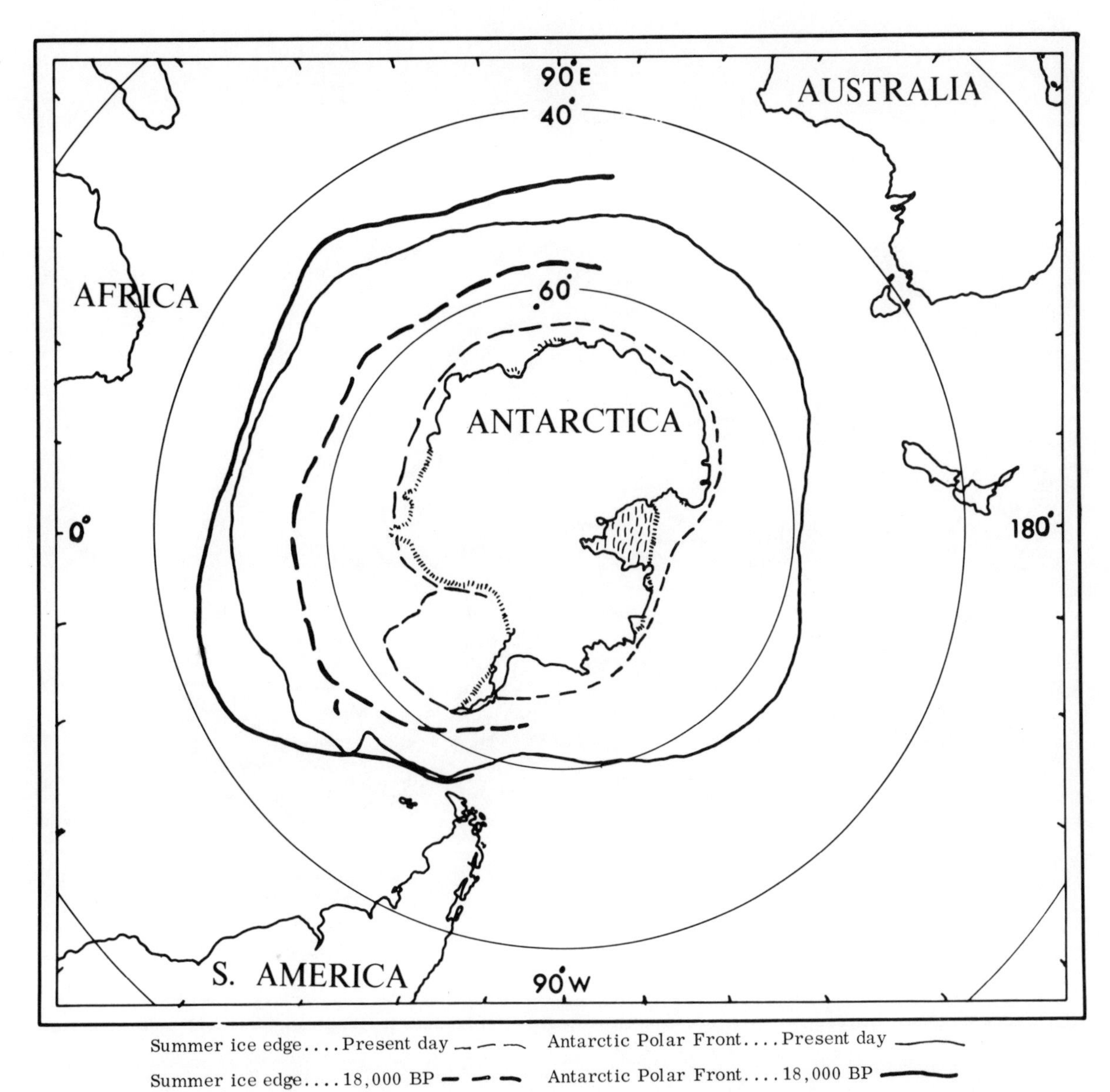

Fig. 1. Antarctica and the Southern Ocean showing the position of the polar front relative to continents of the Southern Hemisphere. Changes in the 18 000 BP position of the polar convergence and Antarctic summer sea ice (after Hays et al., 1976a) would have exerted a strong influence on the pleniglacial climate of southern Australia.

tables. To avoid the connotation of increased rainfall inherent in the term 'pluvial' this specific event is called here the Mungo Lacustral Phase from Lake Mungo where it was first identified.

The second significant hydrologic phase, the onset of widespread aridity, is expressed through three main lines of evidence:

1. drying of lakes;
2. construction of lakeshore gypsum and clay dunes;
3. reactivation of desert longitudinal dunes, indicated by their transgression onto floors of previously full lake basins and by the increased contribution of silt sized desert dust to both lake basin and shoreline sediments.

1. Age and extent of hydrologic changes

The record from the Willandra Lakes (Fig. 2) provides clear and well dated sequences, for which the techniques of facies analysis used to interpret them have been outlined elsewhere (Bowler, 1976).

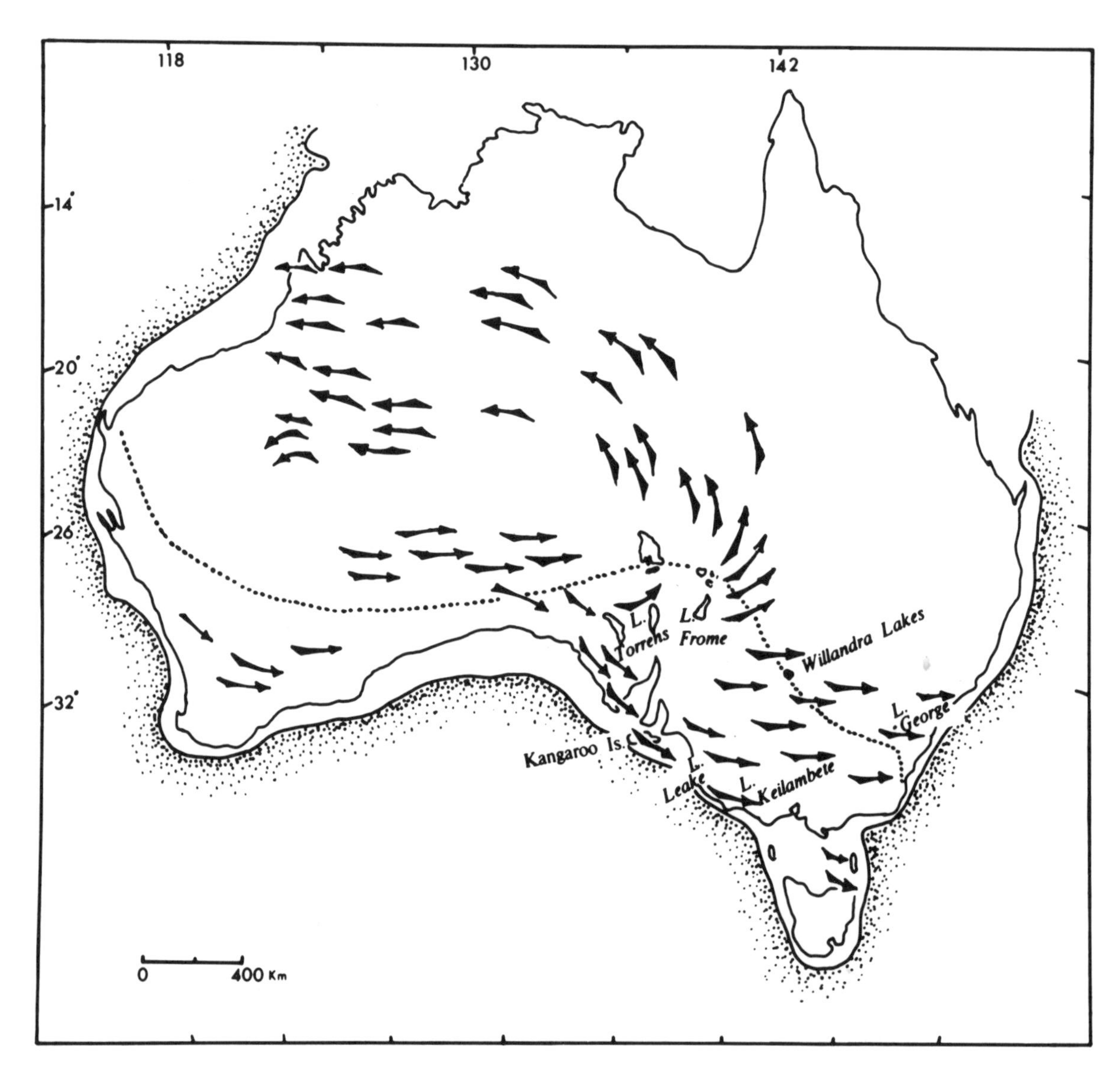

Fig. 2. Sand flow lines representing the trends of dunes believed to have been active during the last glacial maximum. Dotted line represents zone boundary of equal summer–winter rainfall. Note the association of westerly sand flow with winter rainfall region.

These lakes were first filled about 45 000 BP following a long, relatively dry interval. The large freshwater lakes contained aquatic faunas including large fish and freshwater mussels. Early man was attracted to them and continued to frequent the area at least seasonally from 40 000 BP until the lakes finally dried some 25 000 years later. These lakes provide the earliest known record of man's presence on the continent.

The Mungo Lacustral Phase coincided with the oldest described glacial and periglacial environments on the Australian mainland dated between 30 000 and 35 000 BP (Costin 1972; Costin & Polach, 1971). Therefore it is reasonable to assume that both summer and winter temperatures were considerably reduced below present levels. To explain slope deposits near Canberra, Costin & Polach (1971) postulated mean annual temperature reduction to 10° C below present values. Although such estimates from the highlands should not be extended unreservedly to the dry inland plains, a considerable reduction in evaporation loss accompanied the lower temperature regime. This resulted in more effective precipitation and greatly increased runoff which, in turn, was registered as a regional change in lake levels, water-tables and stream discharge.

The available evidence does not permit an unequivocal reconstruction of precipitation-temperature

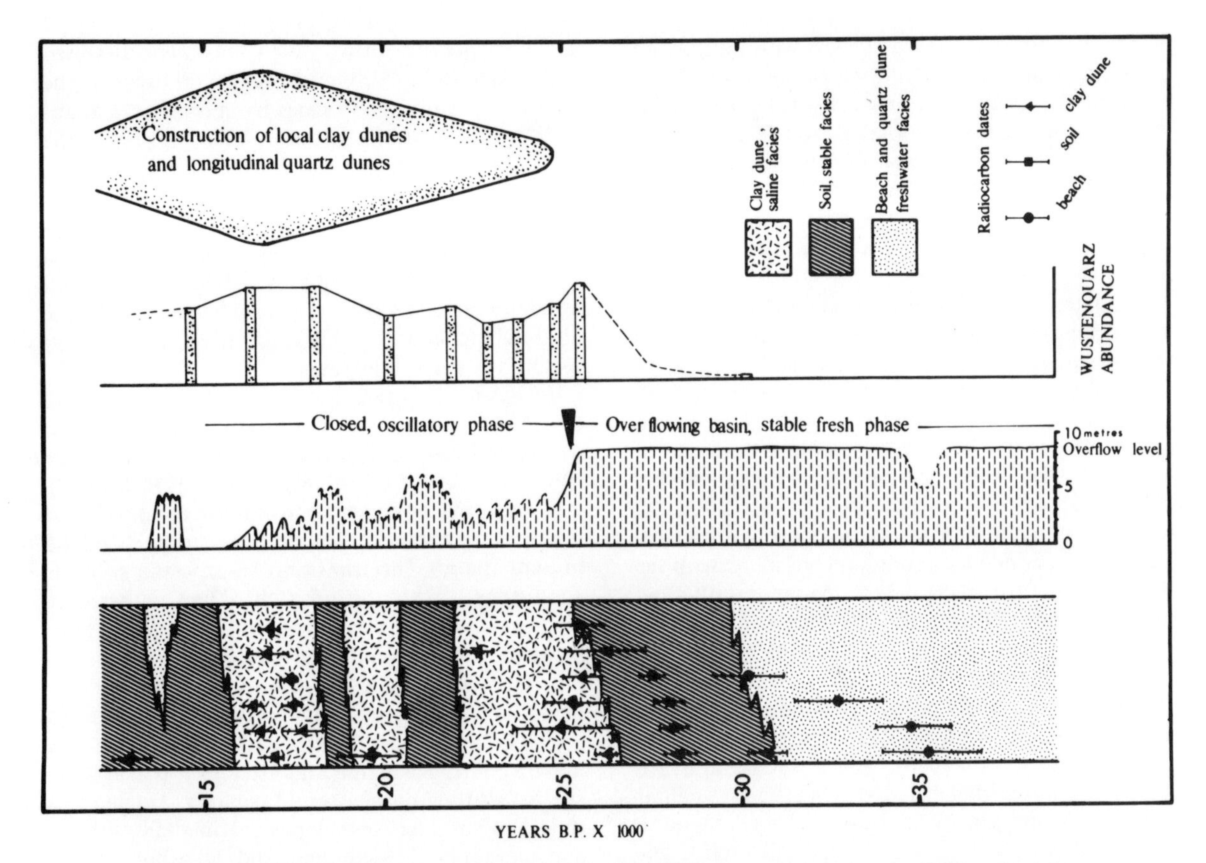

Fig. 3. Lithofacies diagrams of stratigraphic sequence from Lake Mungo, Willandra lakes (Fig. 2), showing radiocarbon dated horizons and their relationship to water level changes. Frequency of relative abundance of desert quartz *(Wüstenquarz)* reflects onset of regional longitudinal dune activity, a phenomenon that occurred synchronously with fall in lake level and with local building of clay-rich lunettes and linear quartz dunes.

relationships. The question of whether precipitation was higher or lower than today's values remains largely unresolved for this period. However, this interval represents the wettest conditions experienced throughout the past 100 000 years. Moreover, the lake full phase (which buried an ancient palaeosol in the floor of the Willandra Lakes about 45 000 BP) began some 30 000 years before the glacial maximum as recorded in the extension of Northern Hemisphere ice masses (Flint, 1971; Driemanis & Karrow, 1972) and in the deep-sea core record of ice volume changes (Shackleton & Opdyke, 1973).

As an example of hydrologic change of great magnitude and extent the Mungo Lacustral Phase is equalled by the events that succeeded it. Studies in the past ten years have demonstrated the nature and widespread affects of aridity that accompanied events of full glacial age, events that reached a climax with the final drying of lakes and the expansion of desert dunes in the period between 18 000 and 16 000 BP, precisely at the time of maximum expansion of global ice (Bowler, 1976). However, the events that heralded this change began much earlier. Indeed, those changes which represent the end of one phase and the initiation of another may be more critical in identifying causes and natural thresholds than the maxima and minima of the major oscillations. Moreover it is now possible to pin-point the onset of these changes in western New South Wales with a high degree of accuracy.

The transition from a positive hydrologic budget when many lakes were overflowing to that of a negative regime with increasing salinity can be identified and dated. The accumulation of many radiocarbon dates from interdisciplinary studies (archaeology by Allen, 1972; palaeomagnetics by Barbetti & McElhinny, 1976; Quaternary geology by Bowler, 1971) provides a tight chronologic control. The transition is first indicated in the shoreline stratigraphy by aeolian gypsum and sandy clays derived by seasonal deflation from saline lake floors. The formation of clay-rich dunes requires a seasonal oscillation of water-tables with efflorescent salts breaking clays into pellet aggregates. Strong winds then transfer the clay pellets with quartz and seed gypsum to the shoreline where they accumulate to form the crescentic transverse dunes so characteristic of late Quaternary lakes over large areas of southwestern and southeastern Australia (Bettenay, 1962; Campbell, 1968; Bowler 1973).

Figure 3 demonstrates the relationship between some of the C-14 sites at Mungo in the Willandra Lakes (Fig. 2) and the lakeshore facies from which these dates derive. The transition from deep, freshwater lacustrine environments represented by well sorted, quartz beach and aeolian sands, to saline calcareous and gypseous aeolian clays is critical. This change was accompanied by a geochemical transition from a calcitic to a dolomitic environment (Bowler, 1971); the transition to the saline, clay dune facies occurred between 26 000 and 25 000 BP.

Further important evidence of major hydrological change is provided by a simultaneous increase in the abundance of red, clay-coated, fine sand to silt-sized quartz grains. With their characteristic red clay coatings (cutans) these grains appear identical to the desert dust grains *(Wüstenquarz)* found in the Atlantic Ocean and derived by wind storms from Saharan Africa (Radczewski, 1939; Seibold et al., 1976; Diester-Haas, 1976). Today such grains are transported over long distances from the continental interior by summer dust storms that may extend to southeast Australia especially during droughts. The increase in the desert quartz grains, coinciding with the transition from high to low lake levels, points to re-activation of quartz sand dunes west of the Willandra Lakes at the same time as those lakes began to dry. Thus the desert sands and the lakes were responding in the same sense at the same time.

The contrast between pre-25 000 and post-25 000 conditions is particularly worthy of attention. Elsewhere in southern Australia the change from wet to comparatively dry conditions, as recorded by Dodson (1975) at Lake Leake, occurred also about this time. At Lake George, Coventry (1976) has estimated the change from an overflowing system to closed basin environments as dating from between 20 000 and 25 000 BP. From the same lake, Singh (in Bowler et al., 1976) records a major vegetational change reflecting a transition to drier environments at a time which, on the basis of present C-14 control, he places near 22 000 BP.

In arid South Australia, Lake Frome supported large populations of calcareous algae *(Chara)* before 20 000 BP which represent relatively low salinity environments. Soon after 20 000 BP clay and gypsum dunes were built on the lake floor providing conclusive evidence of a major hydrologic change (Bowler, 1976). In the nearby Flinders Ranges, the period about 24 000 BP is seen by Williams (1973) as representing a transition from earlier soil forming conditions of the Wilkatana palaeosol to a more unstable and arid period during which fans were dissected and dunes built. Similarly, in the far west of the continent, radiocarbon dates from lakeshore dunes in Western Australia (Bowler, 1976) cluster in the period between 20 000 and 16 000 BP reflecting an intensification of aeolian deposition coincident with comparable events in the southeast. Dune building continued in western N.S.W. from 25 000 until about 14 000 BP. Certainly by 13 000 BP most sand dunes on the eastern desert margin had been stabilised although pockets of mobile dunes may have persisted locally for some time later.

In Tasmania, a region where one expects a strong maritime influence, Colhoun (1975) and Colhoun, Mook and van de Geer (pers. comm.) have recently shown that by 22 130 ± 180 (GrN-7689) seasonally very dry environments were extensively developed over the northern, central and eastern parts of the present island. The transition from wetter to drier conditions, yet to be precisely dated, is seen by Colhoun as the effect of a regional reduction in temperatures and precipitation which occurred simultaneously with renewed glacial advance in the highlands of western Tasmania. Very dry conditions prevailed until after 16 590 ± 110 BP (GrN-7882) in northwestern Tasmania and after 15 740 ± 700 BP (SUA-376) in southeastern Tasmania. During this time aeolian silts were deposited in shallow lakes and parabolic longitudinal and lakeshore dunes were developed. It is significant that the trends of dunes built at this time have a WNW-ENE orientation reflecting in some measure a continuation of trends apparent in the drier and warmer continental environment further north (Fig. 2).

2. *Two-phase regime characteristic of Australian pleniglacial*

When viewed on a continental scale several important conclusions follow from analysis of southern Australian environments. The last glacial period was characterised by the development of two hydrologic regimes drastically different from those of today.

Firstly a late Pleistocene humid phase of major proportions produced stable soil-forming conditions in the Flinders Ranges. Elsewhere lakes now dry stood at high levels, some remaining so for some 20 000 years. Whilst the climatic parameters that controlled this event cannot be specified accurately, runoff was very much higher and seasonal evaporation lower than today.

Secondly, about 25 000 BP a drastic change occurred in the hydrologic regime. Lake levels fell and extensive dune building followed. Several minor hydrologic fluctuations culminated in a phase of maximum aridity recorded in simultaneous dune construction between 20 000 and 15 000 BP across the southern part of the continent.

A strongly developed palaeosol underneath the deposits of the last pleniglacial stage in Lake Mungo implies a period of non-deposition. From the high degree of pedogenesis evident in that soil at least 40 000 years is involved in its formation (Bowler, 1976). Added to the age of the basal sediments of the Mungo Lacustral phase (about 45 000 BP) the origin of lacustrine deposits on which the soil formed dates from more than 85 000 years ago. Sediments of this phase (Golgol unit of Bowler, 1971) may correlate with, or be older than, the previous glacial event (dated to about 140 000 years ago, (Shackleton & Opdyke, 1973); they cannot be younger than 85 000 years.

The dramatic contrast between the two hydrologic events that characterise the last pleniglacial in southern Australia cannot be explained by local events. Although factors such as increased runoff from highland catchments played an important part, the areal extent of the changes point to climatic variations on a regional and intercontinental scale. The search for an explanation must take us to Antarctica and the Southern Ocean, the generating area of much of Australia's climate, as well as to other low latitude deserts and to high latitudes of the Northern Hemisphere to seek comparative data for events of this age.

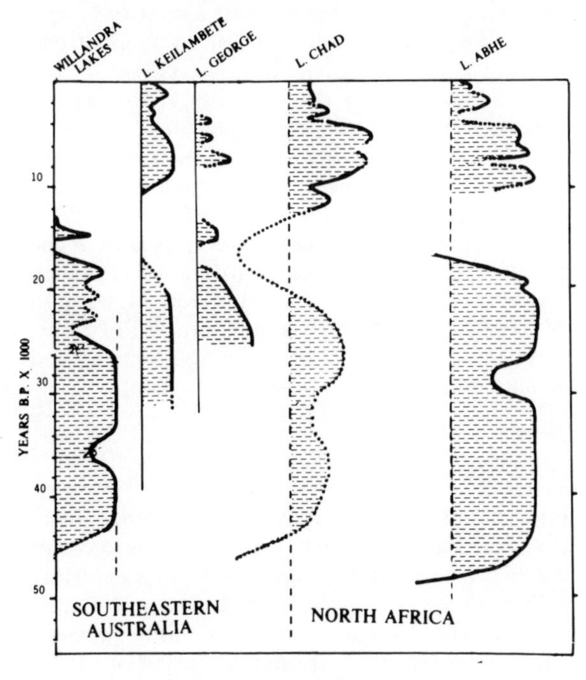

Fig. 4. Sequence of water level curves from lakes in southeastern Australia compared with two lakes from northern Africa. Data from Coventry (1976), Bowler et al. (1976), Servant & Servant (1970) and Gasse (1975; 1977). Broken lines represent limits of reliability.

EVIDENCE FROM OTHER PARTS OF THE WORLD

1. Low latitude deserts

A full review of the relevant literature lies beyond the scope of this paper which must necessarily be selective. Moreover the availability of sites suitable for comparison is limited by two factors. Firstly, few sites from sub-tropical arid or semi-arid areas have yielded comprehensive data in the relevant time range. Secondly, few of those data that are available have the degree of continuity and reliability of dating desirable to permit detailed comparisons to be made. Of those that are available, most are north of the equator, and the majority lie in northern Africa (Rognon, 1976; Rognon & Williams, 1977). Given the different climatic setting of the northern African and southern Australian continent the value of such comparisons must be treated with caution. However, two of the longest and best dated sequences from the equatorial side of the African desert warrant specific attention.

Figure 4 illustrates the sequence of lake level changes recorded from Lake Chad (southern Sahara) and Lake Abhé (Ethiopia) compared with those from southeastern Australia. Three features are worthy of note. Firstly, the beginning and duration of the major wet phase in the period 30–40 000 BP are approximately identical in all cases. Secondly, the peak arid oscillation represented by low lake levels was synchronous. Thirdly, and perhaps most important, the transition from wet to drier climates occurred considerably before the 18 000 BP glacial maximum. At Lake Chad, this transition is placed about 22 000 BP; in the tightly dated Lake Abhé sequence, Gasse (1975, 1977) places the event of major drying about 20 000 BP. However, this was preceded by a transition from a temperate to a tropical diatom assemblage about 25 000 BP, the transition from cool to warm occurring at a time when the record from most other parts of the world would indicate a temperature decline rather than an increase. Thus Lake Abhé became warmer at 25 000 BP when the Willandra lakes suddently began to fall. As in western N.S.W., a phase of renewed continental dune building in the Chad region followed the drying of the lakes (Servant and Servant, 1970). In northwestern Sahara, Alimen (1976, Fig. 4) has described a similar sequence with a wet phase (50 000 to 20 000 BP) followed by an arid interval from about 20 000 until 9 000 BP.

Pursued even further, events of Holocene age between the two continents present further striking similarities. The relatively wet conditions that developed in the early Holocene record of Africa (summarised by Street & Grove, 1976) are paralleled in the high lake levels of Lynchs Crater in North Queensland (Kershaw, 1974, 1975) in Lake George (Coventry, 1976; Singh, in Bowler et al., 1976) and in Lake Keilambete (Dodson, 1974; Bowler & Hamada, 1971). Indeed the water level curve from Lake Keilambete (Fig. 4) shows a remarkable similarity to that from Lake Abhé.

Considering that the African evidence is drawn mainly from the equatorial side, rather than the polar side of the anti-cyclonic belt, we may well question the significance of the similarity with the southern Australian record. Until detailed sequences become available from equivalent latitudes in northern Australia, the meaning of the similarity must remain speculative, but the reality of synchronous wet and dry phases between large areas of southern Australia and equally large areas of northern Africa is beyond doubt.

Perhaps the most significant aspect concerns the onset of aridity. As measured by falling lake levels and intensified dune building in both environments, it began long before the glacial maximum recorded in oceanic and high latitude evidence. It persisted throughout the glacial maximum and ended soon after. By 12000 BP dunes in Africa and Australia were stabilized after the long period of intensified aeolian activity, a chronology also similar to that of the Rajasthan Desert (Singh et al., 1972).

The comparisons set out above have involved making deliberate selections of data. Sequences are known in which the hydrologic budgets were out of phase with those dated from southern Australia. Thus the records from lakes in North America such as Lakes Lahontan and Bonneville (Broecker & Orr, 1958; Broecker & Kaufman, 1965) or Searles Lake (Flint & Gale, 1958; Smith, 1968) do not show the same relationships as those of the southern Sahara. The account of lake oscillations of Lake Alexandersfontein in South Africa (Butzer et al., 1973) may be similarly out of phase although the recent statement of Helgren & Butzer (1976: 34) that 'the + 7 shorelines are referable to the mid-Upper Pleistocene (⩾25 000 BP)' may suggest a closer resemblance than previously indicated. At first sight this may seem to provide a cogent case against attributing any special significance to those sequences that appear in-phase. However, the sites selected for comparison with southern Australia have a number of common climatic characteristics. Firstly, all lie well away from the immediate hydrologic and climatic influence of ice sheets, an influence capable of substantial local climatic modifications. Secondly, all possess climates with a high degree of continentality today, a factor that would have been accentuated in times of glacial low sea levels. These common elements strengthen the value of comparing data from such sites.

2. High latitude semi-arid regions

To test whether the similarities observed between the age and onset of aridity in Africa and Australia, have significance in terms of global synchroneity, they can be compared with glacial age aeolian events at higher latitudes. Unfortunately the chronology of late Quaternary events from the extensive cold deserts of Patagonia and Mongolia are not well known, but the comparison may be usefully pursued by examining the chronologies of extensive loess deposits of the Northern Hemisphere, especially the dry loess of Fink (1969).

As a result of the activities and publication of the INQUA Loess Commission, the regional extent, age and stratigraphical relationships of loess deposits of southeastern Europe are now well documented. Figure 5 summarizes those sequences of the last glaciation that have been dated by radiocarbon analysis. Several features are particularly worthy of note.

Firstly, the interval from 45 000 until after 30 000 BP was characterized by widespread stability and soil formation. This was a stable period represented in Austria by the Stillfried B soil (Fink, 1969) or in southwestern USSR by the Briansk soil of Velitchko (in Ivanova, 1969). More importantly the onset of aeolian deposition as recorded by the C-14 evidence dates from about 24 000 BP over a relatively large region. Deposition continued then to about 10 000 BP in parts of the USSR (Ivanova, 1969).

In North America the similarity in the ages of initiation of the last loess depositional phase is equally striking. The final Wisconsinan loess in the upper Mississippi–Missouri Valley is underlain by the widespread Farmdale palaeosol (Ruhe, 1968; 1969). From detailed radiocarbon studies Ruhe specified the age of the onset and duration of aeolian deposition; it extended from between 24 700 until 14 000 BP. Thus there is a high measure of agreement between the ages of major soil forming and aeolian events on both sides of the Atlantic.

3. Oceanic and Antarctic correlations

For the last glacial maximum, evidence from elsewhere indicates a contribution of dust to ocean sediments considerably above that of today. Most such evidence is derived from the Atlantic Ocean and records increased deflation from Saharan Africa (Biscaye, 1965; Bowles, 1973; Parkin, 1974; Sarntheim & Walger, 1974). Moreover Parkin & Shackleton (1973) have demonstrated the strong correlation between increased continental dust and ^{18}O maxima; the higher contribution of dust is thereby related directly to periods of increased glacial ice volume.

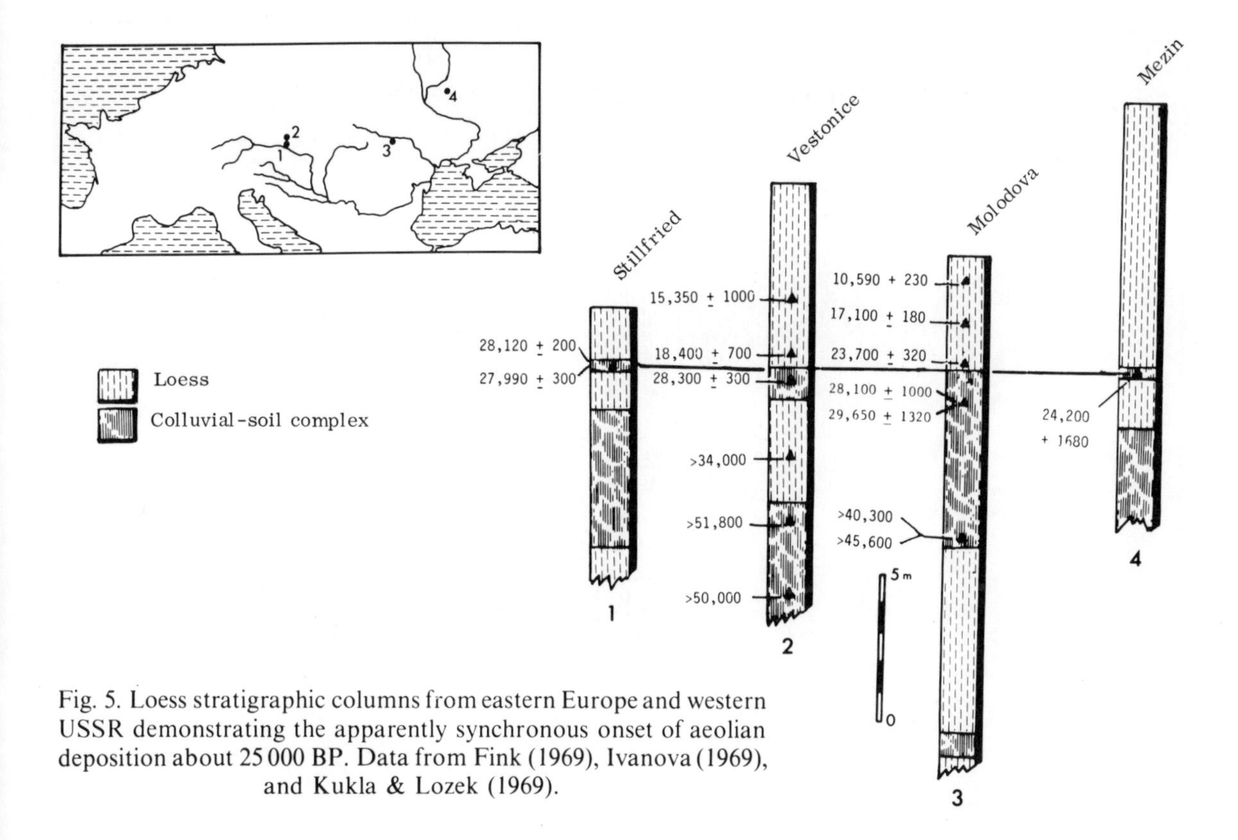

Fig. 5. Loess stratigraphic columns from eastern Europe and western USSR demonstrating the apparently synchronous onset of aeolian deposition about 25 000 BP. Data from Fink (1969), Ivanova (1969), and Kukla & Lozek (1969).

Turning to Antarctica, the abundance of microparticles (0.65–0.82μm) in the Byrd ice core, interpreted by Thompson et al. (1975) to represent a global rather than local dust component, finds its peak in the interval 25 000 to 14 800 BP agreeing almost exactly with the beginning and end of the dune building period determined from southeastern Australia. Moreover, events of this period coincided with the maximum loess deposits of the Northern Hemisphere and with dune building in Saharan Africa.

EARLIER PLEISTOCENE CORRELATIONS

The evidence points to a strong correlation between glacial maxima and increased dust in regions of both high and low latitudes. Some measure of the validity of this correlation throughout earlier Pleistocene cold phases is provided by long sequences available from deep sea cores and continental loess sequences.

1. Atlantic and Czechoslovakian evidence

Correlation of high dust peaks with ^{18}O maxima has been carried by Parkin & Shackleton (1973) back through successive glacial phases to the beginning of the Brunhes epoch. Each successive isotope peak representing increased ice volume is matched by an increase in the quantity of Saharan dust contributed to the Atlantic downwind.

A similar correlation has been established for the cold loess sequence of Czechoslovakia where Kukla (1970) demonstrated the presence of eight depositional cycles above the Matuyama reversal. More recently Fink & Kukla (1977) have extended the loess — ^{18}O correlation to the base of the sequence establishing the presence of some seventeen cycles throughout which each phase of increased ice volume is matched by a period of loess deposition.

2. Evidence from loess of China

No discussion of loess would be adequate without reference to the classical sequence of China, the area that first provided Baron von Richtofen (1882) with persuasive arguments in favour of a definite aeolian origin. A visit to China in 1975 by a group of Australian Quaternary scientists enabled us to see at first hand some typical sections and to discuss with the Chinese research workers responsible, the most recent and very significant advances made in the study of these deposits.

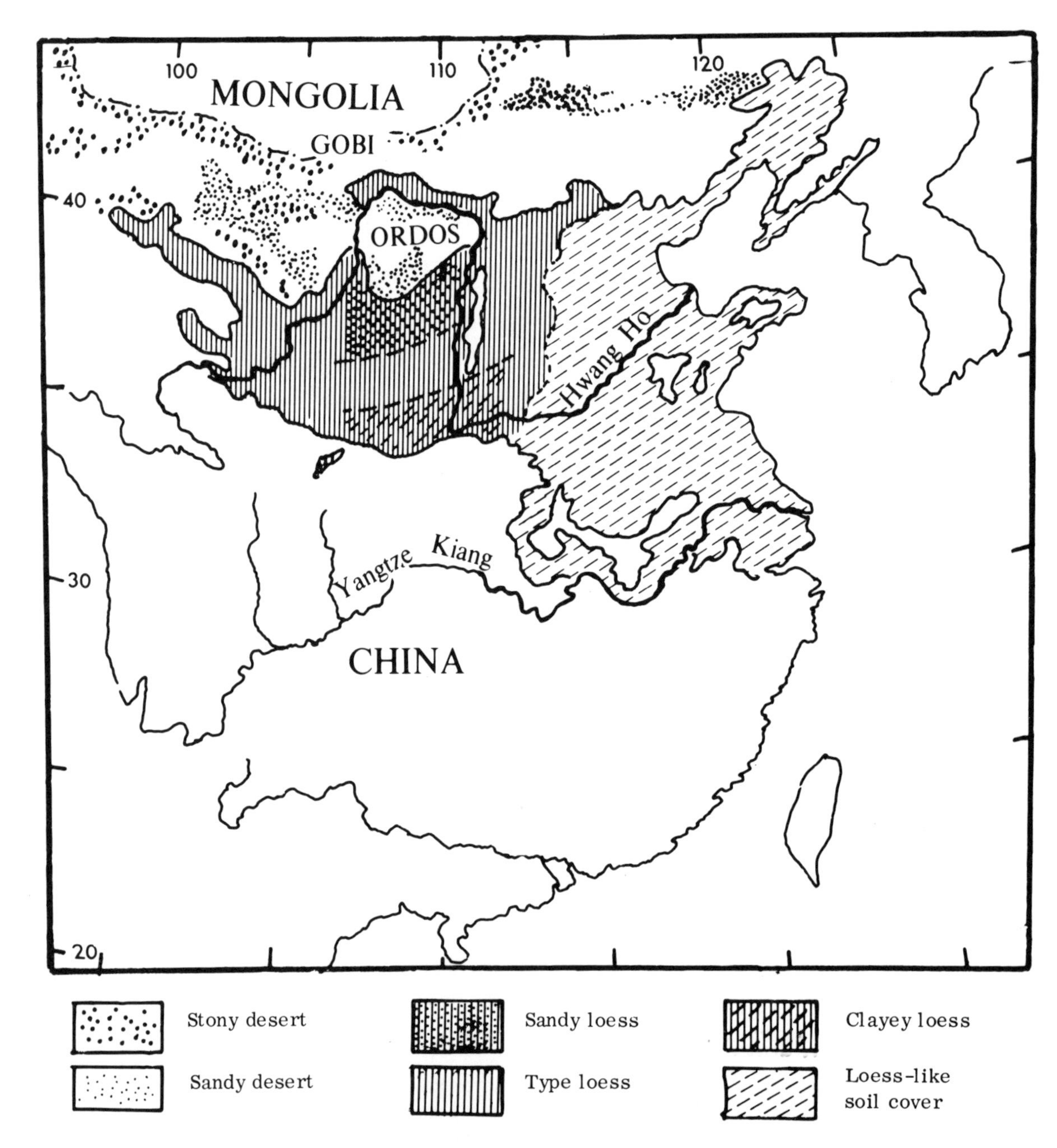

Fig. 6. Map showing the distribution of loess and associated deposits of north China (from Lu Yanchou et al., 1976).

Extending over an area of some 300 000 km^2 centred on the arid and semi-arid region of the Hwang Ho (Fig. 6), the loess deposits reach a maximum thickness of 175 m. Throughout the extensive plateau of Shansi and Shensi provinces, where loess overlies uplifted and horizontally bedded Mesozoic sediments, slope failure on steep valley sides often exposes spectacular vertical sections. Intensive studies of stratigraphy, sedimentology and faunal content initiated under the direction of Professor Liu Tung-Sheng are now being continued by younger workers from the Kweiyang Institute of Geochemistry.

From northwest to southeast, the sediments become progressively finer reflecting distance from source, believed by the Chinese scientists to lie in the glacially and periglacially affected provinces of the Gobi desert and its surrounding highlands with a possible contribution from fluvioglacial outwash from the Hwang Ho floodplain. However, the

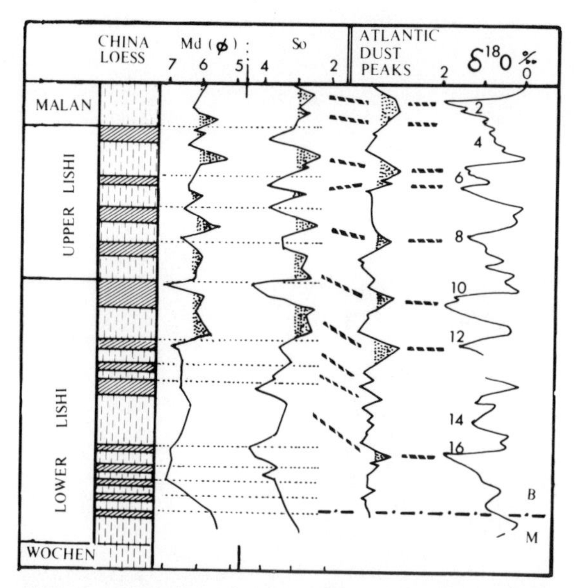

Fig. 7. Stratigraphic sequence of Chinese loess showing multiple episodes of aeolian deposition (vertical hachuring) and soil formation (cross hachuring) above the Brunhes-Matuyama boundary. Note increased coarseness expressed as medium diameters (Mdø) and better sorting (So) near centre of loess units. Loess stratigraphy from Liu Tung-Sheng et al. (1969) and Li Hua-Mei et al. (1974). The loess units are correlated here with the marine oxygen isotope and Atlantic dust curve of Parkin & Shackleton (1973). Even numbers represent cold stages as used by Shackleton & Opdyke (1973).

presence of loess on the north side of the Hwang Ho confirms that at least some of the loess originated from the stony desert regions further north.

The climatic environments that accompanied loess deposition and subsequent soil formation have been reconstructed on the basis of molluscan, pollen and vertebrate assemblages. Aeolian accumulation took place under dry steppe conditions with reddish brown calcareous palaeosols formed during stable periods of increased humidity. Deposition is equated with cold glacial phases while soils formed during interglacials. Thus the association of glacial events with dry aeolian episodes is again substantiated from the Chinese sequence.

The loess sequence has been subdivided by Liu et al. (1966) on the basis of the palaeosols found within it (Fig. 7). Four formations are defined, the boundaries being drawn at major pedogenic horizons. From top to base they are Malan (representing deposits of the last glacial period), Upper Lishi, Lower Lishi and Wocheng overlying red Pliocene *Hipparion* deposits. The Brunhes-Matuyama palaeomagnetic boundary, identified near the Lower Lishi-Wocheng contact (Li Hua Mei et al., 1974), is overlain by at least 12 palaeosols.

From the pattern of sediment and intervening soil thicknesses, the Chinese loess sequence can also be correlated with the marine oxygen isotope curve. In Figure 7 the sequence of Liu et al. (1966) is shown with textural and sorting peaks reflecting different degrees of wind transporting capacity, each appearing to be most effective near the mid-point of the loess units. The correlation of loess peaks, and the Atlantic record of Parkin & Shackleton (1973) is based not only on the number of units but on the recurring relationship between thickness of units and size of each equivalent event as registered in the ^{18}O record. The correlation presented tentatively here (to be discussed more fully elsewhere) demonstrates a sequence which bears a strong resemblance to that of southeastern Europe and reinforces the relationship between ice volume changes and loess depositional episodes over wide regions throughout the past 700 000 years.

GLACIAL AGE WIND REGIMES

The contribution of continental dust to the oceans during glacial times was not only larger in quantity than during interglacials, but the size of the particles was also considerably increased. From detailed analyses of such dusts Parkin (1974) and Parkin & Shackleton (1973) have argued for a regime of increased vigour during glacial episodes, a conclusion that finds independent support in the amount of saltation load contributed by the Sahara to the Atlantic Ocean (Sarntheim & Walger, 1974). Additional studies based on deep sea cores (Biscay, 1965; Parmenter & Folger, 1974; Diester-Haas, 1977) provide evidence consistent with this interpretation. Moreover, in the Chinese loess sequence, the better sorting and coarser textures of the mid-points of the loess units (Fig. 7) suggest stronger winds with more effective carrying and sorting capacity than those that controlled the beginning and end of the loess episodes.

Evidence of changes in glacial age wind regime of the southern hemisphere are restricted mainly to the Namib and Kalahari Deserts of South Africa (van Zinderen Bakker, 1976; and to the desert dunes of Australia (Sprigg, 1965; Brookfield, 1970). From the orientation of dunes of southern Australia (Fig. 2), Bowler (1975) suggested that the summer wind regime of the last glacial episode possessed a stronger westerly and north-westerly component than today. This is perhaps most strikingly demonstrated by the trends of dunes on Kangaroo Island (Figs. 8 and 9). Here a sequence of glacial age short linear dunes and transverse lakeshore dunes (lunettes) developed under the influence of prevailing northwesterlies.

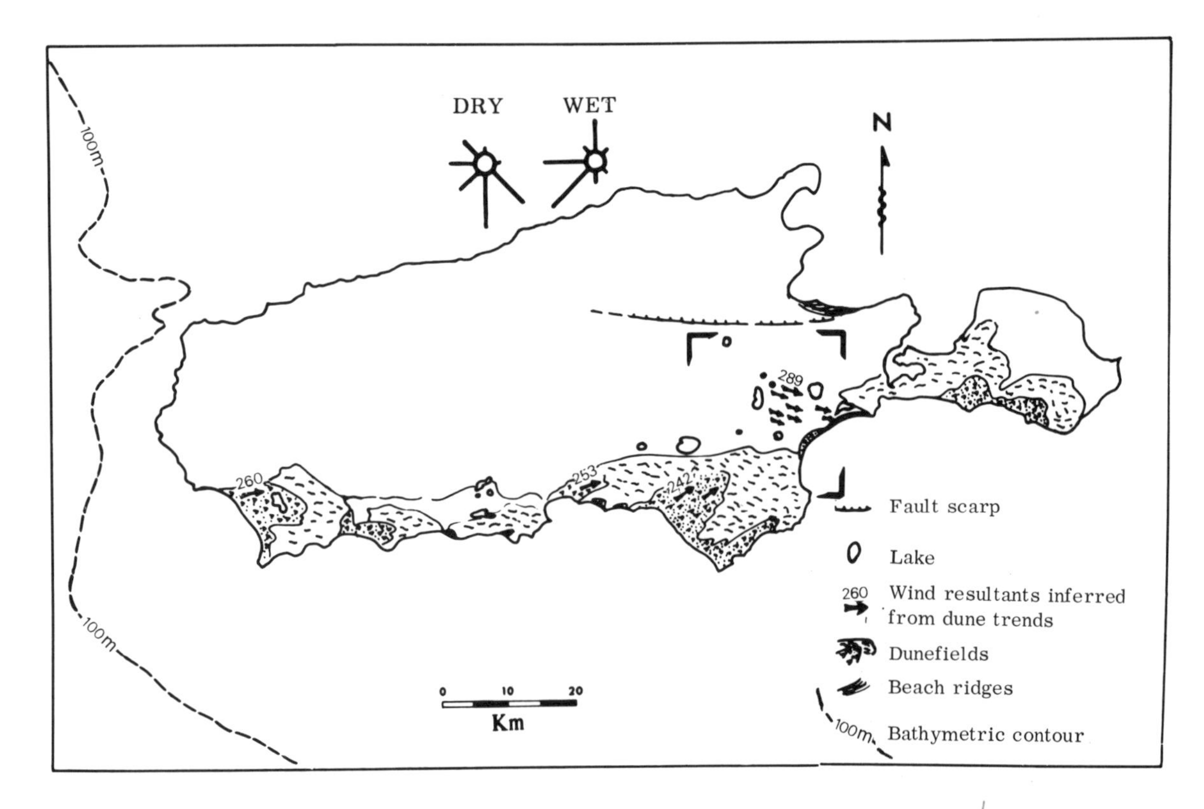

Fig. 8. Regional map of Kangaroo Island (Fig. 2) showing trends of Late Quaternary continental dunes compared to post glacial shoreline dunes. Wind roses for wet and dry seasons after Campbell (1968). For details of inset, see Fig. 9.

Orientation of the linear dunes is at about 290°, their regional trends representing the continuation of the desert dunes that extend from the continental core. At the time of their development, the Pleistocene shoreline would have been located only 70 km south of the island near the present 100 m isobath.

Following the post-glacial sea level rise, a new set of coastal dunes developed, the trends of which lie dominantly between 260° and 240° some 30–50° out of alignment with their Pleistocene predecessors. Moreover, in the post-glacial regime, lunettes now developed on White Lagoon reflect the influence of southwesterly winds whereas northwesterlies dominated lunettes development during full glacial times.

Several important implications follow from these observations. Firstly the glacial age winds that were responsible for construction of longitudinal and transverse dunes (lunettes) over large areas of southern Australia had their origins deep in the continental interior (Bowler, 1975). They originated largely from outbreaks of continental air masses travelling to the southeast from the dry, high isolation areas in Central Australia.

Secondly, these winds were capable of constructing continental dunes to within a few kilometres of the coast in a region of reliable winter rainfall and at a time of globally reduced evaporation losses. The ability of such continental winds to dominate over moister onshore southwesterlies points to a greater increase in the frequency and strength of the northwesterlies.

DISCUSSION

1. Limitations of Australian evidence

The long sequences available from dust peaks in deep sea cores, Northern Hemisphere loess deposits and the ^{18}O record demonstrates repetition of glacial-aeolian environments throughout the last 700 000 years. In southeastern Australia, aeolian deposits of the last glacial age correlate well with those from other parts of the globe. However, the known sequences of episodic deposition and soil formation are limited to four or five cycles. There is no equivalent of the many cycles that characterize the loess of Brunhes age in Czechoslovakia or China. Nor can the Australian aeolian sequence yet match the number of Saharan dust peaks in Atlantic Ocean

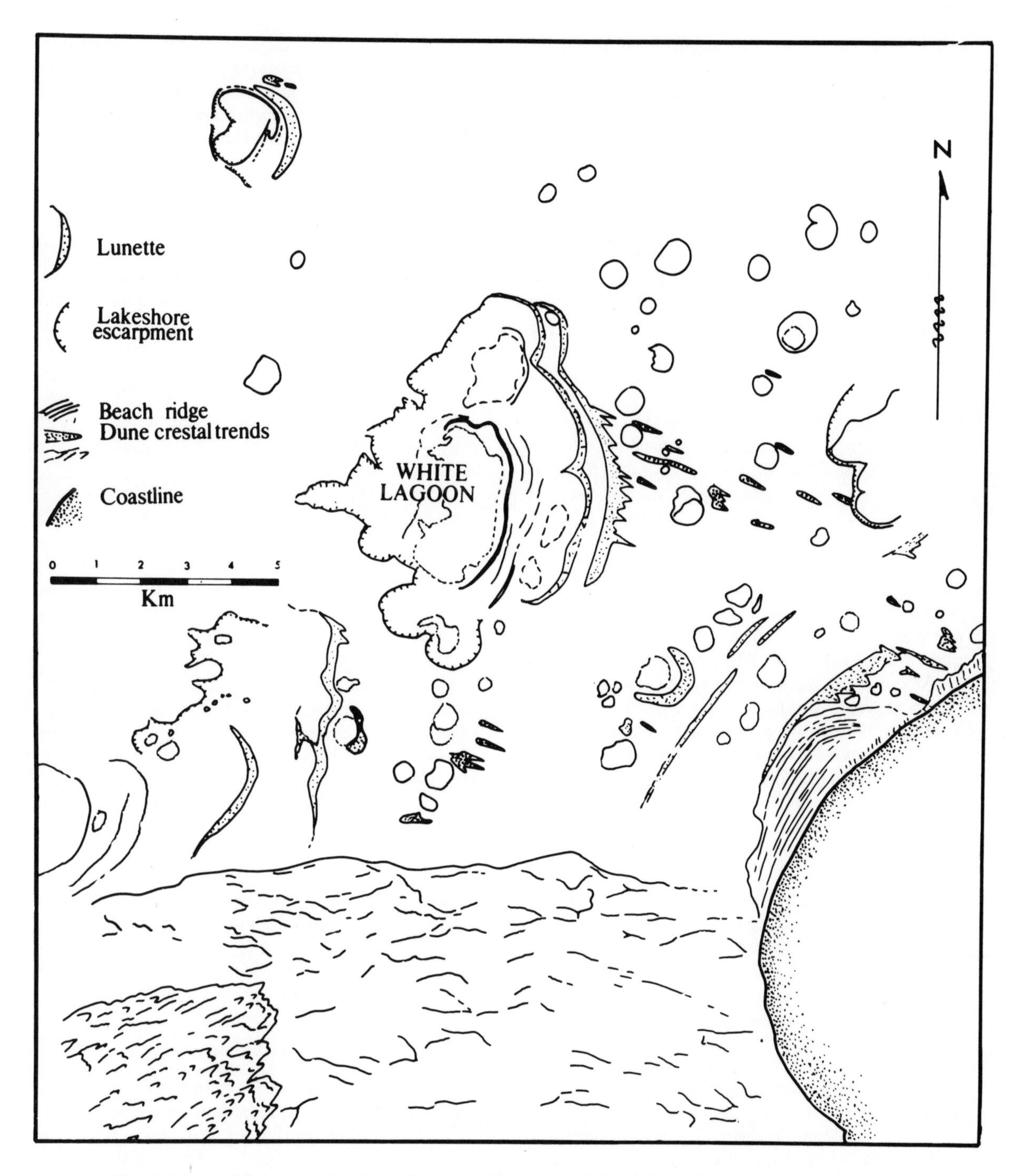

Fig. 9. Map of dune trends of southeastern Kangaroo Island (Fig. 8) showing relationship between transverse dunes (lunettes) and 'tear-drop' longitudinal dunes constructed during last glacial maximum. Trend of longitudinal dunes is nearly 50° out of alignment with that of post-glacial dunes transgressing from southwest.

cores. The reasons for this apparent anomaly are possibly three-fold.

Firstly, it may be a matter of discrimination. The known episodic depositional – soil forming phases have been identified from stratigraphic units within longitudinal dunes (Churchward, 1963) and lunettes (Bowler, 1971) in southeastern Australia. These units are capable only of rather 'coarse' resolution.

Secondly, the data seem to record only the highest intensity events. Whilst aeolian activity peaked about 16–17 000 BP, close to the ^{18}O maximum, no corresponding response is observable during the earlier cold phase between 64–75 000 (Stage 4, Fig. 10), an event clearly registered in the ^{18}O record. In the Atlantic Ocean a drastic drop in sea surface temperature during Stage 4 exceeded any later reduction. Thus both the summer and winter temperatures of the eastern Atlantic equatorial zone as indicated by core A180–73 (Gardiner & Hays, 1976) show greatest temperature reduction during this interval. Similarly, off the west coast of Australia core RC 9–150 (Bé & Duplessy, 1976) indicate an ^{18}O maximum in Stage 4 equivalent to that of Stage 2 although analyses of *Orbulina universa* diameters suggest the subtropical convergence moved further towards the equator during the subsequent cold phase about 18 000 BP. The apparent failure of the lakes and dunes of western N.S.W. to respond to events of Stage 4 indicates that, in this part of the Southern Hemisphere at least, the continental climate of this interval was not as rigorous as in the 18 000 BP glacial maximum.

A third factor also affects aeolian events in the sub-tropical semi-arid zones. Construction of desert margin dunes both in Australia and Africa occurred after a long period of high lake levels and high water-tables. Some dune building was a consequence of that transition from high to low water-tables. This is particularly true of the lunette clay-rich dunes, the major construction and preservation of which is achieved only during such a hydrologic transition (Bowler, 1973). Similarly the reactivation of the longitudinal dunes of western N.S.W. may have been facilitated by salinization of swales following a long period during which the surface supply of water exceeded evapo-transpiration loss. With a reversal of the hydrologic budget, the accumulation of surface salts from evaporating groundwater would provide conditions in which vegetation destruction was accompanied by efflorescence of salts breaking sandy clays into soft dry pelletal layers suitable for deflation. The high clay composition of such dunes and the presence within them of sand-sized clay pellets supports this interpretation.

Thus the aeolian units recognized in southeastern Australia may have developed only during transitional phases when wet conditions were succeeded by very dry ones. In this way the Australian data may be registering a response only to combined events of major proportions. Many of the smaller oscillations may not be recorded. Resolution of this problem awaits tighter chronologic control of pre-Mungo deposits, possibly through palaeomagnetic studies.

2. *Synchroneity and two-stage nature of glacial age events*

Events of the last pleniglacial summarised diagrammatically (Fig. 10) demonstrate similarities in the timing and nature of hydrologic changes from Australia, Africa and European aeolian deposits. The two-stage nature of these environments described from Australia has an obvious parallel in hydrologic changes registered in Lake Chad and Lake Abhé. A major wet phase gave way to dry conditions at about the same time.

In the higher latitude loess sequences the two-stage association is expressed in the transition from stable soil-forming conditions to aeolian depositional environments, the transition again occurring about 25 000 BP as in Australia. Even in Antarctica the coincidence of the global dust peak in the interval 25–14 000 BP emphasises the widespread effects of dry dusty conditions.

The evidence demonstrates that not only were the low latitudes of both hemispheres responding in the same sense at the same time (Williams, 1975) but, perhaps more importantly, the same degree of synchroneity existed in the aeolian deposits between low and high latitudes. Thus re-activated dune building in the hot deserts of Australia, India and Africa occurred simultaneously with aeolian accumulation in the cold semi-arid loessic regions of Europe, USSR, and the Upper Mississippi Valley of North America. Although not yet accurately dated, the Malan loess of China is almost certainly of similar age. The interval between 25 000 and 14 000 BP might properly be called 'a global dust phase'.

3. *Aridity and glacial wind regimes*

Regional aridity may be explained by a worldwide reduction in oceanic evaporation loss (Flohn, 1953), by shifts in the location of rain bearing systems (Rognon & Williams, 1977) or by intensification of anti-cyclones (Wyrwoll & Milton, 1976). While these factors may each have been important, none by itself provides a satisfactory mechanism to explain the synchroneity observed in the data between the different climatic zones.

We have been considering data drawn primarily from seasonal water deficiencies represented by falling lake levels, dune and loess deposition. This may occur in either of two ways; by reducing the local and regional rainfall or by increasing the

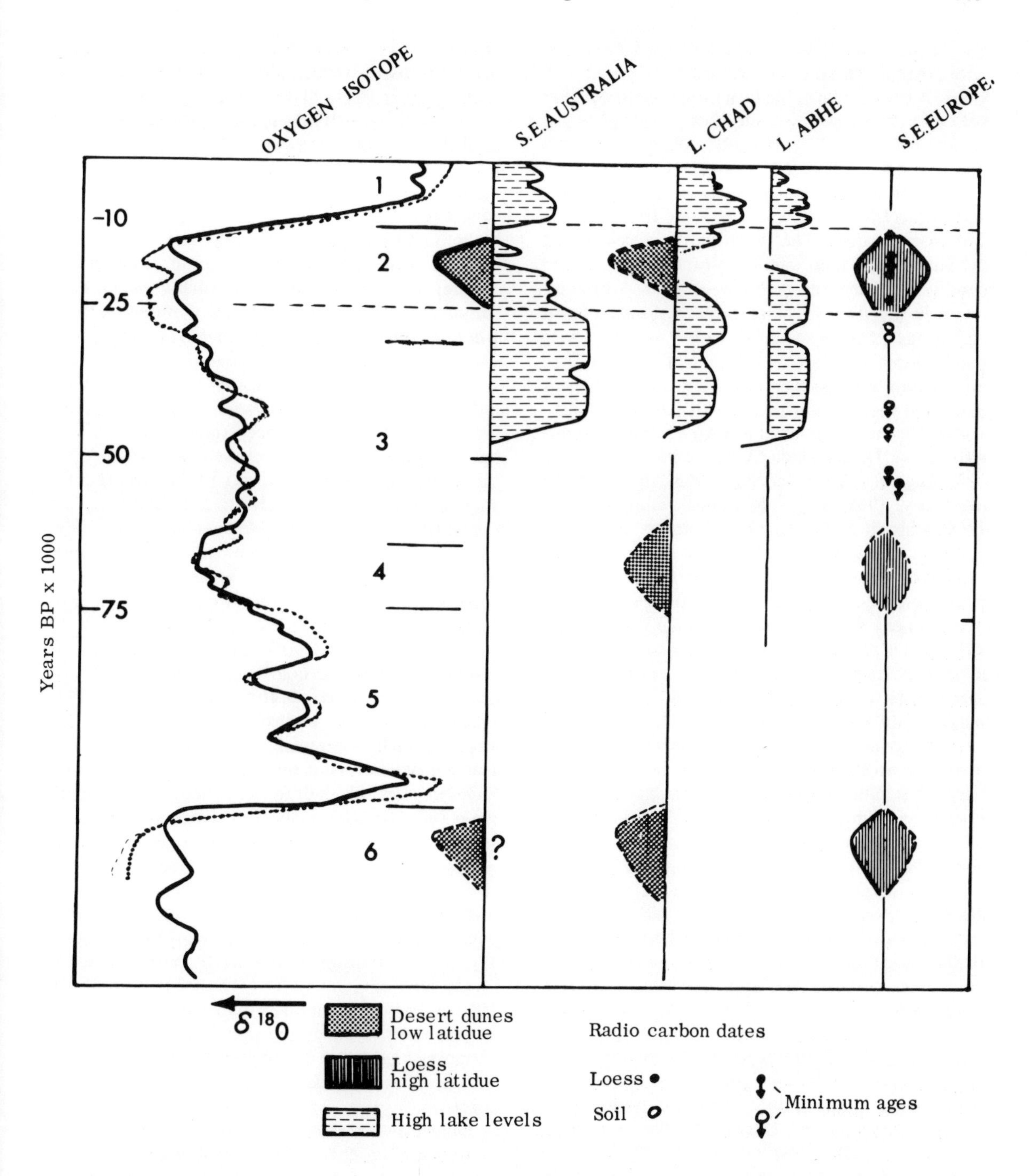

Fig. 10. Summary curves of hydrologic and aeolian events of last glacial age as known from southeastern Australia and northern Africa compared with European loess sequence and marine oxygen isotope curves. Age matching of isotope curves is based on the stage boundaries dated by Shackleton & Opdyke (1973). Broken lines of dust peaks represent limits of reliability. In isotope curves, full line represents record of V19–28 (Ninkovitch & Shackleton, 1975); dotted line RC 11–120 (Hays et al., 1976).

evapotranspiration loss. While the first mechanism is theoretically feasible and is at least consistent with some of the evidence, the latter appears in conflict with a world wide trend towards depressed pleniglacial temperatures at the beginning of the major arid phase. But can it be ruled out on that account?

The fall in lake levels between 25–20 000 BP involved a transition from a positive to a negative hydrologic budget. The synchronous development of dunes and loess points towards dry windy conditions in the absence of vegetation. Although dunes do not always require the development of arid conditions, their association with low lake levels points firmly to a water deficit in the landscape. The mechanism responsible was such that whilst it destroyed vegetation it created extensive dunefields, not only in Australia, northern Africa and India but also in the Orinoco and Amazon basins (Tricart, 1975; Bigarella & de Andrade, 1965), in the Congo (de Ploey, 1965) and in Rhodesia (Flint & Bond, 1968) whilst dust was carried poleward to Antarctica.

The prevalence of glacial age aeolian deposits thus extended from within a few degrees of the equator to polar latitudes. Reduced precipitation assisted the development of such aridity but even a fifty per cent reduction is not likely to have produced such effects in tropical regions at a time when expanded ice sheets were reducing global evaporation rates, thus making rainfall more effective.

A change in the nature of atmospheric circulation involving acceleration of surface winds may have played a significant role in bringing about the transformation from wet to dry regimes. Within Australia, the evidence from Kangaroo Island demonstrates the resultants of sand shifting winds of this time. Compared with the trends of modern dunes of this area, those of glacial age trace paths of winds that originated in the continental interior. However, further north the trend of linear dunes in the arid cores does not diverge significantly from modern sand shifting resultants (Brookfield, 1970). The Pleistocene circulation is consistent with intensification of the general circulation that is evident today with variations between dune trends and modern winds being apparent only on the southern margin of the anticyclonic cell. Here post-glacial relaxation of wind stress has resulted in less unidirectional flow. The Kangaroo Island dunes lend support to the claim that summer dune building winds of pleniglacial time, with a strengthened northwesterly component bringing air from the interior, were hot and dry (Bowler, 1975).

A change in the frequency of occurrence and in velocities of winds flowing from the continental interior helps explain some apparent anomalies. It provides a mechanism whereby increased seasonal evaporation would result from steepened meridional temperature gradients. Moreover, it would produce conditions in which high winter runoff would maintain high stream discharges simultaneously with summer drought conditions on the western plains, a condition that persisted in the fluvial record until about 15 000 BP (Bowler, 1976; 1977). Increased ground-level wind velocities with a consequent increase in near surface turbulence under dry adiabatic conditions would enlarge the vertical heat flux by converting radiant to sensible heat. This would then be advected out of the region towards the moister continental margins. In such a system high summer insolation in the arid continental core would ensure local diurnal temperatures approaching those of today. Any temperature reduction consequent on glaciation at high altitude and high latitude would have been much attenuated in the arid continental core. Temperature depression may have been amplified in winter but would be cancelled out in summer by the efficiency of direct radiation. Indeed, given a more effective mechanism for converting radiant to sensible heat as would follow increased turbulence under a more vigorous regime, pleniglacial summers in such areas may have been capable of producing evaporation as high or even higher than today's.

In today's climate outbreaks of hot dry northwesterlies from the continental interior produce large increases in evaporation rates (Loewe, 1943; Hounam, 1961). In Pleistocene times the evidence suggests such outbreaks were more frequent and reached new levels of intensity. Such winds would have been capable of drastically changing the hydrologic balance by transporting large quantities of sensible heat during the summer season to areas where the vegetation may already have been subject to considerable stress due to the earlier onset of cold and frosty conditions to which the Australian eucalypts are ill-adapted (Gentillii, 1960; Parsons, 1968). Any intensification of northwesterly summer winds such as seems to have occurred about 25 000 BP would impose an additional evaporation stress; it would at the same time have lowered lake levels, increased vegetation destruction and thus facilitated the simultaneous development of various dune forms.

Dune trends and sand flow lines of the Australian continent (Fig. 2) help to demonstrate how longitudinal and transverse dunes (lunettes) were formed. The trends reflect a pattern of long overland fetch for all dune building winds. Even those winds that built longitudinal dunes trending WNW–ESE on the northeast coast of Tasmania would have travelled long distances over dry land during times of low sea level. The entire anti-clockwise whorl of the Australian dunefield therefore reflects the influence of continental air masses.

A similar pattern exists in the sand flow lines of northern Africa where a clockwise orientation is centred on the continental core (Mainguet & Canon, 1976). Thus if the development of high velocity continental winds were responsible for high summer evaporation rates in southern Australia, a similar process may help explain the hydrologic changes that occurred about the same time in northern Africa. Seen in this light, the appearance of tropical diatoms in Lake Abhé at 25 000 BP (Gasse, 1977) may reflect the onset of increased advective heat flow, a change that dried that lake out a few thousand years later. The explanation for the increased aridity advanced here from Australian evidence seems in good accord with the northern African dates. It may also hold true for other continental regions such as tropical Africa and South America.

4. Ice expansion as a controlling mechanism

The mechanism proposed to explain the synchroneity of aeolian events throughout widely different areas involves a substantial increase in the strength of glacial age winds. It is appropriate to consider what may have caused such widespread and simultaneous changes.

Environments in the Northern Hemisphere were likely to have been affected by proximity to continental ice sheets. The general synchroneity between loess depositional events in Europe and North America about 20 000 BP may reflect the southerly ice expansion about that time. Dreimanis & Karrow (1972) have documented the expansion of the Laurentide ice sheet after 30 000 BP. A contemporary expansion of the Scandinavian sheet and alpine glaciers onto the plains of central Europe with the construction of the Brandenberg moraines (Flint, 1971; Cepek, 1965) was associated with the onset there of loess deposition. But were these Northern Hemisphere events sufficient in their magnitude to have produced simultaneous changes from wet to dry environments across northern Africa and from the west to the east coast of southern Australia? Was the latitudinal shift of the ice front of such magnitude as to create virtually new climatic conditions over such large areas of the globe?

Although the change in the Laurentide ice sheet between 30 000 and 20 000 BP was extensive over North America it was not continuous aroung the Arctic. A corresponding change took place in eastern Europe, but the overall latitudinal shift in Arctic circulation would have been considerably less than the 8° latitude involved in the Laurentide advance. Averaged around the Northern Hemisphere the effective area of increased influence was probably less than 3 or 4° latitude and therefore unlikely to represent the primary cause of the drastic changes recorded from the Southern Hemisphere.

A further line of evidence is available from the ^{18}O record of ice volume changes at this time. Examination of the curve of Hays et al. (1976a), together with the high resolution record of Ninkovitch & Shackleton (1975), is particularly instructive (Fig. 10). The period 13 000 to 32 000 BP corresponding to Stage 2 (Shackleton & Opdyke, 1973) spans the end of the wet phase and the entire global dust phase. However the downturn in the ^{18}O curve near 25 000 BP provides little indication of the very large changes that took place on land at this time. Some indication of changes in oceanic thermal structure may be provided by the increased abundance of the radiolarian *Cycladophora davisiana* at this time in core RC11-120 from southwest to Western Australia (Hays et al., 1976a, Fig. 2). But, in general, both oxygen isotope curves and estimates of sea surface temperatures register changes that are disproportionately small when compared with the large climatic variations of middle and low latitudes. So great are the differences in the continental record before and after 25 000 BP that the controlling atmospheric circulation would almost appear to have entered a different mode about that time.

The construction of glacial age dunes in southern Australia represented outbreaks of hot dry continental air rather than the cold, moister maritime southwesterlies that prevail in many southern areas today. When this hypothesis was proposed (Bowler, 1975) the nature of the controlling mechanism was not apparent. Almost simultaneously results from the Southern Ocean provided what appears to be the forcing mechanism. The large northward extension (5° to 7°) of Antarctic summer sea ice at 18 000 BP in the Atlantic and southern Indian Ocean sector (Hays et al., 1976a) would have had a significant effect on the generating areas of southern Australia's climate. If continuous around the Antarctic continent, such changes may have accelerated Southern Hemisphere westerlies and, at the same time, extended the global influence of corresponding changes in the Northern Hemisphere. Moreover, in doing so, conditions of continentality already extended by lowered sea levels would be accentuated. Over large mid-latitude regions, the new wind regimes would assist in producing greater seasonal contrast in which cold stormy winters alternated with hot dry dusty summers. The southern Australian evidence is consistent with such an interpretation.

Synoptic conditions that favour outbreaks of hot dry continental air masses commonly form today when an anticyclone, situated south of the continent, gives way to a westerly travelling cyclonic depression. The frontal situation thus developed produces strong northerlies and northwesterlies which, under drought conditions, produce dust storms that sweep across the southeast, sometimes extending as far as New Zealand (Loewe, 1943). Any intensification of equatorward compression of the sub-Antarctic pressure cells such as may have occurred following the northerly displacement of the Antarctic convergence demonstrated by Hays et al. (1976a) may have accentuated these conditions.

The magnitude and timing of the change recorded about 25 000 BP, involving the transition from wet to dry conditions in southern Australia and from soil formation to loess deposition in the Northern Hemisphere, suggests important modifications at high latitude in the Southern Hemisphere, perhaps equivalent to those known from Europe and North America. It is tempting to suggest that the northward displacement of the Antarctic Convergence dates from 25 000 BP. This would assist in producing the changes in atmospheric circulation at this time that both the Australian and Northern Hemisphere evidence seems to require. Moreover, a relatively large expansion in the area of sea ice would be accomplished without a great addition to the world's total ice volume. Thus a change of this nature would be consistent with the relatively small change in the ^{18}O oceanic ice volume record between 30 000 and 18 000 BP.

5. *Consistency with model simulations*

The observations presented from southern Australia and the explanations proposed may be tested against results obtained from model simulations of ice age conditions. The general trend towards glacial aridity is reproduced in both the NCAR models of Barry & Williams (1975) and the Rand simulation of Gates (1976). Estimates of reduced winter precipitation range from 50% in the NCAR version to about 23% reported as a Southern Hemisphere average by Gates. Additionally both models reproduce a substantial lowering of air temperatures over the continent. July temperatures over inland Australia range 5° to 9° lower than now with a January reduction of 6° to 8° reported by Barry and Williams.

Whilst the models reproduce changes up to 50% in winter rainfall, the temperature reductions postulated involve large reductions in evaporation. Using modern temperature-evaporation relationships from Lake Keilambete in western Victoria, a mean 6° drop in temperatures (conservative by model standards) would reduce present evaporation (99 cm/yr) to about 43 cm/yr. To dry Lake Keilambete under such conditions would require that glacial precipitation was reduced by 65%, a change that is considerably greater than the models suggest.

The explanation advanced here of enhanced evaporation brought about by changes in the wind systems simultaneously with reduced precipitation is neither confirmed nor invalidated by the simulated ice age conditions. Whilst the surface pressure levels reported by Gates are slightly higher than now, due to lower sea level, the meridional pressure gradients are little changed. However the position of the January low pressure cells are displaced 5–10° equatorward in the NCAR model consistent with the northern displacement of anticyclones postulated by Bowler (1975) to help explain the difference between Pleistocene dunes trends and modern wind resultants.

In terms of the 25–18 000 BP climatic paradox recorded from continental environments, when extremes of wet and dry developed in succession, the 18 000 BP simulations cannot provide the answers. We must wait to compare them with further palaeoclimatic simulations that focus on environments that existed before 25 000 BP.

6. *Summary of glacial environments in semi-arid southeastern Australia*

The conditions that prevailed throughout late pleniglacial time in southern Australia were dominantly dry with a great expansion of desert conditions into what is now the sub-humid and even humid margins of the continent. Although mean annual precipitation may have been considerably lower than today, this would have been off-set to a large extent by the colder conditions that prevailed at least for part of the year. Development of arid conditions on such a vast scale was assisted by the nature of the wind regime that was initiated about 25 000 BP. The summer winds of this period are seen as being derived from hot, dry continental air masses passing with accelerated velocities from northwest to southeast.

The overall effect of enhanced summer advective heat flow would have been to produce conditions of greater seasonality coincident with the glacial maximum, conditions more severe than any others experienced within the past 100 000 years. The relevance of high summer evaporation rates proposed here must be considered in the context of the effects that the equatorward shift in the Antarctic Convergence would have produced on Australian winters. Cold maritime air masses would not only have penetrated more deeply into the continent, but their temperatures would have been much lower than in the winters of today, bringing rain and snow further north than in post-glacial winter regimes. Although the moisture carrying capacity of the cold air masses may have been reduced, this effect was probably more than offset by increased run-off. Despite this factor, the seasonal evaporation loss after 25 000 BP was of a magnitude sufficient to lower lake levels and permit deflation from lake floors as well as to maintain active dunes over large regions of predominantly winter rainfall.

7. Implications

The regional evidence in general and the Australian evidence in particular suggest that intensified aeolian activity lasted some 12 000 years from 25 000 until about 13 000 BP. Moreover the repetition of these conditions back through Pleistocene time, as in Atlantic core dust peaks and loess sequences, means that each of the terminations characteristic of cycles in the oxygen isotope record (Broecker & van Donk, 1970) was preceded by a phase of greatly accelerated aeolian activity. Depending on its magnitude and duration each such event would have produced corresponding modifications to both earth surface forms and atmospheric heat balance. Whilst many of the geomorphic effects are obvious, the influence of such changes on local and regional thermal budgets has received scant attention.

We should explore further the implications of the global increase in atmospheric dust and advective heat transfer implied in the evidence from the deserts of Australia and North Africa. The effects of increased atmospheric dust on surface and tropospheric temperatures have received much attention in recent years (Kellog et al., 1975). The 'thermal-blanket' theory (Idso, 1974) and earlier theories of dust-induced heating (Bugaev et al., 1952; El-Fandy, 1949) take on new significance when considering the consequences of the pleniglacial dust episodes and the rapid return to interglacial conditions that always followed them. Could the thermal blanket effect, coupled with increased heat flow from high insolation areas, produce sufficient feed-back to modulate the growth of ice sheets and eventually assist in their swift destruction? In view of the wide synchroneity of glacial age aeolian processes with its implication of increased sensible heat transfer this becomes a real question, answers to which must be sought by further testing the hypothesis against field studies and numeric models.

Viewed over the longer time scale, the repeated coincidence throughout the Quaternary period of dust episodes with ^{18}O maxima, calls for additional analyses of both oceanic and the continental records of the Southern Hemisphere. Systematic changes in dust components through time, apparently related to ice volume thresholds, may provide a most important tool for inter-continental correlations, for identifying regional circulation changes and for understanding atmosphere–ocean heat transfer processes that may help modulate local and regional climatic variations. The time is now appropriate to reaximine in greater detail the occurrence, ages and environmental origins of such aeolian deposits and to compare them between high and low latitudes, between continents and ocean basins.

CONCLUSIONS

The salient points of this paper may be summarised as follows:

1. Pleniglacial environments in southern Australia were characterised by two hydrologic phases, one much wetter than today, followed by one that was much drier. The transition from one to the other, involving a major change in lake levels and the onset of dune building, occurred close to 25 000 BP. The dry phase lasted until about 13 000 BP.
2. A sequence of hydrologic events very similar to those of southern Australia is recorded in the evidence from northern Africa where the transition from wet to dry took place about the same time.
3. Over large areas of continental Europe, USSR and USA a transition from stable soils to loess deposition dates also from about 25 000 BP.
4. Accumulation of macro-particles believed to represent a global dust increment in Antarctic ice reached maximum proportions in the period between 25 000 and 14 000 BP simultaneously with the extension of desert margin dunes in southeastern Australia.
5. The association of major aeolian episodes with events of the last pleniglacial can be extended back through the Pleistocene by correlation of aeolian deposits from Czechoslovakia and China, with oceanic ^{18}O maxima, and with peaks in the Saharan dust contributed to the Atlantic Ocean. This association, demonstrating simultaneous reactivation of hot and cold deserts, holds throughout the latter part of the Pleistocene.
6. The evidence for the last pleniglacial in southern Australia is consistent with the development of enhanced summer evaporation rates probably controlled by increased flow of sensible heat advected from the continental interior. Hot dry summers appear to have alternated with cold stormy winters. Such conditions of increased seasonal contrast were enhanced by low sea level and increased velocities of glacial age winds.
7. The cause of the major change that occurred at 25 000 BP remains to be isolated. It may have been due to a large increase in meridional temperature gradients such as is known to have developed by 18 000 BP when Antarctic sea ice and the Laurentide ice masses achieved their greatest latitudinal extent. The hydrologic transition from wet to dry dated to 25 000 BP over large regions suggests that there must have been an equivalent event in the Southern Hemisphere to help produce such widespread synchroneity. It is possible that the first major expansion of Antarctic sea ice dates from this time.

8. The consistent re-appearance of itensified aeolian activity at both high and low latitudes immediately preceding the end of each full glacial phase strengthens the possibility that increased vigour of atmospheric circulation, either directly or indirectly, stimulated deglacial retreat and final return to interglacial conditions. The effects of greatly increased dust in the global atmosphere and of increased rate of heat flow from low to high latitudes must be considered as factors in modulating glacial-interglacial transitions.

ACKNOWLEDGEMENTS

I am grateful to my Canberra colleagues Dr J.N. Jennings and Dr C.D. Ollier for constructive suggestions on presentation; Dr N.J. Opdyke, visiting from Lamont-Doherty Geological Observatory, offered valuable comments on the arguments presented. John Magee drafted the figures, checked the text and contributed to the discussion.

Typing was in the careful hands of Marilyn Gray.

I am particularly indebted to those Quaternary scientists in China who introduced me to the excitement of their spectacular and most significant loess province, surely the 'Grand Canyon' of Quaternary stratigraphy. The scientific exposition of Professor Liu Tung-Sheng, Wen Qizhong, Lu Yanchou and their colleagues was matched only by the genuine warmth of their hospitality and the epicurean quality of the Chinese cuisine. Future students of Quaternary research will benefit greatly from the quality and extent of the current investigations being pursued so enthusiastically by Chinese scientists.

DISCUSSION

N.A. Mörner: How could you say that there is no correlation between your wet/arid change and the glacial volume record? The ^{18}O curves you refer to may have a much more complex origin. The glacial volume records from the ice caps themselves suggest an increase between about 30 000 and 20 000 BP of about $\frac{1}{3}$ (or even $\frac{1}{2}$) of the total volume of 20 000–18 000 BP.

J.M. Bowler: I did not imply that there is no correlation between the ^{18}O and continental hydrologic curves. My point was that the *magnitude* of the hydrologic change registered between 30 000 and 18 000 BP, a change that occurred about 25 000 BP is not matched at that point by a change of corresponding magnitude in the marine ^{18}O record. The only possible event of such magnitude is recorded in Chappell's sea level reconstructions from New Guinea in which a stand about —30 m at 27 000 BP preceded the full glacial fall to —130 m around 18 000 BP implying more than a doubling of ice volume. However this has yet to be confirmed by ice volume ^{18}O evidence from deep-sea cores.

M. Sartheim: I wonder whether a lot of the disagreement between the ^{18}O curve and the onset of the aridity at 25 000 was not related to the difficulties in dating precisely the Stage 2–3 boundary. The 'official' age of 32 000 could be easily shifted to 24 000 thus resulting in age agreement of both events.

J.M. Bowler: Yes, I agree that some of the apparent inconsistency would be removed if the age of Stage 2–3 boundary was located closer to 25 000 BP.

H.B.S. Cooke: The data presented are impressive but there may be other factors involved. For example, the drop in sea level leads to expansion of the exposed shelf area and to increased continentality. Increased intensity of the zonal circulation of the atmosphere would increase the intensity of the anticyclone cells. It seems possibly that the associated cooling would extend the duration of the continental winter anticyclone and hence the clear skies, evaporation and wind system activity. Cooled waters would also provide less moisture to the cyclonic systems and starvation may well have aided the death of the glaciers.

J.M. Bowler: I fully agree that the glacial age environments in low to mid-latitudes probably experienced an intensification of anticyclonicity of the type you describe. However the passive retreat of ice masses (by starvation) and subsequent warming does not explain the hydrologic events observed in Australia and other low latitude regions. The building of large gypsum and clay dunes, a summer-autumn event, was controlled by air masses derived from the continental interior. These would have been hot and dry, capable of producing high evaporation rates. Glacial age intensification of such processes implies the seasonal production in low latitudes of large quantities of advective heat. If all low latitude continental regions were responding in the same way, as the Australian and African evidence suggests, we have a powerful seasonal heat source developed simultaneously with the climax of glacial conditions.

H. Flohn: The evidence of increasing aridity in so many parts of the world during the last Würm—Wisconsin maximum glaciation poses several problems from the climatological point of view. Strong and persistent upwelling in both Pacific and Atlantic (nearly 150° long.) needs a great part of incoming radiation to warm upwelling water – there energy is lost and certainly responsible for decreasing evaporation of oceans between 5° N and 10° S. Other effects (eustatic lowering of shelves, increase of drift-ice both in N and S hemispheres) add to this diminution. Both equatorial continents (Australia and Africa) have been semi-arid at reduced temperatures; increasing flux of sensible heat can only be possible at few arid regions in the sub-tropics. It seems difficult to interpret these losses of the surface heat-budget quantitatively under the (tacit) assumption of a really constant solar constant. This will give food for serious future discussion.

J.M. Bowler: There certainly is a climatological problem in the timing of low latitude aritidy synchronously with maximum extension of ice sheets. Even if precipitation was substantially below today's levels, glacially reduced temperatures would result in rain being much more effective than today. To evaluate this factor we need to be able to reconstruct glacial age climatologic parameters on land more precisely than present data allow.

D.A. Livingstone: Temperature data that are becoming available for altitudes around 2 000 m from tropical East Africa remove some of this embarrassment. Bada's racemization analysis of amino acids from bone and C4–C3 interpretation of grass cuticles agree in suggesting a temperature lowering of about 5° C during the last ice age.

J.M. Bowler: This may indeed be so but I'm not sure it solves the problem. Until we can obtain estimates of seasonal climates and especially changes in summer-winter thermal regimes we can but speculate on the factors that controlled the annual continental hydrologic budgets. To provide the climatologists with the data they require we must develop seasonally discriminating parameters just as CLIMAP has done in the marine environment.

REFERENCES

Alimen, H. 1976. Alternances "pluvial-aride" et "érosion-sédimentation" au Sahara nord-occidental. *Rev. Géogr. Phys. Géol. Dyn.* 18: 301–312.

Allen, H. 1972. *Where the crow flies backwards: man and land in the Darling Basin.* Thesis, Australian National University, Canberra. 382 pp.

Barbetti, M. & Polach, H. 1973. ANU Radiocarbon datelist V. *Radiocarbon* 15: 241–251.

Barbetti, M. & McElhinny, M. W. 1976. The Lake Mungo geomagnetic excursion. *Phil. Trans. Roy. Soc. London* A, 281: 515–542.

Barry, R.G. & Chorley, R.J. 1971. *Atmosphere, weather and climate,* Methuen, London. 379 pp.

Barry, R.G. & Williams, Jill. 1975. Experiments with the NCAR global circulation model using glacial maximum boundary conditions: southern hemisphere results and interhemispheric comparison. In: R.P. Suggate & M.M. Creswell (eds.), *Quaternary Studies*, Roy. Soc. N.Z., Wellington: 57–66.

Bé, A.W.H. & Duplessy, J. 1976. Subtropical convergence fluctuations and Quaternary climates in the middle latitudes of the Indian Ocean. *Science* 194: 419–422.

Bellair, P. 1965. Un exemple de glaciation aberrante. Les Iles Kerguelen. *Com. Nat. Français Rech. Antarctiques* 11.

Bettenay, E. 1962. The salt lake systems and their associated aeolian features in the semi-arid regions of Western Australia. *J. Soil Sci.* 13: 10–17.

Bigarella, J.J. & Andrade, G.D. de. 1965. Contributions to the study of the Brazilian Quaternary. In: H.E. Wright & D.G. Frey (eds.), *International studies on the Quaternary*, Geol. Soc. Am., Spec. Paper 84: 433–451.

Biscaye, P.E. 1965. Mineralogy and sedimentation of recent deep sea clay in the Atlantic Ocean and adjacent seas and oceans. *Bull. Geol. Soc. Am.* 76: 803–831.

Bowler, J.M. 1966. Aridity in Australia: Age, origins and expression in aeolian landforms and sediments. *Earth-Sci. Rev.* 12: 279–310.

Bowler, J.M. 1971. Pleistocene salinities and climatic change evidence from lakes and lunettes in southeastern Australia. In: D.J. Mulvaney & J. Golson (eds.), *Aboriginal man and environment in Australia,* ANU Press, Canberra, A.C.T.: 47–65.

Bowler, J.M. 1973. Clay dunes: their occurrence, formation and environmental significance. *Earth-Sci. Rev.* 9: 315–338.

Bowler, J.M. 1975. Deglacial events in southern Australia: their age, nature and palaeoclimatic significane. In: R.P. Suggage & M.M. Creswell (eds.), *Quaternary Studies*, Roy. Soc. N.Z.: 75–82.

Bowler, J.M. 1976. Quaternary chronology of Goulburn Valley sediments and their correlation in southeastern Australia. *J. Geol. Soc. Aust.* 14: 287–292.

Bowler, J.M. 1977. Quaternary climate and tectonics in the evolution of the Riverine Plain, southeastern Australia, In: J.L. Davies & M.A.J. Williams (eds.), *Time, pace and landforms in Australia*, ANU Press, Canberra (in press).

Bowler, J.M. & Hamada, J. 1971. Late Quaternary stratigraphy and radiocarbon chronology of water level fluctuations in Lake Keilambete, Victoria. *Nature* 232: 330–332.

Bowler, J.M., Hope, G.S., Jennings, J.N., Singh, G. & Walker, D. 1976. Late Quaternary climates of Australia and New Guinea. *Quat. Res.* 6: 359–394.

Bowles, F.A. 1973. Climatic significance on quartz content in sediments of the eastern equatorial Atlantic. *Trans. Am. Geophys. Un.* 54: 328 pp.

Broecker, W.S. & Orr, P.C. 1958. Radiocarbon chronology on Lake Lahontan and Lake Bonneville. *Bull. geol. Soc. Am.* 69: 1009–32.

Broecker, W.S. & Kauffman, A. 1965. Radiocarbon chronology of Lake Lahontan and Lake Bonneville II, Great Basin. *Bull. geol. Soc. Am.* 76: 537–566.

Broecker, W.S. & Donk, J. van. 1970. Insolation changes, ice volumes and ^{18}O record in deep-sea cores. *Rev. Geophys. Space Phys.* 8: 169–198.

Brookfield, M. 1970. Dune trends and wind regime in Central Australia. *Zeitsch. Geomorphol. N.F.* 10: 121–153.

Bugaev, V.A., Dzhordzhio, V.A. & Dubentsov, V. R. 1952. *O termicheskom effekte pyli pri pyl'nkh i peschanykh buriakh* (The thermal effect of dust during dust and stand storms). Akademia Nauk, SSSR, Isvesita, Ser Geograficheskaia, 3: 44–45 (in Russian). English Abst. in: *Meteorol. Geoastrophys. Abstracts* 1954, 5: 1277.

Butzer K.W., Isaac, G.L., Richardson, J.L. & Washbourne-Kamau, C. 1972. Radiocarbon dating of East Africa lake levels. *Science* 175: 1069–1076.

Butzer K.W., Fock, G.J., Stuckenrath, R. & Zilch, A. 1973. Palaeohydrology of late Pleistocene Lake Alexandersfontein, Kimberley, South Africa. *Nature* 243: 328–330.

Campbell, E.M. 1968. Lunettes in southern South Australia. *Trans. Roy. Soc. S. Aust.* 92: 85–109.

Cepek, A.G. 1965. Geologische Ergebnisse der ersten Radiokarbondatierung von Interstadialen im Lausitzer Urstromtal. *Geologie* 14: 625–657.

Churchward, H.M. 1963. Soil studies at Swan Hill, Victoria, Australia, IV: Ground-surface history and its expression in the array of soils. *Aust. J. Soil Res.* 1: 242–255.

Colhoun, E.A. 1975. A Quaternary climatic curve for Tasmania. *Australasian conference on climate and climatic change,* Royal Metereological Society.

Costin, A.B. 1972. Carbon-14 dates from the Snowy Mountains area, southeastern Australia and their interpretation. *Quat. Res.* 2: 579–590.

Costin, A.B. & Polach, H.A. 1971. Slope deposits in the Snowy Mountains, southeastern Australia. *Quat. Res.* 1: 228–235.

Costin, A.B. & Polach, H.A. 1973. Age and significance of slope deposits, Black Mountain, Canberra. *Aust. J. Soil Res.* 11: 13–25.

Coventry, R.J. 1976. Abandoned shorelines and the Late Quaternary history of Lake George, New South Wales. *J. Geol. Soc. Aust.* 23: 249–273.

de Ploey, J. 1965. Position géomorphologique, génèse et chronologie de certains dépôts superficiels au Congo occidental. *Quaternaria* 7: 131–154.

Diester-Haas, L. 1976. Late Quaternary climatic variation in northwest Africa deduced from East Atlantic sediment cores. *Quat. Res.* 6: 299–314.

Dodson, J.R. 1974. Vegetation and climatic history near Lake Keilambete, Western Victoria. *Aust. J. Botany* 22: 709–717.

Dodson, J.R. 1975. Vegetation history and water fluctuations at Lake Leake, southeastern Australia. II. 50,000 to 10,000 B.P. *Aust. J. Botany* 23: 815–831.

Dreimanis, A. & Karrow, P.F. 1972. Glacial history in the Great Lakes – St. Lawrence region, the classification of the Wisconsin(an) Stage and its correlatives. In: *Proc. Internat. Geol. Congress, Montreal, Section 12, Quat. Geol.* 5–15.

El-Fandy, M.G. 1949. Dust – an active meteorological factor in the atmosphere of Northern Africa. *J. Appl. Physics* 20: 660–666.

Fink, J. 1969. Le loess en Autriche. In: La stratigraphie des Loess d'Europe, *Bull. Ass. Française Étude Quaternaire,* Supplement 3–12.

Fink, J. & Kukla, G.J. 1977. Pleistocene climates in Central Europe: At least 17 interglacials after the Olduvai Event. *Quat. Res.* 7: 363–371.

Flint, R.F. 1971. *Glacial and Quaternary geology*, Wiley, New York.

Flint, R.F. & Gale, W.A. 1958. Stratigraphy and radiocarbon dates at Searles Lake, California. *Am. J. Sci.* 249: 257–300.

Flint, F.R. & Bond, G. 1968. Pleistocene sand ridges and pans in western Rhodesia. *Bull. Geol. Soc. Am.* 79: 299–314.

Flohn, H. 1953. Studien über die atmosphärische Zirkulation in der letzten Eiszeit, *Erdkunde* 7: 266–275.

Flohn, H. 1973. Antarctica and the global Cenozoic evolution: a geophysical model: *Palaeoecology of Africa* 8: 37–53.

Galloway, R.W. 1965. Late Quaternary climates in Australia. *J. Geol.* 73: 603–618.

Gardiner, J.V. & Hays, J.D. 1976. Responses of sea-surface temperature and circulation to global climatic change during the past 200,000 years in the eastern equatorial Atlantic Ocean. *Geol. Soc. Am. Mem.* 145: 221–246.

Gasse, F. 1975. *L'évolution des lacs de l'Afar central du Plio-Pléistocène à l'Actuel. Reconstitution des paléomilieux lacustres à partir de l'étude des Diatomées.* Thesis, University of Paris. 3 vols, 383 pp.

Gasse, F. 1977. Evolution of Lake Abhé (Ethiopia and TFAI), from 70,000 B.P. *Nature* 265: 42–45.

Gates, W.L. 1976. Modeling the ice-age climate. *Science* 191: 1138–1144.

Gentillii, J. 1960. Il fattore termico nell'ecologia deglia eucalitti. *Centro di Sperimentazione Agricola e Forestale*, Rome, 4:

Gentilli, J. (ed.). 1971. *Climates of Australia and New Zealand. World Survey of Climatology*, 13, Elsevier, Amsterdam, 405 pp.

Hays, J.D., Lozano, J. Shackleton, N. & Irving, G. 1976a. Reconstruction of the Atlantic Ocean and western Indian Ocean sectors of the 18,000 B.P. Antarctic Ocean, In: R.M. Cline & J.D. Hays (eds.), *Investigation of late Quaternary paleoceanography and paleoclimatology Geol. Soc. Mem.* 145: 337–372.

Hays, J.D., Imbrie, J. & Shackleton, N.J. 1976b. Variations in the earth's orbit: pacemaer of the Ice Ages. *Science* 194: 1121–1132.

Helgren, D.M. & Butzer, K.W. 1976. Historical geomorphology of the pans in the lower Vaal Basin, South Africa. *Am. Quat. Assoc. Abstracts*, Arizona State University, Tempe: 33–36.

Hounam, C.E. 1961. *Evaporation in Australia: A critical survey of the network and methods of observations together with a tabulation of the results of observations.* Bull. 44, Comm. Bur. Met. Aust. 88 pp.

Idso, S.B. 1974. Thermal blanketing: a case for aerosol-induced climatic alteration. *Science* 186: 50–51.

Ivanova, I.K. 1969. Les loess de la partie sud-ouest du territoire Européen de l' U.R.S.S. et leur stratigraphie. In: La Stratigraphie des Loess d'Europe, *Bull. Ass. Française Étude Quaternaire*, Supplement 151–159.

Kellogg, W.J. Coakley Jr, J.A. & Grams, G.W. 1975. Effect of anthropogenic aerosols on the glocal climate. In: *Proc. WMO/LAMAP symposium on long-term climatic fluctuations, Norwich*: 323–330.

Kemp, E. 1977. Tertiary climatic evolution and a vegetation history in the southeast Indian Ocean region. *Palaeogeography, Palaeoclimatology, Palaeoecology* (in press).

Kershaw, A.P. 1974. A long continuous pollen sequence from north-eastern Australia. *Nature* 251: 222–223.

Kershaw, A.P. 1975. Late Quaternary vegetation and climate in north-eastern Australia. *Bull. Roy. Soc. N.Z.* 13: 181–188.

Kukla, G.J. 1970. Correlations between loesses and deep-sea sediments. *Geol. Föreningen Stockholm Förhandlingar* 92: 148–180.

Kukla, J. & Lozek, V. 1969. Trois profils caractéristiques de la Bohème Centrale et de la Moravie du Sud. In: La Stratigraphie des Loess d'Europe. *Bull. Ass. Française Étude Quaternaire*, Supplement 53–56.

Kuo Shieh-Tung & Lu Yanchou. 1966. The grain size cycles of loess: A discussion of the grain size distribution. In: Liu Tung-sheng (ed.), *Composition and texture of the loess of China*, Academia Sinica, Peking: 23–25 (in Chinese).

Lamb, H.H. 1972. *Climate: present, past and future*, V.1. Fundamentals and climate now, Methuen, London. 613 pp.

Liu Tung-sheng, Wang Ke-lu & Chu Hai-tze. 1966. Features of the regional distribution of the Malan loess. In: Liu Tung-sheng (ed.), *Composition and texture of the loess of China*, Academia Sinica, Peking: 1–8 (in Chinese).

Liu Tung-sheng et al. 1966. *Composition and textures of the loess of China.* The Science Press, Academia Sinica, Peking. 132 pp. (in Chinese).

Li Hua-mei, An chih-sheng & Wang Chun-ta. 1974. Preliminary paleomagnetic study of loess from the Wucheng section, northern China. *Geochimica* 6, Academia Sinica: 93–104 (in Chinese with English abstract).

Loewe, F. 1943. *Duststorms in Australia.* Bull. 28m, Comm. Bur. Met. Aust. 16 pp.

Lu Yanchou, Wen Qizhong, Huang Baijun, Min Yushun & Denh Huazing 1976. A preliminary study on the source of loessic materials in China – a study of the surface textures of silt quartz grains by transmission electron microscope. *Geochimica* 1, Academia Sinica: 47–53 (in Chinese with English abstract).

Mainguet, M. & Canon, L. 1976. Vents et paléovents au Sahara. Tentative d'approche paléoclimatique. *Rev. Géogr. Phys. Géol. Dyn.* 18: 241–250.

Mercer, J.H. 1973. Cenozoic temperature trends in the southern hemisphere: Antarctic and Andean glacial evidence. *Palaeoecology of Africa* 8: 85–114.

Ninkovitch, D. & Shackleton, N.J. 1975. Distribution, stratigraphic position and age of ash layer "L", in the Panama Basin region. *Earth and Planetary Sci. Letters* 27: 20–34.

Parkin, D.W. 1974. Trade winds during the glacial cycles. *Proc. Roy. Soc. London*, A 337: 73–100.

Parkin, D.W. & Shackleton, N.J. 1973. Trade-wind and temperature correlation down a deep-sea core off the Saharan coast. *Nature* 245: 455–457.

Parmenter, C. & Folger, D.W. 1974. Eolian biogenic detritus in deep-sea sediments. *Science* 185: 695–697.

Parsons, R.F. 1968. An introduction to the regeneration of mallee eucalypts. *Proc. Roy. Soc. Vic.* 81: 59–68.

Radczewski, O.E. 1939. Eolian deposits in marine sediments. In: P.D. Trask (ed.), *Recent marine sediments*, Soc. Econ. Paleontol. Mineral. Spec. Pub. 4: 496–502.

Richthofen, F. von. 1882. On the mode of origin of the loess. Reprinted from Geol. Mag. 9: 292–305. In: I.J. Smalley (ed.), *Loess, lithology and genesis*, Benchmark Papers in Geology 26, Dowden, Hutchinson & Ross: 24–36.

Rognon, P. 1976. Essai d'interprétation des variations climatiques au Sahara depuis 40 000 ans. *Rev. Géogr. Phys. Géol. Dyn.* 18: 251–282.

Rognon, P. & Williams, M.A.J. 1977. Late Quaternary climatic changes in Australia and North Africa: a preliminary interpretation. *Palaeogeo-*

graphy, Palaeoclimatology, Palaeoecology 21: 285–327.

Ruhe, R.V. 1968. Identification of paleosols in loess deposits of the United States. In: C.B. Schultz & J.C. Frye (eds.), *Loess and related Eolian deposits of the world*, Vol. 12, Proc. INQUA Congress, U.S.A.: 49–65.

Ruhe, R.V. 1969. Paleosols and soil stratigraphy. In: M. Ters (ed.), *Etudes sur le Quaternaire dans le Monde*, 8th INQUA Congress, Paris: 335–340.

Sarntheim, M. & Walger, E. 1974. Der äolische Sandstrom aus W-Sahara zur Atlantikküste. *Geol. Rundschau* 63: 1065–1087.

Seibold, R., Diester-Haas, L., Fütterer, D., Hartmann, M., Kögler, F.C., Lange, H., Müller, P.J., Pflaumann, U., Schrader, H.J. & Suess, E. 1976. Late Quaternary sedimentation of the western Sahara. *An. Acad. Bras. Cienc.*, Supplement 48: 287–296.

Servant, M. & Servant, S. 1970. Les formations lacustres et les diatomées du Quaternaire récent du fond de la cuvette Tchadienne. *Rev. Géogr. Phys. Géol. Dyn.* 12: 63–76.

Shackleton, N.J. & Opdyke, N.D. 1973. Oxygen isotope and paleomagnetic stratigraphy of equatorial Pacific core V28–238: Oxygen isotope temperatures and ice volumes on a 10^5 and 10^6 year scale. *Quat. Res.* 3: 39–55.

Singh, G., Joshi, R.D. & Singh, A.B. 1972. Stratigraphic and radiocarbon evidence for the age and development of three salt-lake deposits in Rajasthan, India. *Quat. Res.* 2: 496–505.

Smith, G.I. 1968. Late-Quaternary geologic and climatic history of Searles Lake, southeastern California. In: *Means of correlation of Quaternary successions*, Vol. 8, VII INQUA Congress, Univ. Utah Press: 293–310.

Sprigg, R.C. 1965. The nature and origin of modern deserts. *Aust. Mus. Mag.* 71: 207–211.

Street, F.A. & Grove, A.T. 1976. Environmental and climatic implications of late Quaternary lake-level fluctuations in Africa. *Nature* 261: 385–390.

Sugden, D.E. & Clapperton, C.M. 1977. The maximum ice extent on island groups in the Scotia Sea, Antarctica. *Quat. Res.* 7: 268–282.

Thompson, L.G., Hamilton, W.L. & Bull, C. 1975. Climatological implications of microparticle concentrations in the ice core from "Byrd" station, western Antarctica. *J. Glaciology* 14: 433–444.

Tricart, J. 1975. Existence de période sèche au Quaternaire en Amazonie et dans les régions voisins. *Rev. Géomorphol. Dyn.* 4: 145–158.

van Zinderen Bakker, E.M. 1973. The glaciation(s) of Marion Island (sub-Antarctica). *Palaeoecology of Africa* 8: 161–178.

van Zinderen Bakker, E.M. 1976. The evolution of late Quaternary palaeoclimates of Southern Africa. *Palaeoecology of Africa* 8: 160–202.

Williams, G.E. 1973. Late Quaternary piedmont sedimentation, soil formation and palaeoclimates in arid South Australia. *Zeitsch. Geomorphol.* N.F. 17: 102–125.

Williams, M.A.J. 1975. Late Pleistocene tropical aridity synchronous in both hemispheres? *Nature* 253: 617–618.

Wyrwoll, K. & Milton, D. 1976. Widespread late Quaternary aridity in Western Australia. *Nature* 264: 429–430.